HISTORY OF TANK WARFARE

HISTORY OF TANK WARFARE

FROM WORLD WAR I TO THE PRESENT DAY

GENERAL EDITOR: STEPHEN HART

amber
BOOKS

This Amber edition first published in 2019

Previously published as *Atlas of Tank Warfare* in 2012

Published by
Amber Books Ltd
United House
North Road
London
N7 9DP
United Kingdom
www.amberbooks.co.uk
Instagram: amberbooksltd
Facebook: www.facebook.com/amberbooks
Twitter: @amberbooks

ISBN: 978-1-78274-720-8

Project Editor: Sarah Uttridge
Design: Andrew Easton
Picture Research: Terry Forshaw
Maps: Courtesy of Red Lion Media

Printed in China

CONTENTS

INTRODUCTION

MOBILITY AND FIREPOWER
Early armoured forces were limited by the mobility of their artillery support. Self-propelled guns like the M109A6 Paladin can stay with rapidly advancing armoured forces, supporting them all the way to the objective.

The first tanks to enter combat in 1917, great armoured machines armed with deadly guns, created utter panic among the defending German troops, and changed the face of land warfare forever. A generation later the Germans had adopted this new technology and used it efficiently and with devastating effect in the Blitzkrieg of 1940 when they came within a hair's breadth of winning World War II. This book traces the development and deployment of tanks from their introduction during World War I, to the present day.

World War II is remembered for some of the most devastating tank battles in history, such as the innovative Operation Uranus, the Soviets' surprise attack on the German Sixth Army at Stalingrad in November 1942. The greatest tank battle of them all, the battle of Kursk on 12 July 1943, raged around the tiny, unremarkable village of Provhorovka in southwestern Russia.

Although both sides in the Cold War developed their armoured forces, the most significant tank engagements occurred in the Middle East, notably during the Six-Day War of 1967, when Israeli Centurion tanks infiltrated Egyptian defences and laid a deadly ambush that all but wiped out the Arab force. In 1991 the tanks of the U.S. VII Corps unleashed a rapid, unstoppable left hook as part of Operation Desert Storm to halt the forces of Saddam Hussein.

The introduction to the battlefield of armoured vehicles powered by the internal combustion engine gradually transformed conventional operations during the twentieth century. This revolution saw the creation of mobile formations, such as armoured, mechanized and motorized divisions and brigades. Each of these formations deployed hundreds of fully-tracked, partially-tracked and wheeled armoured vehicles, including tanks, armoured cars, reconnaissance vehicles, half-tracked personnel carriers, and lorries. The deployment of such force structures allowed armies to wage strategic wars of mobility and manoeuvre at a greater tempo, physically advancing across the terrain many times faster than had been possible on horseback or by foot. The age of strategic armoured warfare was born, where armies employed their armoured forces to break through, outflank, or infiltrate the enemy's defences, before unleashing these forces in audacious advances deep through the enemy's rear areas, thus bringing the enemy to swift defeat. Consequently, many of the wars that have raged across the world since 1916 have unfolded in arenas in which armoured forces have clashed in a titanic struggle of wills, pitching machines, and the humans that operate them, into a bitter life-and-death struggle for survival.

The Great War

During World War I, the belligerents took the first tentative steps to deploy tanks to break the stalemate on the trenches of the Western Front, notably during the latter stages of the Battle of the Somme, at Cambrai in November 1917, and at Amiens on 8 August 1918. Despite massive demobilization after the war's end in 1918, European armies continued to experiment with these capabilities. In this period, the theory of armoured warfare was developed further, notably by J.F.C. Fuller in his *Plan 1919* and Heinz Guderian in *Achtung – Panzer!* The German Blitzkrieg doctrine, as well as the Soviet concepts of Deep Battle and Deep Operations also evolved between the wars. Armies experimented with force structures, with the British Experimental Mechanized Force – the world's first all-arms mobile brigade group – undertaking manoeuvres in the late 1920s. Next, during 1939-42, Germany unleashed its Blitzkrieg methods as its armoured forces gradually conquered most of the European continent. The roll call of bitter armoured encounters during these years is well-known – the 1939 Battle for Poland; the assault on Sedan in May 1940; Operation Barbarossa, the

MOBILE FORTRESS
Germany's first tank design, the A7V, was a large, slow target. Few were built and those that saw action proved to be largely ineffective.

German invasion of the Soviet Union in June 1941; and the siege of Stalingrad, to name but a few. Gradually the Western Allies and the Soviets turned the tide, as their armoured forces slowly drove the enemy back into the Reich itself by early 1945. A sequence of cataclysmic armoured clashes in this period remain famous for their intensity, including the destruction of Sixth Army at Stalingrad, the struggle at Kursk (July 1943), the Soviet summer 1944 Bagration offensive, and the Western Allied Market Garden attacks of mid-September 1944. By May 1945, Allied armour had finally defeated Hitler's Germany, ridding Europe of the terrible scourge of Nazism.

Post-war

But peace was to elude the post-1945 world, and clashes between armoured forces continued to shape international politics into the twenty-first century. From May 1948 until at least 1982, for example, the Israelis found themselves locked into a struggle for national survival,which erupted in the 1967 Six-Day War and the 1973 Yom Kippur War. Similarly, western Allied forces battled Communist armoured forces in Korea, in French Indochina and in Vietnam. In 1990–91 a US-led multinational alliance ousted Iraqi forces from occupied Kuwait, and in 2003 a narrower U.S.-led alliance overran Iraq and overthrew Saddam Hussein's Ba'athist regime. Even in the counter-insurgency campaigns that Western forces have fought in Iraq (2003–08) and Afghanistan (2001–), armoured vehicles have played crucial roles. And while the death of the tank has been pronounced many times by commentators, its ability to move troops under armoured protection around a hostile battlefield and bring lethal fire to bear will continue to ensure that advanced armoured vehicles will play a crucial role in a variety of military operations in the future.

A MATURE WEAPON SYSTEM
Armoured vehicles underwent a rapid evolution during World War II, becoming both a graphic symbol of occupation or liberation and the tool by which these were achieved.

EARLY DEVELOPMENT

HOLT TRACTOR
The Holt agricultural tractor was used as the basis of several military vehicles, from artillery tractors to prototype tank designs.

The armoured fighting vehicle concept originated before World War I, but serious development did not take place until spurred by the needs of the war. Prior to 1914, it was generally assumed that any future war would be fought in the traditional manner by cavalry, infantry and artillery, and within such a force structure there was no perceived need for a cumbersome, unreliable and outlandish weapon such as a self-powered, armoured gun platform.

The necessary components and concepts of the tank were, however, becoming available by the end of the nineteenth century. The caterpillar track had been in (albeit limited) use since the late eighteenth century on steam-powered vehicles, and suitable heavy weapons had been produced for naval use. Similarly, armour plate was readily available, ever since the rise of armoured metal ships from the mid-nineteenth century onwards. All that was missing was a suitable powerplant, and by the outbreak of World War I advances in technology and engineering meant that it was possible to produce an internal combustion engine capable of driving a large, heavy vehicle across rough terrain. Putting all these elements together, however, required some experimental trial and error and, more significantly, a perceived need on the part of those required to authorise and support the project.

Tractors and 'war cars'

Self-powered vehicles had been in military service since the Crimean War (1854–56). The first were steam-powered traction engines used to pull artillery pieces, and were not successful. Nevertheless, by the time of the Boer War (1899–1902), steam-powered traction engines were in use with British engineering units and performing some logistics tasks. While capable of providing motive power and crossing fairly rough ground, these vehicles were entirely unsuitable for use as weapons platforms.

Steam power was not an effective option for a combat vehicle, though some experiments were conducted. As late as 1918, the Holt Company offered a wheeled, steam-powered tank design to the US government. Holt had been manufacturing tracked agricultural tractors for some time, producing a design powered by an internal combustion engine in 1894. The Holt Tractor was adopted for military use and served in World War I as an artillery tractor and logistics vehicle.

Holts were also used to support the early tank units, towing mobile workshops mounted on trailers. They were also used as primitive armoured recovery vehicles, rescuing ditched or broken-down tanks after an attack. The demonstrated capability of the Holt Tractor to cross soft mud and broken ground was a key factor in the adoption of track-laying systems for the first tanks.

Meanwhile, the first armoured cars, or 'war cars' had begun to appear. These were based on commercially available cars or light trucks, fitted with armour plate and a weapon system. The earliest appeared at the end of the nineteenth century and, driven by engines ranging from 15-25hp, were seriously underpowered.

Some vehicles, such as the French Charron-Giradot et Voigt, were mobile weapons platforms rather than true combat vehicles. Lacking any armour protection other than a gun shield and, later, a steel drum surrounding the gun and its operator, such a vehicle could still perform useful service by quickly redeploying to place its weapon in a suitable firing position. However, although the concept was successfully demonstrated to the French military, no orders were placed.

Other designs were over-ambitious or excessively fanciful. The Sizaire-Berwick 'Wind Wagon', designed by airmen, had a conventional two-wheel-drive transmission but for speed over flat terrain could switch to its alternate propulsion system; an aero engine driving a propeller. Armed with a front-facing machine gun with a very limited firing arc and lacking protection for either its radiator or the aero engine, it was not a practical combat vehicle.

Where idle aircrew came up with what was essentially a wheeled fighter aircraft, the British Admiralty tried to field a land-based warship. The Seabrook armoured car had an armoured body with hinged sides that could be lowered to increase the arc of fire of its weapons. These included a 3-pdr (using a 47mm/1¾in shell) gun and up to four machine guns, requiring a crew of six to operate. The Seabrook was built on a 5-ton truck chassis but was still overloaded and lacked mobility. A handful were deployed but failed to make any impression.

Most armoured car designs were somewhat more conventional, and by 1914 a general type had emerged. Built on a light truck or car chassis, most mounted a single or perhaps two weapons in a revolving turret and were protected by light armour plate. Although still overloaded, these vehicles proved useful in several theatres of war, though once the trench deadlock began on the Western Front they were of little use there.

Most armoured cars carried a machine gun as armament, though some mounted light guns up to 37mm (1½in) calibre. Germany experimented

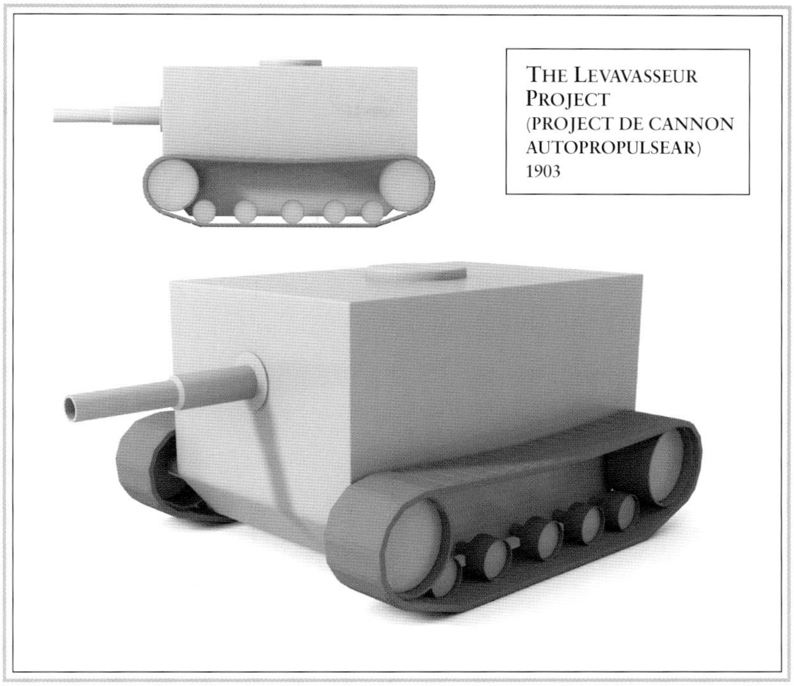

LEVAVASSEUR PROJECT
The Levavasseur project was essentially an armoured housing for a 75mm (3in) gun, transported on a tracked chassis. The project fell by the wayside for lack of interest, but the St Chamond tank which emerged later may have been influenced by its general concepts.

EXPERIMENTAL VEHICLES
Not all early experiments were a success. The Seabrook Armoured Car was too heavily loaded for its chassis and consequently lacked the mobility necessary for a successful armoured vehicle.

'LITTLE WILLIE'
The prototype tank 'Little Willie' was little more than an armoured box of boiler plate mounted atop a tracked chassis. It was not put into production but demonstrated that an armoured combat vehicle was a workable concept.

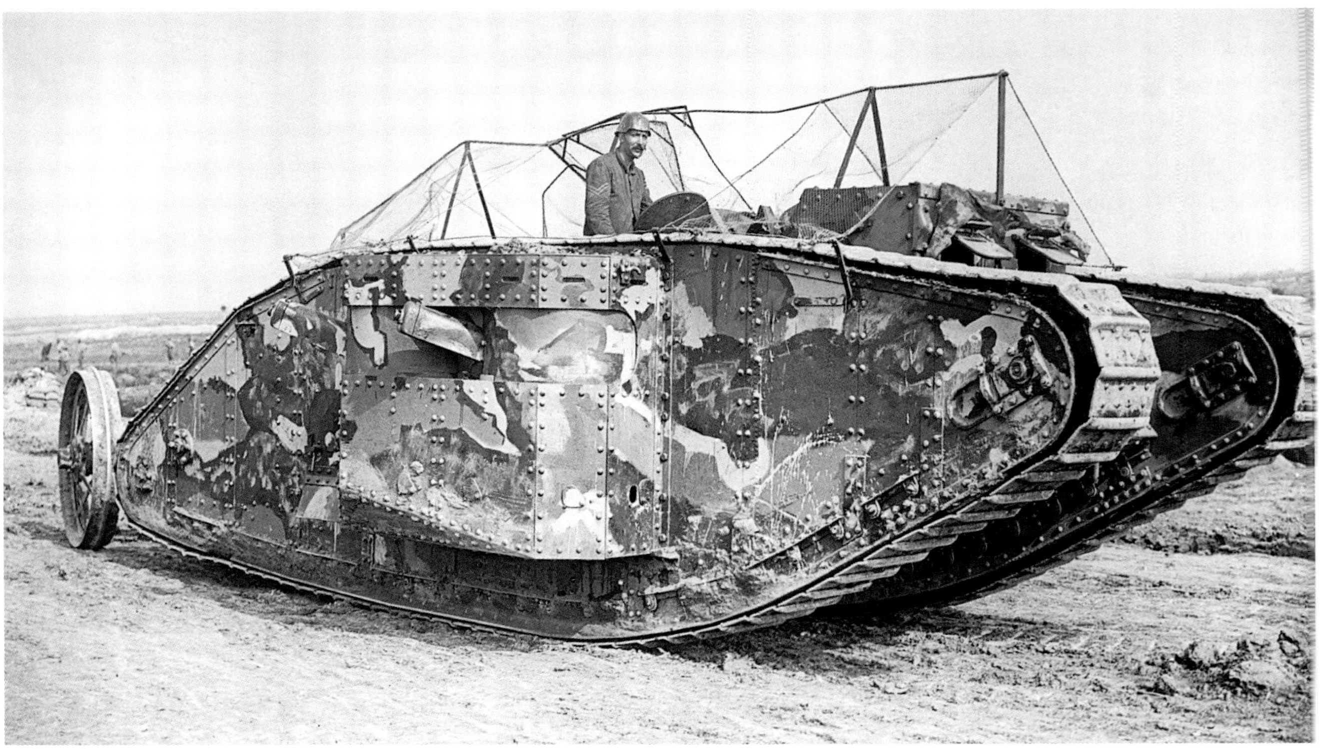

MK I TANK
The second British tank design was designated 'Mother' or 'Big Willie'. It was promising enough to be put into production as the Mk I tank.

with a mobile air defence vehicle, built in 1906 and intended for use against observation balloons. The concept was not followed up at the time, but anti-aircraft vehicles would eventually become an essential component of armoured forces.

The armoured car did perform a number of useful tasks during World War I, notably as a security vehicle and in the reconnaissance and strike role in theatres other than the Western Front. However, it was not the weapon needed to win the trench war. That would require a combination of features from two concepts – the light armoured combat vehicle and the track-laying support and logistics tractor.

Early experiments

Various designs for heavy armoured vehicles were put forward before the outbreak of war, but these attracted relatively little interest. Some were highly impractical, using walking beams or wheels to provide mobility, and it soon became apparent that the caterpillar track offered the only workable solution. Steam power was inadequate and internal combustion was the only powerplant that offered the necessary ratio of power to weight and size.

An Austrian officer named Gunther Bursztyn proposed a workable tank design in 1912. Mounting its weapons in a turret, the Bursztyn armoured vehicle used tracks for mobility, aided by rollers on long beams to assist in crossing rough ground. The idea did not catch on and was largely forgotten, as was a similarly usable design put forward to the British military by Lancelot de Mole. This vehicle was in some ways superior to the design fielded in 1916, but it was proposed at a time when no need for such a vehicle was perceived. Similarly, a French project developed by Captain Levavasseur had found no favour a few years earlier.

The early years of World War I created an urgent need for a weapon that could smash through enemy positions. Various concepts were developed and some deployed, including poison gas and giant flame projectors. However, none restored mobility to the battlefield nor permitted a decisive breakthrough

MK IV TANK
An improvement on the Mk I design, the Mk IV had very impressive rough-ground performance, improved by the lack of rear steering wheels.

Despite resistance from traditionalists, work began on an armoured combat vehicle which might provide the answer.

The new weapon was required to provide three things – mobility, firepower and protection. It had to carry weapons sufficient to eliminate an enemy strong point or a group of infantry, and had to resist at least small-arms fire. Most critically, it had to do these things while moving across shell-cratered, muddy ground and crossing trenches.

'Little Willie' and 'Mother'

Britain took the lead in tank experimentation, giving the main role to the Navy as its personnel had more experience with armoured cars than the Army. The Landship Committee, as the development group was named, developed a track-laying vehicle nicknamed 'Little Willie'. Little Willie mounted a 2-pdr gun (with 40mm/1½in ammunition) and at least one machine gun; various additional armament was tested. Steering was normally performed using a pair of wheels at the rear of the vehicle, but a tighter turn could be accomplished by braking one track and slewing around using the other.

'Little Willie' was protected by armour made from boiler plate and could climb a 0.3m (1ft) high obstacle. Trench crossing was demonstrated over a gap of 1.5m (5ft), all of which was promising enough to commission a more advanced machine .

Sometimes called 'Big Willie' and more often 'Mother', this vehicle was sufficiently effective that it went into production as the Mk I tank. That name originated due to an attempt to maintain secrecy by claiming that the large metal structures under construction were water tanks; the name stuck and has been used ever since.

'Mother' was given all-round tracks and a rhomboidal shape to assist in trench crossing, allowing a much wider gap to be crossed. Driving was a complex business requiring four men, of whom two operated gearboxes. Steering was assisted by wheels at the rear, but these proved to be unnecessary and were later removed.

Although 'Mother' could cross a trench of up to 2.7m (9ft), it had no suspension and frequently caused crew injuries. Crew members were tumbled from their scanty seats or were even rendered unconscious by the impact of crossing an obstacle, often falling against internal projections or hot engine components. The noise of the unsilenced 105-hp engine made communication difficult among the crew, necessitating the use of hand signals.

'Mother' (and the Mk I tank that followed) was an extremely primitive design. It was unreliable and broke down frequently, and was prone to 'ditching', i.e. becoming stuck when crossing an obstacle or trench. One low-tech solution to this problem was the use of an unditching beam carried atop the

BURSZTYN'S TANK DESIGN
The armoured vehicle designed by Gunther Bursztyn was arguably more advanced than the designs that were actually implemented. It had a rotating turret and wheels on hinged beams to assist with trench-crossing. Like many projects of the time it was influenced by the Holt agricultural tractor.

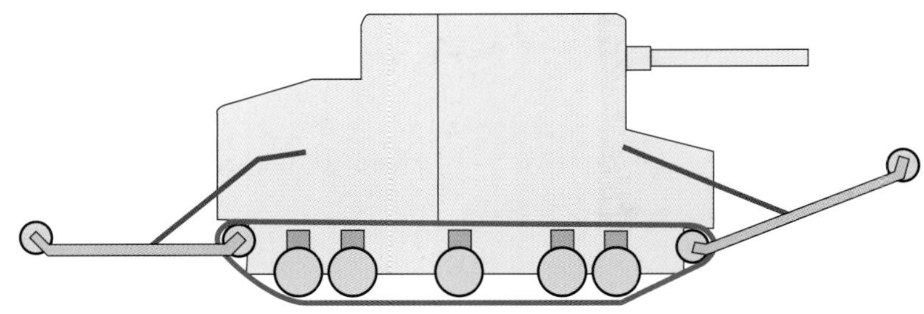

BURSZTYN 'TANK' 1912
MOTORGESHUTZ

vehicle, which could be placed under the tracks for additional purchase in soft ground or where there was nothing under the tracks at that point.

Although turrets had been successfully used in armoured cars, it was decided to mount 'Mother's weapons in sponsons on the sides. A turret-mounted weapon, high atop the superstructure, might not be able to fire into enemy trenches, whereas sponson-mounted guns could do so more easily. This was an inefficient layout, requiring more weapons and thus a larger crew to ensure all-round firepower, but it was an acceptable solution.

Fears that a tank armed with 6-pdr guns (using 57mm/2¼in shells) in its sponsons might be vulnerable to infantry attack from the flanks resulted in the creation of variant designs. 'Males' received a main armament of two 6-pdr guns which had originally been developed for naval use, while 'females' carried machine guns in their sponsons. All tanks carried additional machine guns for anti-personnel work, and some tanks were built as 'hermaphrodites' with a 6-pdr gun in one sponson and machine guns in the other.

'Mother' impressed observers during her trials, or perhaps by that time the British military was desperate enough to give anything a go, and 100 were ordered, with others following on. Some were completed as variants, such as command post/wireless vehicles, which had their armament removed to make room for command and control equipment.

GERMAN ARMOUR
Germany's A7V tank was a critically flawed design. Its huge size made it an easy target, and its ground clearance was too small to permit trench-crossing.

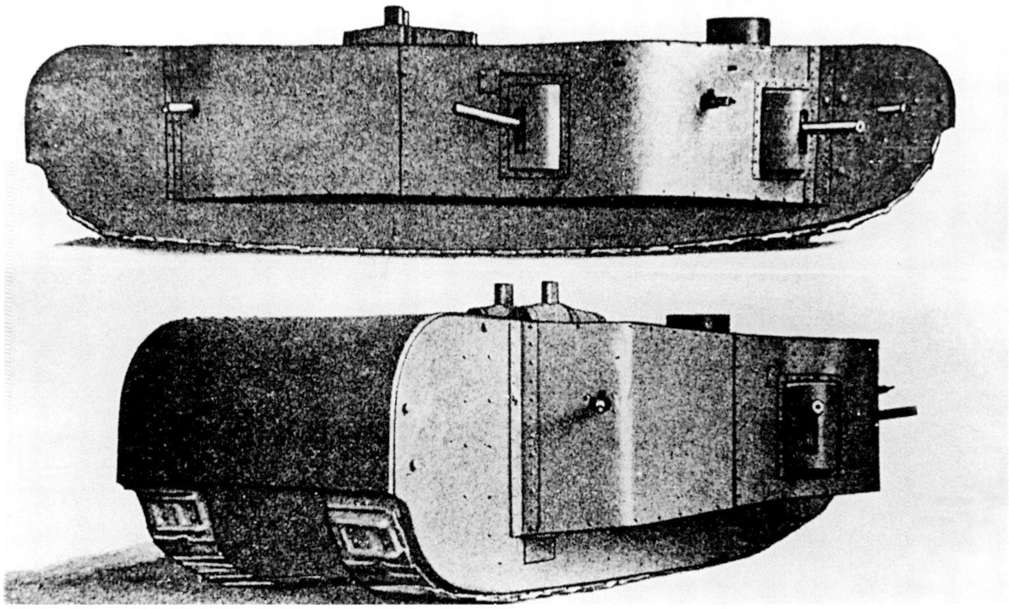

LAND COLOSSUS
The German K-Wagen was a truly gigantic vehicle intended to smash through the Allied trench lines. It never got past the prototype phase.

It was the Mk I tank that made the world's first tank assaults, and despite an inauspicious showing resulting largely from the boggy terrain over which the attack was made, enough was achieved to inspire further development. Among the changes that were made during the transition from the Mk I to the Mk IV, which became the main British tank of the war, was a move to a shorter (23 cal.) 6-pdr gun rather than the original 40-cal. version.

The longer gun had a tendency to ground in rough terrain; it was reintroduced as a front-mounted weapon in the Mk VIII 'International' tank, where its extra length helped protect the driver from the blast of its firing. Armour was also improved, along with the powerplant, but no suspension system was fitted.

Other projects

German tank design was restarted in 1916, but the emphasis was on anti-tank defence rather than the acquisition of an armoured warfare capability. The resulting vehicle, designated A7V, had all of the faults of the Mk I and few of its virtues. Requiring a huge crew of 18 or sometimes more personnel, the A7V was more of a mobile land fortress than an armoured assault vehicle, and had very poor cross-country performance.

Late in the war, German designers came up with another gigantic tank, this one somewhat reminiscent of British designs. The K-Wagen, as it was known, mounted no less than four 77mm (3in) guns and seven machine guns. It required a crew of 22 and would have been difficult to control had it got past the design stage. This colossal device foreshadowed the World War II-era German obsession with gargantuan armoured vehicles, and probably could not have been deployed in sufficient numbers to be effective.

More practical was the LKII design, which was essentially an armoured car fitted with tracks. Designed to overwhelm enemy forces by the use of numbers and intense firepower, it was lightly protected and mounted a 57mm (2¼in) gun or a pair of machine guns. Such a vehicle was unlikely to be able to make a breakthrough, but if one could be achieved it might have made an excellent exploitation vehicle.

French tank design was also beset by problems. The Renault FT-17 light tank was an effective combat vehicle (and remained in use after the war), but the Schneider and St Chamond designs which preceded it were badly flawed. Both suffered from poor cross-country performance, especially the St Chamond, whose body was much longer than its tracks. It also used an electric drive system powered by a petrol engine, which added weight to an already overloaded vehicle.

Internal politics within the French military procurement system resulted in both these similar designs being ordered. Each one demonstrated serious flaws when they first went into action: every one of the St Chamond tanks committed to their first attack ditched at the first trench they had to cross. The result was that work on these designs was limited to finding other uses for them, such as supply carriers, or at least reducing some of their shortcomings, and few remained in action for long.

EARLY ARMOURED FORCES

DEFENSIVE FIREPOWER
The cumbersome machine guns of the World War I era were not mobile enough to be effective in fluid combat. Once defensive positions were established, machine guns and wire made them all but impregnable.

The armoured forces of World War I were created as a counter to the defensive strength of barbed wire, machine guns and rifle-armed infantry firing from entrenchments. This combination changed the face of warfare by making it all but impossible to drive an enemy from his positions without incurring immense casualties.

Two key factors were at play here: firepower and time. In the wars of the eighteenth century and much of the nineteenth, infantry weapons were short-ranged, inaccurate and slow to load and fire. In 1815 an infantry soldier 200m (218 yards) from an enemy who was firing at him was in relatively little danger; massed volleys exchanged between solid formations were the only way to guarantee any degree of effect. Volume of fire was also low, unless a large body of men discharged their weapons all at once.

This meant that the 'threat distance' in front of an infantry formation was short, and could be crossed fairly quickly on foot or very rapidly on horseback. The number of – highly inaccurate – shots that could be fired during the approach was very limited, so it was possible to make an infantry bayonet assault or come to hand strokes with a cavalry sabre without taking enormous casualties.

The advent of percussion-cap rifle-muskets and, later, all-in-one cartridge firearms improved both the number of shots that could be fired and their accuracy, while the time required to cover a given distance remained the same. The effect of this increase in firepower was enhanced by the greater accurate range of infantry weapons, meaning that an attacker had more ground to cover and therefore gave his opponent more time to shoot at him. By the late nineteenth century it was not possible to make a cavalry charge at organized infantry without being shot to pieces. The presence of rapid-fire machine guns simply added to the volume of defensive fire that could be put out in a given time.

In the colonial wars of the late nineteenth century, well-equipped European troops routinely defeated large numbers of warriors armed with hand weapons or black powder firearms using traditional line and square formations in the open. Against an enemy equipped with bolt-action rifles and machine guns, this was suicide, as the early actions of World War I demonstrated. The obvious solution was to dig in, taking cover behind ramparts of earth which protected from bullets, artillery and grenades used to blast troops out of their defences.

Troops firing from an entrenchment made a difficult target at best. Massive artillery bombardments and even poison gas failed to dislodge them. The only effective way to engage an entrenched enemy seemed to be to get close and fire into the trench or to hurl explosives in. However, this required the attacking force to cross the ground in between, moving in the open while troops protected by hard cover fired at them. The situation for the attacker was worse than ever, since the creation of barbed-wire entanglements further increased the time that attacking troops were exposed to fire before they could make any impression on the enemy positions.

An artillery bombardment dropped on to enemy positions was an effective means of reducing the volume of fire the defenders put out, but it was difficult to coordinate and never totally effective. No matter how optimistic commanders might be that enemy wire would be cut and the troops killed or driven from their positions, it was never possible

ARMOURED TRENCH WARFARE
Early tanks were designed with trench warfare in mind. Their sponson-mounted guns could deliver fire into a trench from close by, preventing enemy troops from finding shelter by getting into the guns' 'dead zone'.

to simply 'walk across and take possession' of the enemy trench line, no matter how many guns were brought to bear, nor how big those guns were.

Assault support

If the volume of enemy fire could not be diminished and the ground in front of the trenches could not be crossed quickly enough to make the journey survivable, the only remaining answer was to somehow get weapons to a position where they could fire into the enemy trenches by protecting them from gunfire. This was the mission of the early tank. It needed to be able to carry a reasonable armament and to protect its weapons, systems and crew from enemy fire for as long as it took for the weapons to do their work.

Speed was not a major factor so long as the tank could withstand enemy fire for a sufficient time to do its job, and while tank designers would have been pleased with any increase in speed, the critical factor was the ability to cross shell-cratered ground and trenches. Thus early tank design was a balancing act between armour protection, firepower and all-round ground performance. Speed became more of a factor later, but since the early tanks were designed to lead or support an infantry assault it was not necessary for them to be able to move faster than an infantryman could advance.

For the destruction of enemy machine gun posts, an artillery armament was desirable. Guns did not need a long range nor sophisticated fire control. So long as the tank could rumble up to close range and blast the target over open sights, that would do. Guns would, however, need to traverse and elevate in order to engage their targets. Turret mountings were not considered effective for most early tank designs, partly due to their technical complexity, but mostly because lower-mounted guns could fire into a trench at close range more readily. Thus hull- or sponson-mounted guns became common.

All early tanks carried machine guns for anti-personnel work, but British designers considered

FIELD OF VISION
The field of vision for the driver of an early tank was very limited. The forward horns, carrying the tracks, further restricted what little could be seen. Many tanks drove into obstacles that might have seemed obvious to a nearby infantryman.

ARCS OF FIRE
Although the Schneider's heavier weapons could fire in several directions at once, the FT-17's single weapon offered all-round firepower, which was overall much more efficient. This pattern became the norm for all tanks after the end of the multi-turret phase.

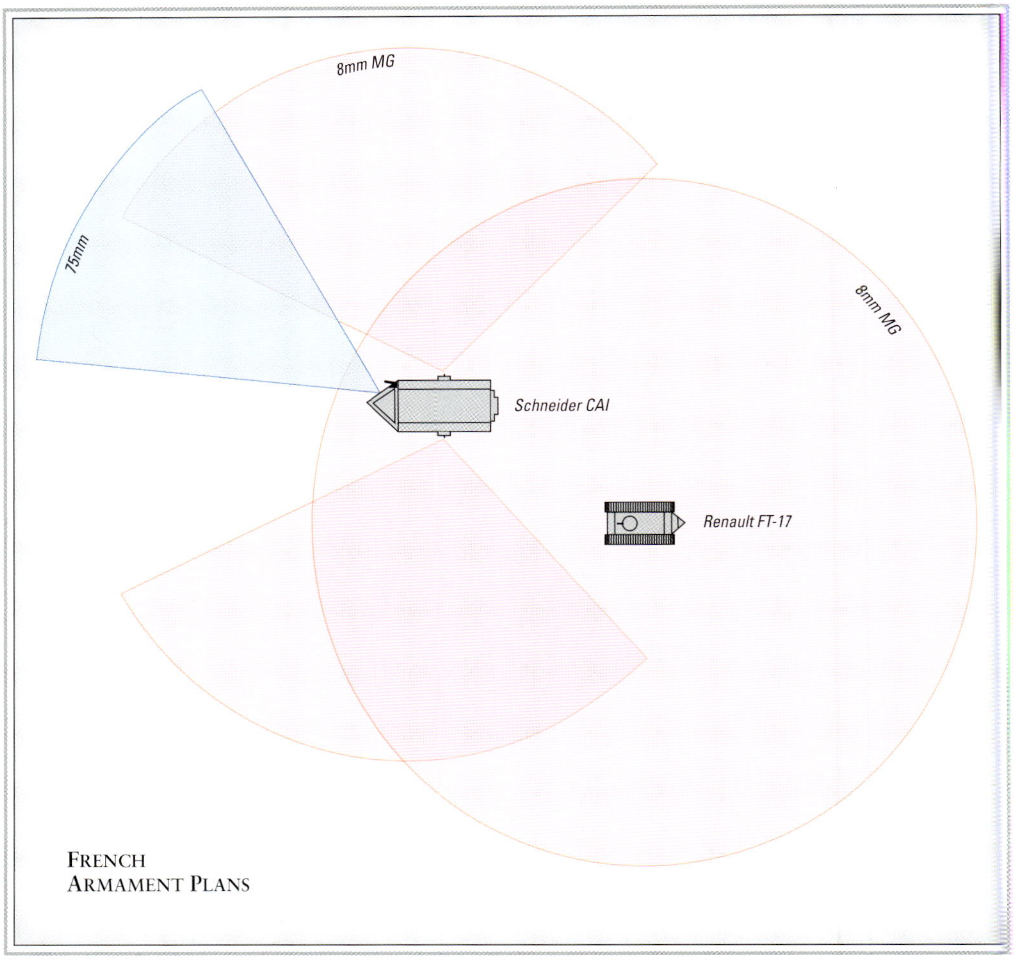

that a gun-armed tank might be vulnerable to infantry attack. The answer was to create a 'male' tank equipped with artillery as its main armament, and a 'female' version with machine guns in its sponsons. The smaller weapons had a much greater field of fire, allowing the 'females' to escort the 'males', while repelling any attempt by infantry to assault the tank or use grenades to disable it.

Early tank design

Early British tank tactics made use of this mutually-supporting partnership, and were tailored to making best use of the tanks' armament. The standard tactic was to advance to the enemy trench line and then drive along it, firing into the trench with sponson-mounted weapons from a vantage point above any troops seeking shelter in the bottom of the trench. In this, the tank fulfilled exactly the role it needed to – it brought weapons close to the enemy trenches where they could be effective.

French tank designers seem to have had a less clear idea of what they were trying to achieve. They did manage to create an armoured vehicle that could carry heavy weapons and protect them, but mobility was lacking in both the Schneider and St Chamond designs. These were very much 'first attempt' projects, so technical difficulties were only to be expected. However, given that the paramount requirement of the early tanks was the ability to reach the enemy's trench system across rough ground, poor cross-country performance was a major deficiency. The extremely limited arcs of fire available to the Schneider's armament were a serious drawback; the best that could be said about it was that it was better than the St Chamond.

Weighing less than a third as much as a St Chamond and requiring a crew of just two, the Renault FT-17 was a much better concept, though it, too, was flawed in some ways. Its turret-mounted armament allowed all arcs of fire to be covered by a single weapon, reducing the number of guns needed and thus the amount of space needed for crew and mountings. Since the

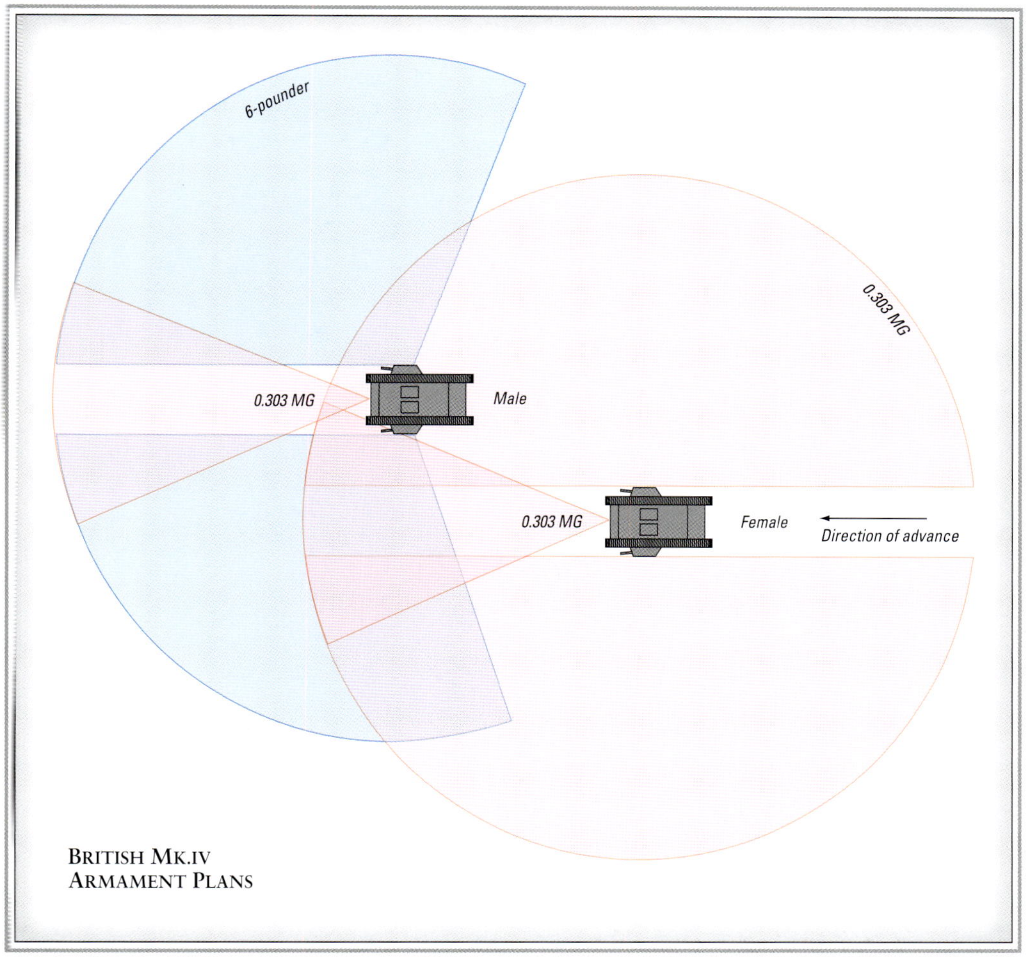

commander had to man the weapon in a very tight space, efficiency was not as high as it might have been.

In many ways the FT-17 was the closest any World War I tank came to a modern design, but it was still very much a product of its time. Designed for infantry support, it was slow and had a short operational range. It was sufficiently successful that it was used as the basis of early vehicles designed in the USA, Italy and the Soviet Union and served adequately in colonial conflicts during the interwar period.

To concentrate or to disperse?

Even after armoured vehicles had proven their worth in massed action, there was strong pressure to disperse them in 'penny packets' or even solo as defensive assets. In this role, tanks were to be dug in as bunkers, retaining the ability to leave their positions to make a local counterattack or to shore up a threatened sector. While this defensive deployment in what became known as 'savage rabbits' might have had some merit, it was not a war-winning strategy and was extremely inefficient. Tanks required significant maintenance even when standing idle, and of course spares and fuel had to be distributed.

By concentrating the tanks, logistical issues were eased and they could be used en masse. This was particularly important as early tanks were very

MUTUAL SUPPORT
British tanks were intended to work together, with the heavily armed 'males' protected by 'females' whose weapons offered greater all-round firepower. Infantry assaulting a given tank were also exposed to fire from other nearby vehicles, provided the tanks remained concentrated.

OBSTACLE CROSSING
Early British tanks could climb over large obstacles, but crewmembers were often injured when the vehicle came crashing back down. These vehicles had no suspension system; shock was transmitted directly to the occupants.

prone to breakdown. A tank that broke down in a successful attack could be salvaged; if the attack failed the enemy might be able to capture the vehicle. It was possible that a small tank unit might be entirely disabled by breakdowns and ditching, whereas a larger force would retain at least some combat capability. The 'savage rabbits' strategy would have rendered the tank force almost useless, and done away with any chance for it to prove its worth. Fortunately, it was decided to remain true to the military principle of concentration of force.

Expansion of the tank forces

The first tank units were small and could be supported on an ad-hoc basis, but as soon as tanks were fielded in numbers it became necessary to create a proper unit structure for them. A certain amount of trial and error was required to create a suitable formation, and there were other influences too. Some officers wanted the tanks dispersed under local command; some thought they were best organized as small units and integrated into infantry formations. Proponents of the tank, of course, wanted to create a specialist armoured formation with its own organization and command structure.

British tanks were originally assigned to the Heavy Section (later Heavy Branch) of the Machine Gun Corps, which was at least in part a measure to maintain secrecy about their nature. As the Heavy Section was expanded, it became known as the Tank Corps, an identity it retained from July 1917 until October 1923 when it became the Royal Tank Corps. In 1939 this name was changed to Royal Tank Regiment.

Early British tank force organization was based on the section of three tanks, often reduced to two by breakdowns. Sections were organized into companies, and companies into battalions which had their own workshops and support capabilities. This concept was not very different from the organization of the other combat arms, though tank forces needed more specialist personnel and supporting 'tail' than other combat arms.

Before and after an assault, organization was important in order to ensure that the tank force was properly supplied and maintained. Tanks required constant maintenance, making the availability of tools and spares a necessity. Perhaps just as vital was the sharing of experience; this was an entirely new weapon system and every action potentially revealed new knowledge. The internal combustion engine was in its infancy, too, so personnel with experience in its operation were by no means common.

LATE-WAR REFINEMENTS
Three of these tanks carry Trench-Crossing Cribs atop their hull. First appearing in 1918, Cribs were used for the same purpose as fascines, i.e. to support a tank as it crossed a wide trench, but weighed less than half as much.

The British Tank Corps drew its personnel from various sources, notably the Motor Branch of the Machine Gun Corps, which operated armoured cars. Others came from the Army Service Corps. Many personnel were never formally transferred from whichever units they came from, but the Tank Corps itself gradually adopted an identity of its own and working practices suitable for its needs.

Expansion of the Tank Corps was rapid, requiring that companies be turned quickly into battalions. This inevitably meant that many personnel did not have time to properly learn the requirements of one level of command before being advanced to the next. The Tank Corps had to learn its trade 'on the job', and an efficient spread of information contributed to its effectiveness. This was especially true in the US tank force, where a private soldier came up with the idea of converting some tanks into mobile workshops to support the others. In a more established force, an enlisted man might not have been able to put forward his ideas and the concept might not have been implemented.

Among the techniques that had to be developed was the control and handling of tank forces and their sub-units. Tanks within a section were intended to support one another and each section was assigned a section of the enemy line to attack. In practice, with limited visibility from within a tank, organization often broke down once action began, with tank commanders acting on their own initiative.

Attempts were made to improve co-operation and communications, ranging from placing carrier pigeons aboard the tanks, to the deployment of supporting officers who were to run between tanks and convey messages. This was an enormously dangerous undertaking at best, and even if a tank crew received new orders they might be unable to comply. Able to see little through their vision slits, tank crews were sometimes in a similar position to medieval knights, whose helmets similarly limited their perception of the battlefield.

Thus in practice, once an attack began many tanks followed the general direction of advance and attacked whatever targets presented themselves. This was often sufficient to inflict heavy damage on the enemy's defences. Early tank forces were to some extent a battering ram rather than a rapier. A great mass of tanks was pointed in the general direction of the enemy and then released. While inefficient, such a heavily armoured mass of firepower was still difficult to stop. This approach sufficed for the remainder of World War I, and in the interwar period there was considerable debate about how best to organize and handle the armoured forces.

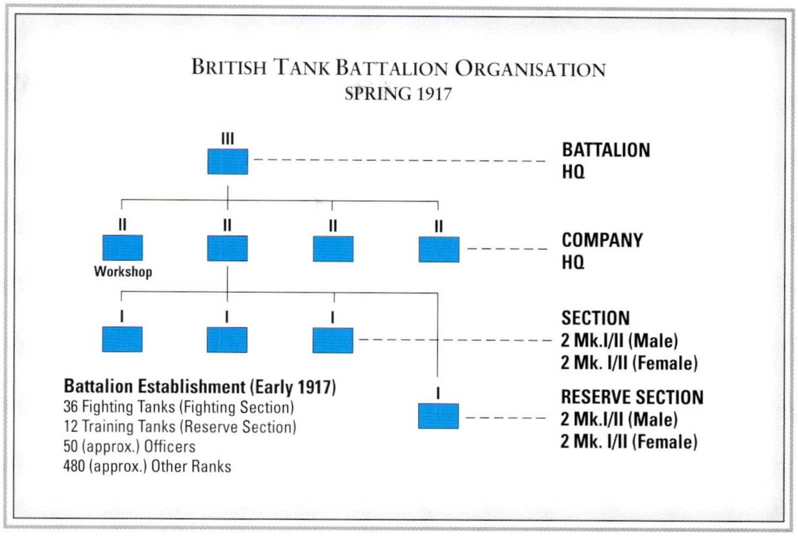

EARLY BRITISH TANK BATTALION
Early tank force organisation had more in common with the structure of an artillery regiment than an infantry battalion. Losses and breakdowns meant that few units ever matched their paper strength.

STRATEGIC MOBILITY
Tanks were moved as close to the battle area as possible by rail. Driving any distance, even over good roads, resulted in losses due to breakdowns and increased the maintenance burden on the unit.

THE BATTLE FRONTS 1914–16

At the outbreak of World War I, there was an expectation among planners that it would be similar to previous conflicts in Europe in recent years, such as the Franco-Prussian War of 1870. Although modern rifles had increased the lethal range of infantry, it was still reasonable to expect that a traditional advance and assault would be successful, and that cavalry could play a useful part on the battlefield.

The machine gun, in particular, had not really shown its potential in large-scale warfare, and so planners had not truly perceived how massively the advantage had swung in the direction of the defender. Nor was this instantly apparent in the first few weeks of the war. Traditional tactics resulted in heavier casualties than previously, but they still worked. However, once the situation began to stabilize, stalemate inevitably set in.

Advances on the Western Front

The initial clashes on the Western Front took a traditional form. Cavalry conducted reconnaissance and attempted to flank enemy formations, and artillery was deployed wheel-to-wheel in the Napoleonic style for the last time. Infantry took cover where they could, but the emphasis was on offensive action, even to the point of the bayonet.

The German plan called for a rapid advance in an arc sweeping across to Paris and the Channel ports. Although the distance to be covered lay at the upper edge of capability for the units on the outer edge of the sweep, it seemed achievable, even in the face of opposition. This theory was apparently borne out as the British and French were driven into a disorganized retreat after the first clashes.

The retreat was characterized by rearguard actions, brief stands made to buy time, and the occasional full-scale counterattack. Unable to form a solid defensive line due to incessant German pressure, those British and French units that made a stand were usually forced into further retreat by defeats elsewhere.

The stalemate begins

Armoured forces played no real part in these events. Various armoured cars did exist at the time, but

PEUGEOT AUTOBLINDE
Built on a civilian chassis, the Peugeot armoured car was crewed by five men and armed with a 37mm (1½in) gun. It could operate only on roads or very firm, flat terrain.

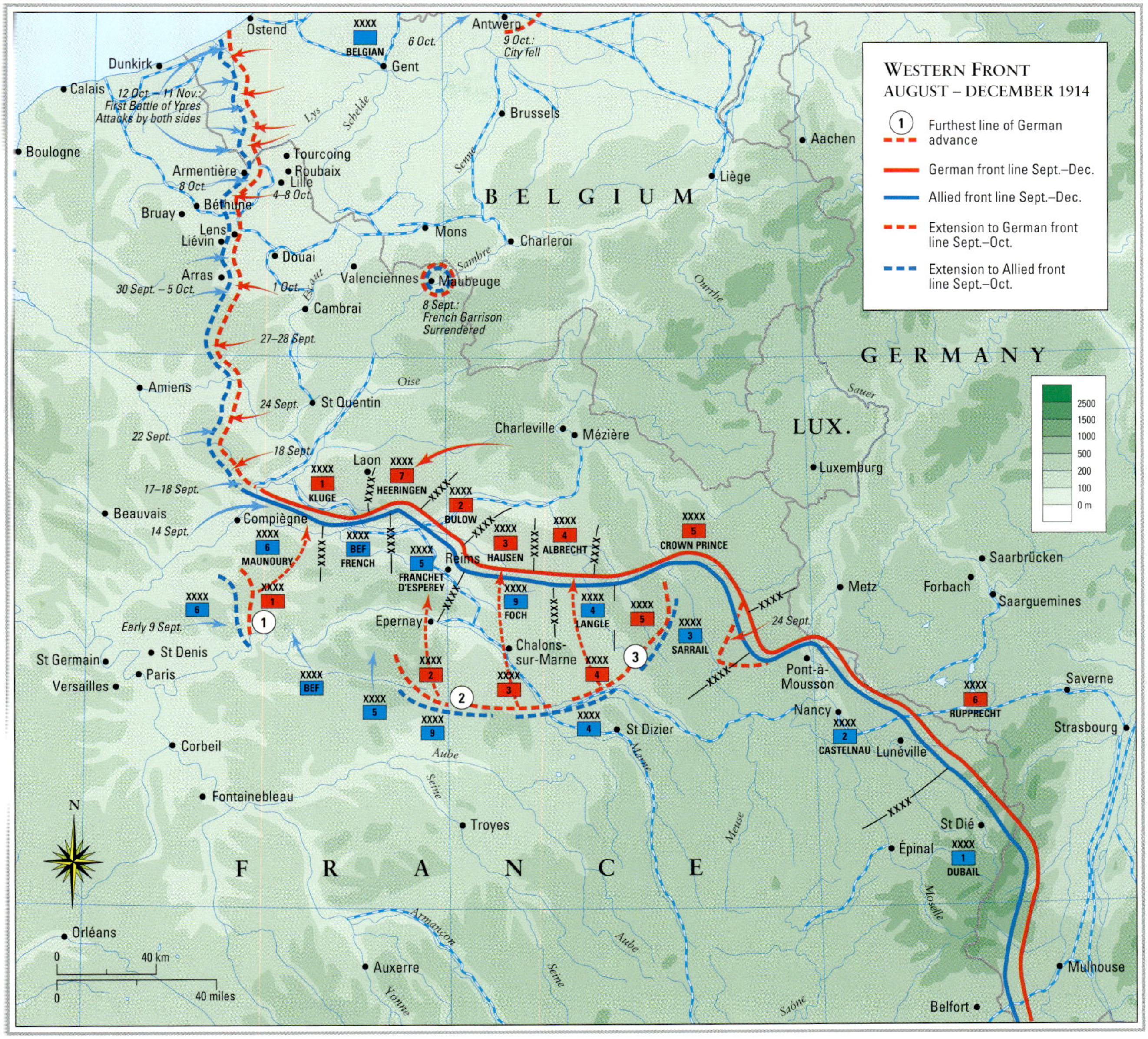

these were primarily suited to security operations along good roads and had extremely poor off-road capability. This was largely due to their primitive nature. Where today the term 'armoured car' refers to a custom-designed light armoured vehicle, the armoured cars of 1914 were exactly that – a touring car or light truck to which weapons and armour plate had been added. Some were so overloaded and underpowered that they could break their own axles on a bumpy road. A few of these vehicles were involved in the early fighting, but not in sufficient numbers to make any real impression.

Thus the early fighting on the Western Front was traditional in nature, and relied on muscle power for mobility. Infantry marched carrying their combat equipment, while cavalry and artillery relied on horses. Inevitably, muscle power gave out. Exhausted men and horses moved more slowly, and at the same time resistance firmed up.

The German plan came close to succeeding. The Anglo-French forces were in real danger of being enveloped or pushed apart by their different needs. The British needed to protect the Channel ports, which were their line of supply and retreat. The French, on the other hand, were primarily concerned with the threat to their capital.

From mobility to trench warfare

The retreat was necessary in order to prevent encirclement and destruction, but eventually a stand

THE WESTERN FRONT, 1914
After the initial, mobile, phase the war settled down into more or less static trench lines. Early attempts to restore mobility were made with massed artillery and infantry attacks, with little effect.

LANCHESTER ARMOURED CAR
Originally developed for the Royal Naval Air Service, the Lanchester armoured car served well in the Eastern theatres of war and achieved good export sales.

RENAULT 1915
The Renault 1915 armoured car used a civilian chassis, which was modified to carry the weight of armour and weapons. Armament was usually a machine gun but a 37mm (1½in) gun could be mounted.

had to be made. The German offensive towards Paris was halted at the Battle of the Marne, and a successful counterattack pushed the German forces away from the French capital. Finally, the British and French were able to throw together defensive positions that resisted attack at enough points to form the basis of a defensive line. Troops dug in wherever they could, and attempts to outflank these positions were countered by an ever-lengthening line of field fortifications.

As it became apparent that a breakthrough was not imminent, German forces also dug in, and so began a stalemate that lasted for most of the war. Both sides maintained forces of cavalry and armoured cars close to the front in case an opportunity arose to use them, but none ever came. Instead, ever-increasing amounts of artillery were deployed to pound the enemy positions, while offensives were carried out using brute-force methods involving vast numbers of infantry.

Although some small gains were made by these methods, any breakthrough was quickly sealed by reserves and the trench lines stabilized. It was apparent that what was needed was a means to not only get across no-man's land and into the enemy positions, but to be able to carry on an advance afterwards, returning to a fluid war where decisive results were possible. Weapons such as poison gas and flamethrowers offered some possibilities, but ultimately failed to deliver.

While the Western Front remained essentially static, engineers worked on a new concept. This mated the idea of the armoured car – a gun-carrying vehicle which could protect its crew from enemy fire – with the tracked agricultural tractor which could reliably carry a heavy weight across rough ground. Not until such vehicles appeared would the trench deadlock be broken.

The Eastern Front

The situation was somewhat different on the Eastern Front. There, the war never became static as it did in the West. Defensive positions were set up and trenches dug, but offensive and counteroffensive moved the front lines on a frequent basis. Cavalry operated in the traditional manner, at least some of the time, and there were engagements between cavalry brigades fighting with sabre and revolver.

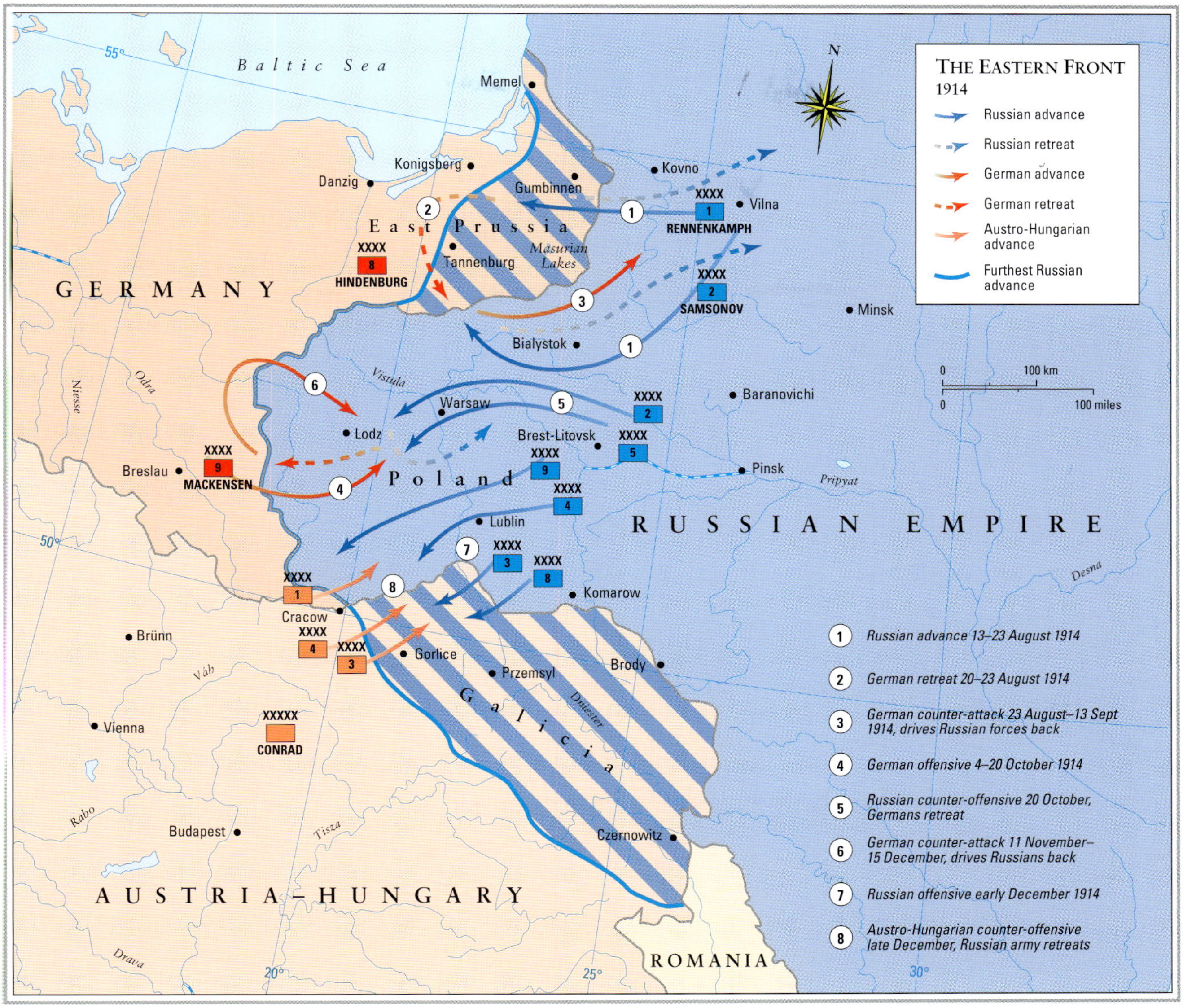

THE EASTERN FRONT, 1914
Russian forces mobilised faster than expected and pushed into Galicia and East Prussia, bringing about fluid combat in which armoured cars and cavalry had an important part to play.

The intended German strategy was to stand on the defensive in the East, launching an offensive to put France out of the fight, then moving East to concentrate on the Russians. Only one of Germany's eight armies was initially positioned in the East, a disposition influenced by the belief that it would take 40 days from the outbreak of war for Russia to ready herself for an offensive.

In the event, Russian forces were ready much sooner than expected and the French were not taken out of the war. However, the Russian advance in the East was hampered by both difficult terrain and very cautious leadership. Despite early setbacks, German forces were able to make a successful counterattack and halt the advance.

Much of the fighting on the Eastern front took place in Poland and Galicia, with the Russians engaged by both German and Austro-Hungarian forces. Dramatic reversals of fortune on both sides necessitated rapid retreats and offensives elsewhere to take some of the pressure off the endangered area. Thus rather than settle down into static trench warfare, the Eastern Front remained relatively fluid throughout the war.

Armoured car operations in the East

In the East, armoured car forces played a significant role. Although limited to roads, they could make a rapid advance to strike or seize an objective, and brought heavy firepower to the field. Rather than functioning in the manner of a modern armoured force, armoured cars often operated as a sort of mobile gun emplacement, driving to a designated

ROLLS-ROYCE ARMOURED CAR
Rolls-Royce armoured cars saw extensive service recovering downed pilots and in security applications. An experimental version was created with a wireless set, acting as a mobile communications post.

firing position, then stopping to fight.

Many of the vehicles used in this manner were converted touring cars mounting a single machine gun in a lightly armoured turret or behind a gun shield. Some carried heavier weapons. These included light commercial trucks which could carry more armour and additional weapons. Multiple machine guns were not uncommon aboard these vehicles, and some carried heavier weapons such as light naval guns.

As with other designs, these weapons carriers were overloaded and totally unable to make a decisive breakthrough and cross-country exploitation into the enemy rear. However, they could move their weapons and ammunition supply quickly to a new firing point to support an attack or block an enemy breakthrough. The fact that these vehicles were resistant to small-arms fire gave them a useful psychological effect on the enemy, causing troops to sometimes fall back in the face of a tiny armoured car force.

Handled with dash and audacity, armoured cars were more effective on the Eastern Front than their numbers might suggest. When the lines were stable, they were useful mainly as mobile reserves, using roads behind the lines to take up positions in case an enemy attack succeeded. On the offense, they moved fast and appeared unexpectedly, sowing alarm and distress among enemy personnel in a manner that foreshadowed future armoured operations.

Palestine

Turkey's entry into the war on the side of the Central Powers resulted in a British campaign in Palestine and the Middle East. An amphibious operation to force passage of the Dardanelles resulted in failure, but success was to be had elsewhere. An advance up the River Euphrates eventually resulted in the capture of Baghdad, and in the meantime

MINERVA ARMOURED CAR
Aggressively-handled Minerva armoured cars played an important part in the defence of Belgium at the beginning of the World War I.

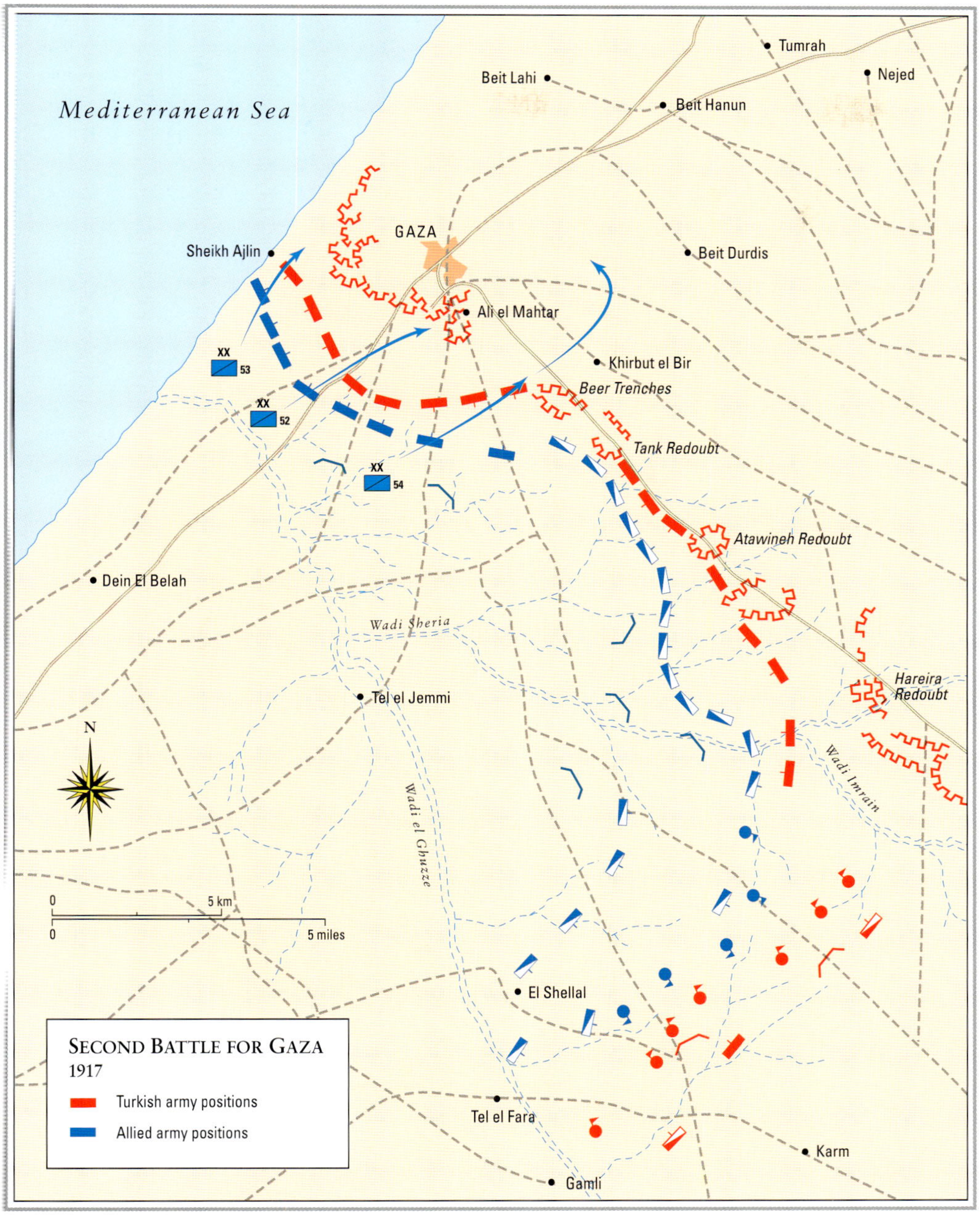

BATTLE OF GAZA, 1917
British tanks were deployed to the Middle East in time for the Second Battle of Gaza. Eight Mk I tanks took part, though these were deployed piecemeal and made little impression on the strong Turkish defences.

British forces defeated a Turkish attempt to capture the Suez Canal and then began advancing north through Palestine. Working in conjunction with local Arab leaders who wanted independence from the Ottoman Empire, forces from Britain and her Empire gradually pushed the Turks northward out of the region.

Operations in Palestine were extremely different to those on the Western Front. Gunboats on the rivers, aircraft, cavalry and armoured car forces were all used alongside the traditional infantry and artillery. The campaign was a long one, and was not without setbacks. The Turks, often with German advisors and support, were highly effective in combat, but were gradually pushed out of one defensive position after another.

The ability of armoured car forces and cavalry, assisted by aircraft, to make long-distance raids as well as to conduct reconnaissance was well proven during the Palestine campaign. Light vehicles carried the firepower of an infantry company and could bring it to bear in places that infantry could not have reached in time to make a difference. However, it was the infantry that captured key targets such as Baghdad, Gaza and Jerusalem, and ensured the Turks were forced to the negotiating table.

THE SOMME JULY 1916

TANKS ON THE SOMME
The stalled Somme offensive prompted the deployment of the few British tanks then available. Their psychological effect was greater than their physical achievements.

To a great extent, the movement of reserve forces dictated the course of World War I. At a tactical level, any breach in the line could be plugged by reserves faster than the attacker could bring up his own reinforcements to consolidate the gains or continue the advance. At the strategic level, this made decisive breakthroughs impossible as the advantage rested so heavily with the defender.

Thus, the German high command set about drawing in and destroying the French reserves by an attack on the fortress of Verdun. The loss of the city would be a terrible blow to French morale and might even force a surrender. Even if the fortress could not be taken, the French army would be drawn into a 'meat grinder' which would eliminate its reserves of manpower and its will to fight.

With Verdun under constant attack, it was necessary for the Allies to take the pressure off in any way possible. This meant launching an offensive to draw in German reserves. Troops committed to preventing an Allied breakthrough could obviously not be employed at Verdun, which would prevent disaster even if the city held out.

GRENADE PROTECTION
Angled wire mesh screens made it unlikely that a grenade could be thrown on to the top of the tank. Even without this protection, grenades were prone to simply roll off a moving tank.

Offensive on the Somme

The ground chosen for a British and French offensive was on the Somme, a region that had seen little action since the beginning of the war. The wisdom of this choice is questionable, as it meant attacking an enemy firmly dug in on high ground, who had benefited from many months of inactivity during which deep bunkers had been constructed.

The Allies had constructed defensive positions of their own, of course, but there was no comprehensively developed infrastructure for the transport of munitions and supplies on a scale necessary for a major offensive. This had to be put in place, and the requisite concentrations of artillery and manpower assembled, before the attack could be launched.

The plan for the Somme offensive was simple in concept: to force the enemy positions by launching successive waves of infantry at them. Survivors of a failed assault would be gathered up by those following on, and would join the next attack until victory was ultimately achieved. This brute force approach was applied to the artillery component as well; a week-long bombardment was made in the hope of demolishing barbed-wire defences and enemy positions, as well as wearing down the defenders.

In the event, the bombardment did little more than move the wire around, and the German dugouts proved largely impervious to shellfire. As the assaulting divisions rolled forward on 1 July 1916, they came under fire from largely intact enemy positions, and suffered immense casualties for little gain. Co-ordination between infantry and artillery was poor, exposing the infantry to the full weight of enemy firepower. There were over 57,000 British casualties within the first 24 hours of battle.

Despite the disastrous first day of the offensive,

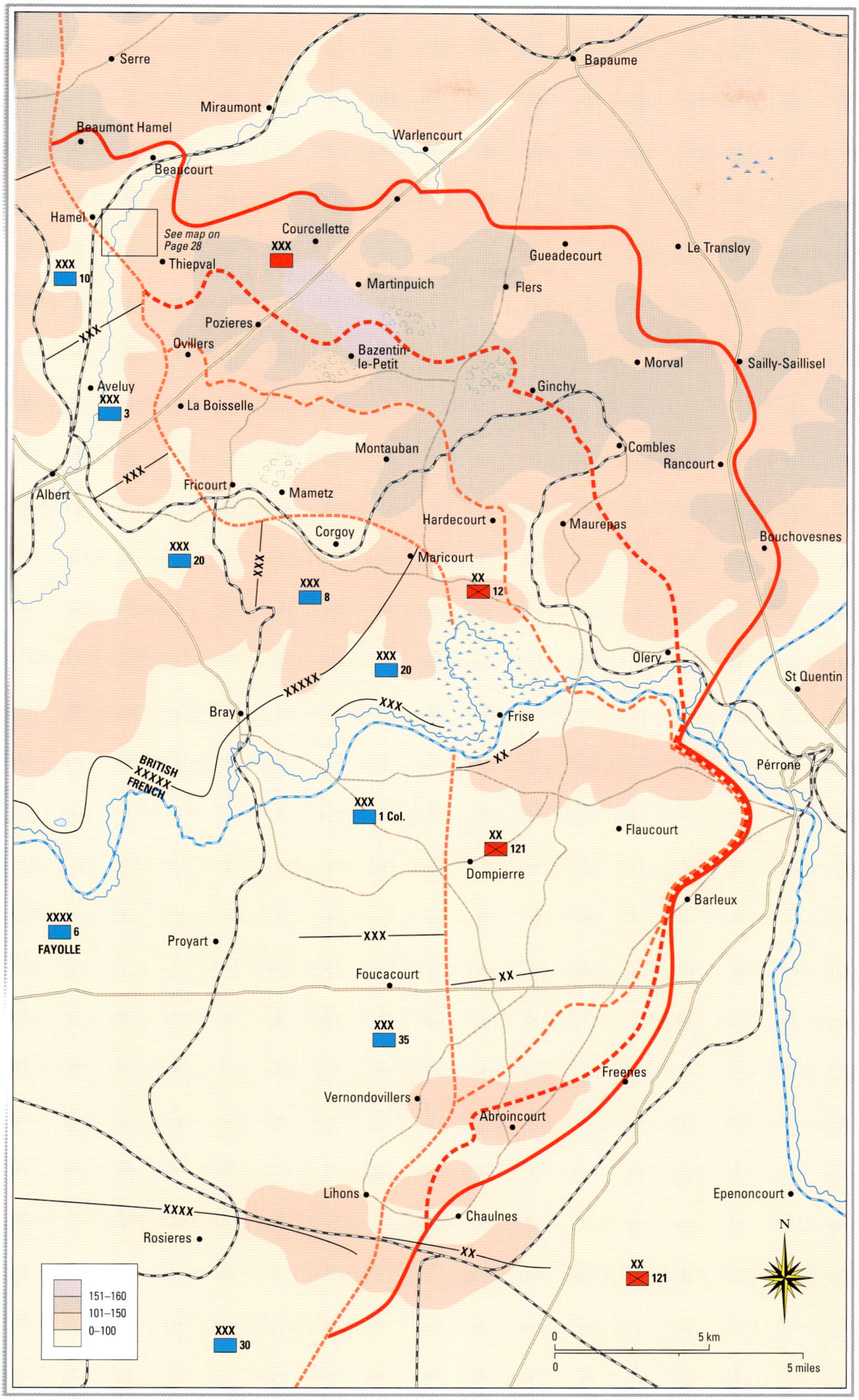

THE 1916 SOMME OFFENSIVE
Once committed to the Somme offensive, the Allies were forced to maintain the pressure, even at a murderous cost in casualties. Ground was gained, but no significant breakthrough was achieved.

OPERATIONAL LOSSES
More tanks were lost, like this one at the Somme, to mechanical failures or irremediable ditching than to enemy action. It was necessary to field large numbers just to ensure that some tanks reached the enemy lines.

the Allies needed to maintain the pressure. Switching to a different axis of attack would require months of logistical buildup, during which Verdun would surely fall. With no choice but to repeat the effort, the Allies fed in fresh divisions and attacked again and again over the same ground.

A change of tactics

The Somme became a battle of attrition, drawing in the reserves of both sides. In this, it succeeded in taking some of the pressure off Verdun, but only at immense cost in casualties. Slight variations in tactics contributed to minor gains made by the Allies, but it was clear that their armies would be exhausted long before any decisive victory could be achieved.

The Allies decided to deploy their new weapon, tanks, in the hope of breaking the Somme deadlock. These Mk I vehicles were extremely primitive, lacking silencers for the exhaust, and using trailing wheels to assist with steering. They broke down frequently and were so slow that merely getting them into position to begin the assault was a difficult and lengthy process.

The 49 available tanks were assigned to the Heavy Section of the Machine Gun Corps, as Britain's tank force was then known. This represented every tank available to the Allies at that time, a force reduced by breakdowns so that only 22 reached their start positions in time. Of these, 15 managed to cross

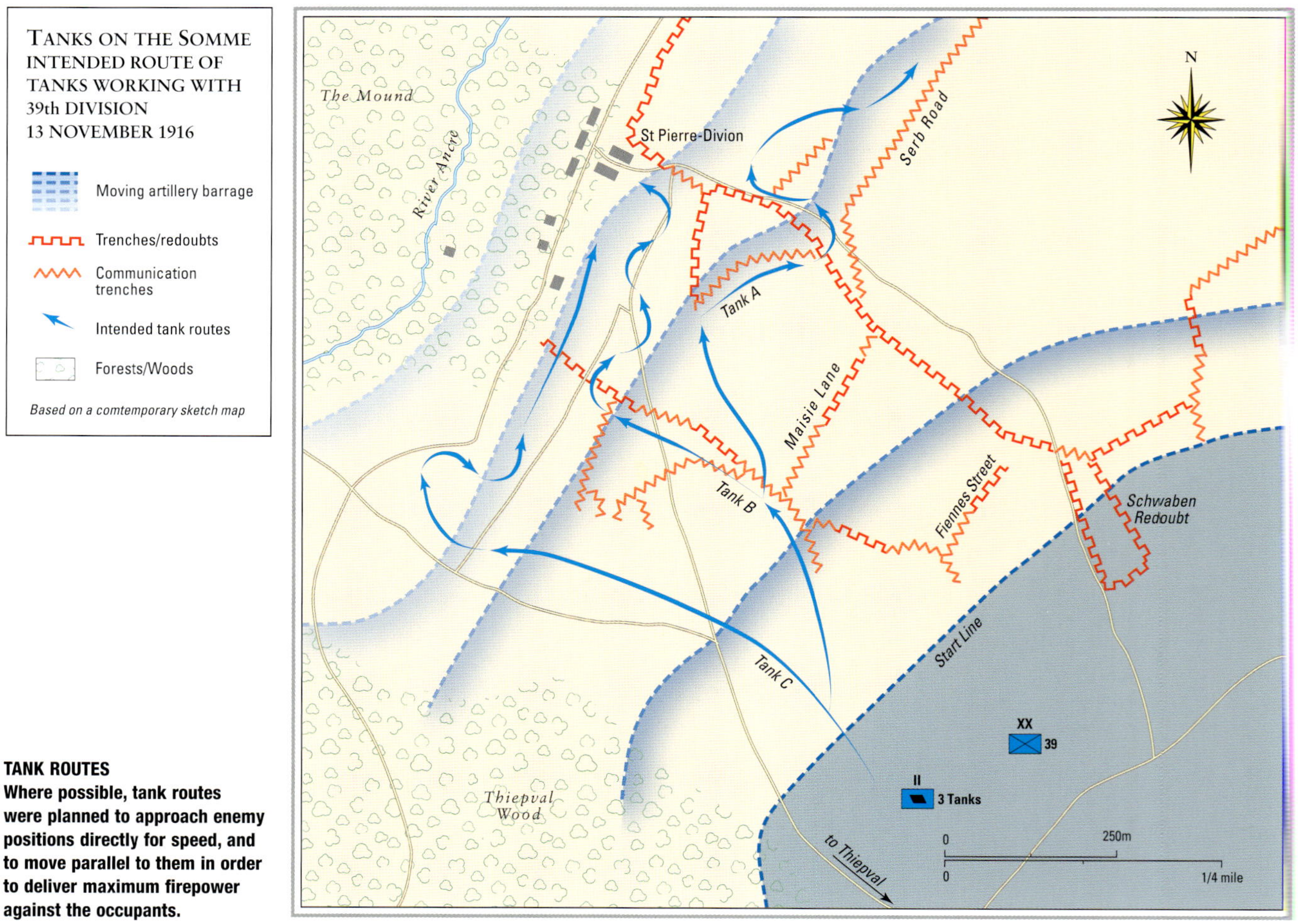

TANK ROUTES
Where possible, tank routes were planned to approach enemy positions directly for speed, and to move parallel to them in order to deliver maximum firepower against the occupants.

their start line on the day of the assault, with some others following on when they could.

The first tank attack

The attack was launched at Fleurs-Courcelette on 15 September 1916. Despite their slow progress and tendency to break down, the tanks caused panic among the enemy infantry. Troops abandoned their positions in what came to be known as 'Tank Terror', even though the Mk I tank could be penetrated by machine gun fire.

In some ways the operation was over-optimistic; many planners had been convinced by demonstrations that the tank was an unstoppable monster which could demolish the enemy lines all on its own. However, some thought had gone into tank operations. The preparatory bombardment had left lanes where there was no shelling, reducing the number of fresh craters the tanks would have to climb in and out of.

Some of the attacking tanks became ditched and one was disabled by an artillery shell, making it the first tank to be knocked out by enemy action. Nine got into the enemy positions, with a similar number coming up later and helping to reduce pockets of resistance. To the enemy, the tanks did indeed seem to be unstoppable, and the effect on Allied morale of seeing them shrug off machine gun fire was considerable.

However, small arms and machine guns could damage the tanks. Some were penetrated by machine gun fire and suffered crew casualties which were not apparent to anyone outside the vehicle. Although 'male' and 'female' versions of the Mk I tank were nominally the same, other than their armament, reports of armour penetration were more common from the crews of 'female' tanks.

Even non-penetrating hits could endanger a tank. Bullets striking the drivers' vision prisms shattered them, and rounds striking the armour could cause hot fragments to fly off inside the cabin and injure the crew. In an attempt to remedy this, 'splash' masks of chainmail were issued to tank crews. Few wore them as they were heavy and uncomfortable.

Despite crew casualties, ditchings and breakdowns, the first-ever armoured assault was a local success. Gains of around 2km (1 mile) were made in three days of fighting, during which the armoured force was involved in beating off counterattacks as well as making assaults of its own. Losses and breakdowns meant that in most cases tanks operated alone, or a handful of vehicles supported an infantry attack. Further successes were prevented by deteriorating weather and the redeployment of German reserves.

MK I TANK
The rear steering wheels of the Mk I tank were of little use in practice and were not used on later models, which steered adequately by use of the tracks alone.

Lasting results

The Fleurs-Courcelette attack achieved respectable gains, especially compared to other attacks in the Somme area in 1916, but it did not change the strategic picture at all. Tanks were used in 'penny packets' over the next few months, generally in support of local infantry actions. No decisive results were possible with such small forces, but the vehicles proved their worth as close support for the infantry. As a result, an order was placed for hundreds more tanks, and in October it was announced that the Heavy Section of the Machine Gun Corps was to be reorganized and expanded.

German commanders drew different conclusions from the battle at Fleurs-Courcelette. They recognized that the tank was a threat, but decided that it was not as dangerous as it had first seemed. Although it seemed to offer much it was problematical in use and prone to taking itself out of action due to breakdowns. Thus although Germany did acquire some tanks by repairing captured examples, its own tank-building programme was half-hearted and produced an ineffective design. Far greater emphasis was placed upon preventing 'tank terror' and creating an anti-tank capability for the infantry and artillery forces.

THE NIVELLE OFFENSIVE APRIL 1917

FRENCH INFANTRY, 1917
By 1917, the French Army was exhausted. Nivelle's grand plan raised hopes of victory which were dashed, leading to the mutiny of 1917.

General Robert Nivelle, appointed as commander of French forces on the Western Front in December 1916, was a graduate of the cavalry school, but had chosen instead a career with the artillery. He was, however, a strong advocate of offensive action. At the Battle of the Marne, Nivelle advanced his artillery through the wavering infantry forces and engaged the advancing Germans over open sights, and his innovative tactics at Verdun in October 1916 brought the French Army its first significant victory in many months. Nivelle thus had good reason to suppose that his proposed offensive would succeed, and managed to convince the Allied political leadership to authorise it.

Nivelle's plan

Nivelle's' plan of attack was a classic pincer movement, directed against a large salient in the German line created by the earlier Somme offensive. An initial blow would be made by the British against the northern shoulder of the salient, pulling in German reserves. French troops would then assault the southern shoulder and break through, leapfrogging fresh units over tired ones, who would consolidate, rest, then move up.

Nivelle repeatedly emphasized the need for aggression and rapid forward movement to overwhelm the defenders, who were known to be weak after the massive losses of the Somme campaign. Infantry were trained in rapid assault techniques of the sort used at Verdun, and formations were supplemented with light field artillery to help them deal with enemy strongpoints. Meanwhile, tank units practised co-operation with the artillery, with the central concept being 'keep moving'.

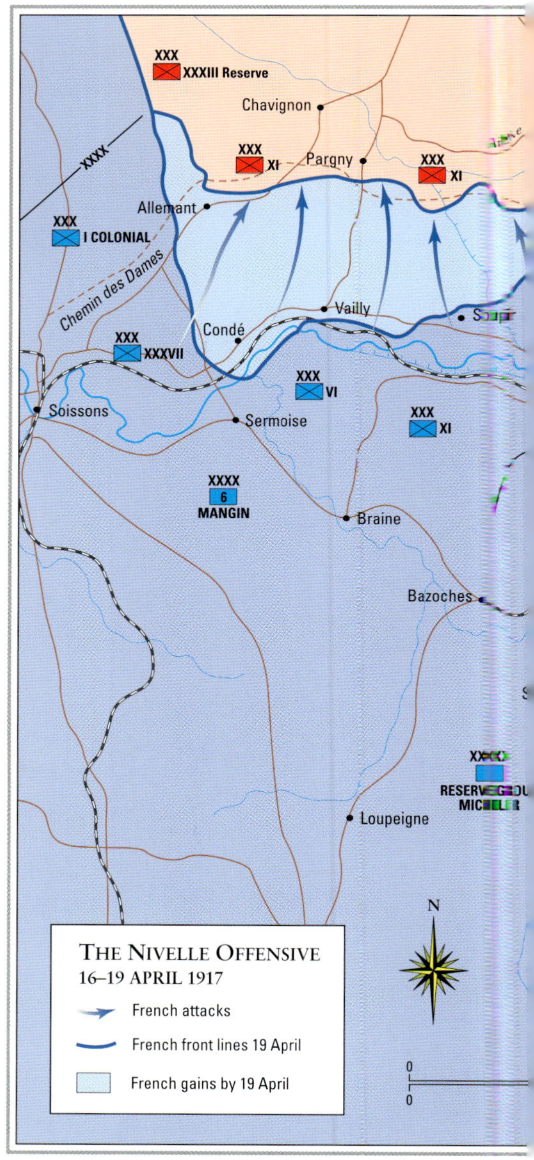

Knowing that morale was low in the French Army, and that many soldiers had completely lost sight of the bigger picture – everything beyond their dugout, in some cases – Nivelle decided to 'sell' his plan to the troops by a campaign of internal propaganda which, unwisely, included documents containing the entire operational plan. Copies of these were captured by German troops, and soon their commanders knew in detail what Nivelle planned to do.

German responses

Realizing that they could not defend the salient, German commanders decided to pull back their troops and establish a shorter, straighter defensive line to the rear. This increased troop density even before any reinforcements were brought in. After sabotaging anything that might be of use to the Allies, German forces retired to the Hindenburg Line, beginning from 9 February. This move created the strongest defensive position of the entire war, taking advantage of high ground both for defence and to shield the rear area from artillery observation.

By the middle of March, the salient had been quietly abandoned, making Nivelle's plan to bite it off obsolete. French troops moved forward to take possession once it became apparent that the Germans had gone, and were opposed only by booby-traps and the hazards of a wrecked countryside. The 'capture' of the salient was hailed as a victory, and Nivelle shifted the aim of his planned offensive to the new German line. In doing

THE NIVELLE OFFENSIVE
Nivelle's double-envelopment plan might have been workable had the German Army not pulled back from the targeted salient and strengthened the defences along a shorter line.

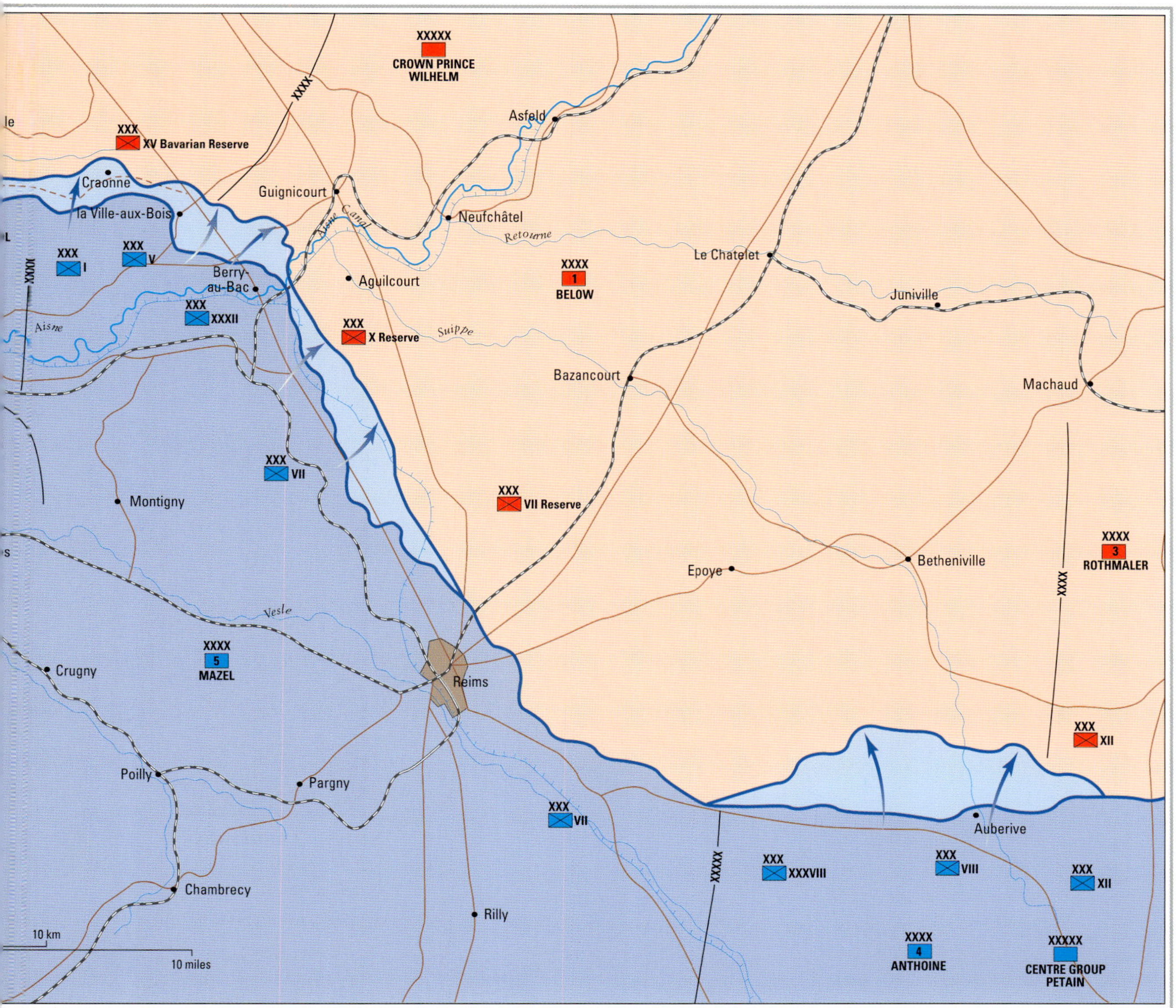

so he disregarded the fact that the strategic picture had changed entirely, and not just in the local area.

The collapse of the Tsar's regime in Russia and subsequent peace overtures meant that dozens of German divisions were freed from the Eastern Front to reinforce those fighting in the West. On the other hand, it seemed likely that the U.S.A. would soon enter the war and provide enormous manpower to the Allied cause. Delay or a new plan seemed to be in order, but Nivelle decided to press ahead with only minor changes to his original offensive.

The offensive opens

The attack was preceded by a heavy bombardment, and at dawn on 16 April 1917 the first units moved forward. Despite a cold night waiting in forward positions, the infantry were confident at first, and 128 tanks were assigned to the offensive. These were Schneider designs, equipped with a 75mm (3in) gun and machine guns, and theoretically well protected against small arms fire. However, their fuel tanks proved vulnerable and, more seriously, they were prone to becoming stuck in rough ground.

After the initial shock of encountering British tanks in 1916, German planners had evolved ways of dealing with an armoured assault. These included 'tank traps', which were simply deep and wide ditches, as well as artillery sited to fire over open sights at the slow-moving vehicles. Infantry were equipped with large-calibre anti-tank rifles which could penetrate a tank on a solid hit, and they also had some light guns pressed into the anti-tank role.

Infantry defences, too, were good. The defenders had massed hundreds of machine guns and benefited from deep entrenchments. As the French infantry moved forward, their progress was illuminated by flares. German artillery pounded both tanks and infantry, and the French creeping barrage failed to protect them. The barrage, using a precalculated timetable based on optimistic calculations, moved forward faster than the assault force, passing over defenders who could then fight unimpeded by incoming shellfire. An attempt to revise the bombardment timetable resulted in the assault troops, stalled between their own trenches and those of the enemy, coming under artillery fire from both sides.

Nor were the tanks much help. The majority became bogged down or ditched well short of the enemy's forward positions, where they were brought under direct fire from light and medium guns, as well as bombardment by heavier artillery. Most were

SCHNEIDER TANK
An inadequate design committed to action for lack of anything better, the Schneider tank performed poorly in its debut action and was never built in large numbers.

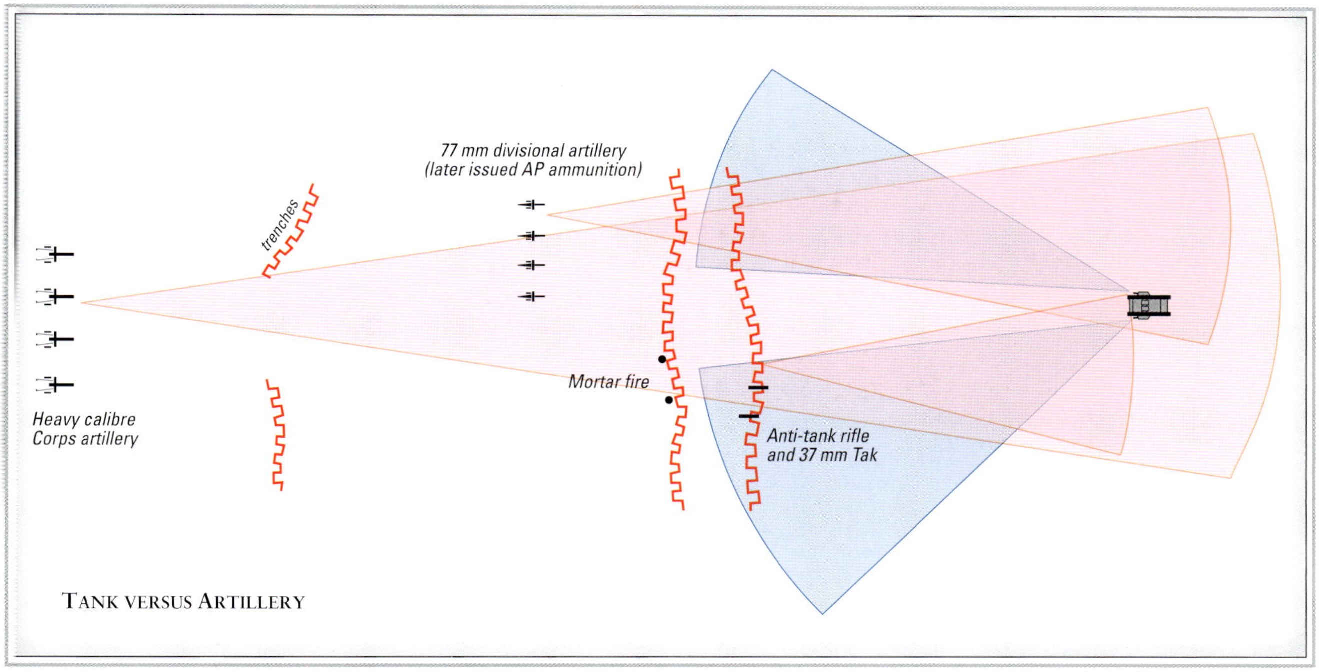

soon disabled. By mid-afternoon the assault had completely stalled, and a German counterattack began. Some units broke and fled, while others fought stubbornly to prevent a general collapse. At the end of the first day, the French Army had suffered 90,000 casualties rather than the projected 10,000, and had advanced about 550m (600 yards) rather than the expected 10km (6 miles).

The offensive continues

Despite this disastrous opening the offensive continued, reaching its high point the next day. This was more due to a German redeployment than a break-in to their positions. The campaign then became more or less static, with virtually no gains made despite intense fighting. Far from a victory achieved by innovative new tactics, this was slaughter in the old style, and by 29 April the campaign was called off. By then the French tank force had been virtually wiped out and the infantry had suffered over 134,000 casualties. Some small gains had been made, and significant casualties had been inflicted on the German Army, but the consequences for the French were far-reaching.

Insubordination and outright mutiny broke out in about half of French divisions, though in most cases it was of a limited sort. Most units were willing to defend their sector and even to undertake minor offensive actions that could be seen to benefit their position. However, there were many cases of 'combat refusal' where units flat-out declined orders to engage in offensive action. Although the French were extremely vulnerable during this time, the extent of the mutiny was concealed as far as possible from the enemy.

Nivelle was replaced by General Henri Philippe Petain, who took measures to restore morale, and by the end of June the situation had improved. However, the French Army had made it clear that it would no longer consent to murderous frontal assaults against heavily defended positions. Better support, and more of it, was needed before the French infantry was willing to undertake any more large-scale attacks.

LAYERED DEFENCES
Advancing tanks came under direct fire from light weapons in the front trenches and heavier guns sited out of range of the tanks' weapons.

DISABLED BRITISH TANK
Disabled tanks could be recovered if the assault was successful, but if not then they might well remain in no-man's land as obstacles for any renewed assault to contend with.

CAMBRAI 1917

PREPARATIONS FOR ASSAULT
A British Mk IV tank moves up to its start position before the battle of Cambrai, 1917.

TANK RAID (RIGHT)
By 1917, British commanders had become conditioned not to expect decisive results from an offensive. Thus provision was not made to exploit the Cambrai 'tank raid' to its fullest extent.

FULLY COMMITTED
A number of specialist vehicles were created to support the Cambrai attack. Wire-pulling tanks were the forerunners of armoured engineering vehicles.

As with any new weapon system, doctrine had to be evolved for the effective use of tanks, which required a certain amount of trial and error. Some considered armoured vehicles to be a waste of resources, or at best an interesting novelty. Others felt that they might best be used in small numbers to support infantry in local attacks and counterattacks. Many tank proponents felt that they should be held back until sufficient numbers were available, then unleashed en masse.

There was other opposition to the use of tanks. Many traditionalists within the army were not merely suspicious of innovation; they were concerned about the implications for their own arm of service. Cavalry had for many years been the 'arm of decision' on the battlefield, and its officers were at times too concerned about their prestige to embrace the new mode of warfare offered by tanks. Similarly, there was concern among senior commanders that, not knowing how to best use the armour under their command, they might end up presiding over a fiasco.

One question that had yet to be answered about tanks was: how did they fit into the existing structure of military operations? With their guns they might be considered to be artillery platforms, but artillery did not lead the assault. Mobile, and capable of breaking an enemy line, they had much in common with cavalry, though many cavalrymen resented the comparison. Many thought that they should be integrated into infantry formations as a support system; others pushed for an independent tank arm.

Preparations

Even if these questions could be adequately answered, it still remained to be seen how best to employ the capabilities of armoured vehicles. Thus the decision that tanks would lead a major offensive was a difficult one to make. Staking the success of an operation that would require months of planning and logistical build-up on an unproven weapon system was a big risk, but as little had been achieved by any other methods, a massed tank attack was authorized. The area chosen lay close to Cambrai, where the ground seemed suitable for vehicle operations.

Many aspects of the operation and its preparations were tailored to the tanks. Rather than a prolonged bombardment which would crater the ground and make even reaching the enemy trenches difficult, it was decided to make a short, heavy bombardment without warning and then launch the assault. Guns were thus registered using map co-ordinates rather than by firing and observing the fall of shot. This was not a tank-specific tactic, of course. Surprise assaults had achieved good results elsewhere, whereas the massive bombardments previously used served mainly to warn the enemy that an attack was imminent.

Infantry were trained in tank co-operation as far as was possible, but it was within the tank force that the most extensive preparations were

Number of Tanks for Cambrai Attack
20 November 1917

378	Fighting Tanks (nine battalions at 42 tanks each)
54	Supply Tanks and Gun Carriers (7)
32	Wire Pulling
2	Bridging Tanks
9	Wireless Tanks (one per battalion)
1	Transport Tank (to carry telephone cable for Army HQ)
476	TOTAL

Source: *Official History*, 1917, Vol. 3, p. 28

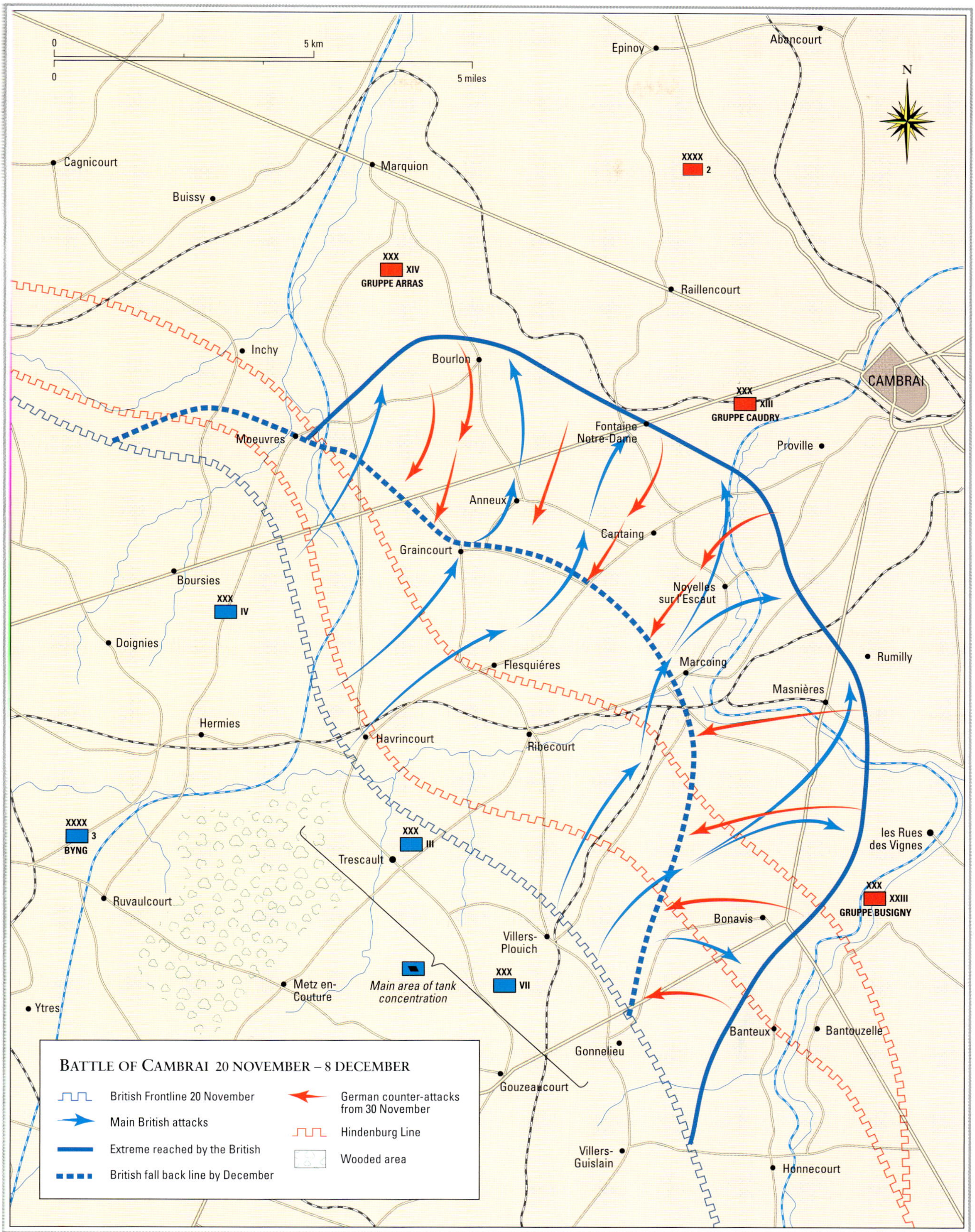
0
5 km
0
5 miles
N
Epinoy
Abancourt
Cagnicourt
Marquion
Buissy
XXXX
2
XXX
XIV
GRUPPE ARRAS
Raillencourt
Inchy
Bourlon
CAMBRAI
XXX
XIII
GRUPPE CAUDRY
Fontaine Notre-Dame
Moeuvres
Proville
Anneux
Cantaing
Graincourt
Boursies
XXX
IV
Noyelles sur l'Escaut
Doignies
Rumilly
Flesquiéres
Marcoing
Masnières
Hermies
Havrincourt
Ribecourt
les Rues des Vignes
XXXX
3
BYNG
XXX
III
Trescault
XXX
XXIII
GRUPPE BUSIGNY
Ruvaulcourt
Bonavis
Villers-Plouich
Metz en-Couture
Main area of tank concentration
XXX
VII
Ytres
Banteux
Bantouzelle
Gonnelieu
Gouzeaucourt
Villers-Guislain
Honnecourt
BATTLE OF CAMBRAI 20 NOVEMBER – 8 DECEMBER
British Frontline 20 November
Main British attacks
Extreme reached by the British
British fall back line by December
German counter-attacks from 30 November
Hindenburg Line
Wooded area

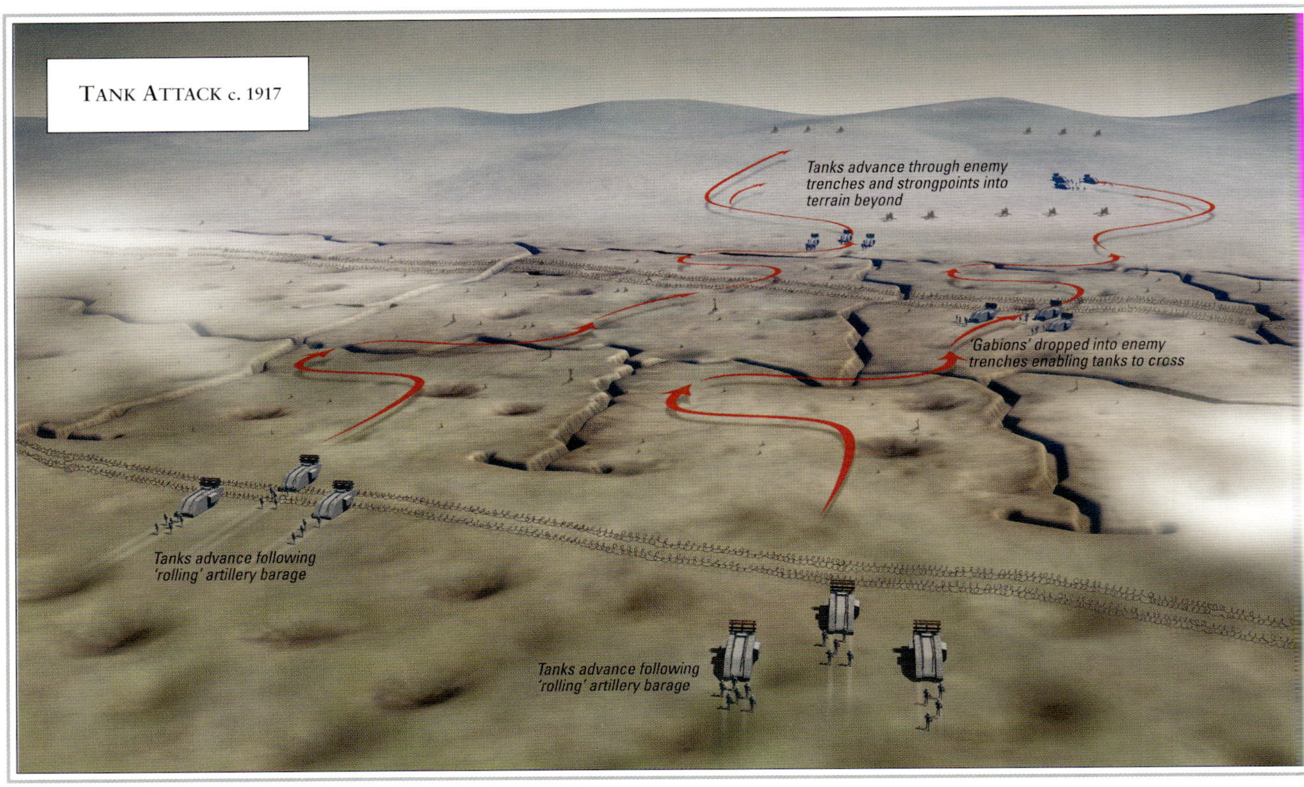

TANK/INFANTRY COOPERATION
Each section of three tanks was followed by a force of infantry, who were sheltered from enemy fire behind the armoured bulk of the tanks. The tanks would also crush wire obstacles or drag them aside.

made. Tanks were organized into sections of three vehicles, two 'females' armed only with machine guns and one 'male' with 6-pounder guns. Each tank was to carry a fascine, a bundle of brushwood, atop the hull, which was rolled into the enemy trenches to facilitate crossing.

The standard tactic was for each section to advance in a triangular formation until the enemy trench line was reached. There, the lead tank was to roll its fascine into the trench but not cross. Instead, it would turn left and proceed along the trench line, firing its starboard guns on the occupants. Meanwhile the other tanks of the section would cross the front trench using their fascines. One would immediately turn left and proceed between the front and support trenches, firing into both.

The third tank was to cross the support trench and move along it, firing its port guns into the trench and engaging targets of opportunity with its starboard weaponry. Each tank section would thus clear a stretch of trench which would then be secured by infantry. The tanks were then to re-form behind the enemy trench line and continue the advance towards the second line of trenches.

Such was the role of the fighting tanks, but within the formation there were specialist vehicles as well. These included some fitted with wire-dragging hooks and tasked with pulling aside the enemy's wire defences to permit infantry to pass. Supply and communications tanks, the latter fitted with a wireless set, also accompanied the advance.

DIRECT HIT
A direct hit by artillery was unlikely, but usually catastrophic. Even if the crew were not killed instantly, they would likely be trapped inside a burning tank hull.

Tank raid

The attack at Cambrai was conceived as a 'tank raid' rather than a full-scale attempt to create a decisive breakthrough. The concept had been expanded somewhat in the planning stage, but other than some vague hope there was no real expectation of massive success. Thus although cavalry were assigned to stand ready to exploit a breakthrough if one were made, they were deployed too far back to be able to respond quickly.

In other ways, however, the attack at Cambrai was an all-out effort. The entire British Tank Corps (as it had recently become known) was committed to the operation. Some 476 armoured vehicles were deployed, most of which were Mk IV models. These contributed most of the 'fighting' tanks, while the more lightly protected Mk I and Mk II vehicles were converted to specialist roles.

The assault began on 20 November 1917 with a sudden bombardment by around 1,000 artillery pieces, which arrived as a complete surprise. After pounding the enemy's positions for a short time, the artillery switched to a 'creeping barrage' ahead of the assault force, and the tanks rolled forward. Resistance was at first sporadic and light, with the assault force reaching the German forward trenches without undue difficulty.

As the shock of the attack wore off and the defenders collected their wits, resistance increased. This was largely on a piecemeal basis, with some units making a stand, while others fell back. Counterattacks were launched, but were repelled by the firepower of the British tanks and their supporting infantry. Where the tank-infantry tactics that had been trained were adhered, success was generally achieved. However, in some areas the tanks and infantry became separated, with the result that each was held up by something that the other could have handled.

Tactical success

By the end of the day, British infantry had managed to advance as much as 8km (5 miles) at some points, which was a spectacular achievement by the standards of the Great War. However, there was no breakthrough that could be exploited into a strategic victory. Some elements of the supporting cavalry were able to contribute to the advance, but after years of waiting for a suitable opportunity, the cavalry was not ready when the time came and failed to exploit the other arms' achievements.

Nor did the tanks and infantry succeed in winning more than a tactical victory. The advance was resumed the following day, but this time the enemy was forewarned and determined to stand and fight. The assault lost momentum and stability returned for a few days before a German counterattack reversed the situation and re-took most of the lost ground.

The attack at Cambrai cost 65 tanks disabled by enemy action and an additional 114 either ditched or broken down, although many of these were recovered and returned to service. The 'tank raid' turned out to be just that, albeit a raid that captured 150 enemy guns and proved decisively that massed-tank tactics could be highly effective.

Although Cambrai did not achieve decisive strategic results, and most of its tactical gains were quickly reversed, the battle had a profound effect on the course of the war. With the capabilities of tanks well proven, the Allies committed to building them in vast numbers and using them en masse. This led to strategic breakthroughs later in the war and ultimately contributed to the collapse of the German Army. Had the tanks failed at Cambrai, the Allies might have reverted to massed infantry assaults, which would likely have prolonged the war considerably.

The action at Cambrai seemed to prove that tanks should be viewed as infantry support weapons, moving at the speed of men on foot and thus constrained to their speed of advance. This view would colour the thinking of Allied planners for some years, though there were some who came to believe that the tank should lead and everyone else must find a way to keep up.

TANK TACTICS
Early tank tactics were sophisticated, with individual vehicles assigned to clear sections of trench and infantry following close behind to take possession as the tanks moved forward.

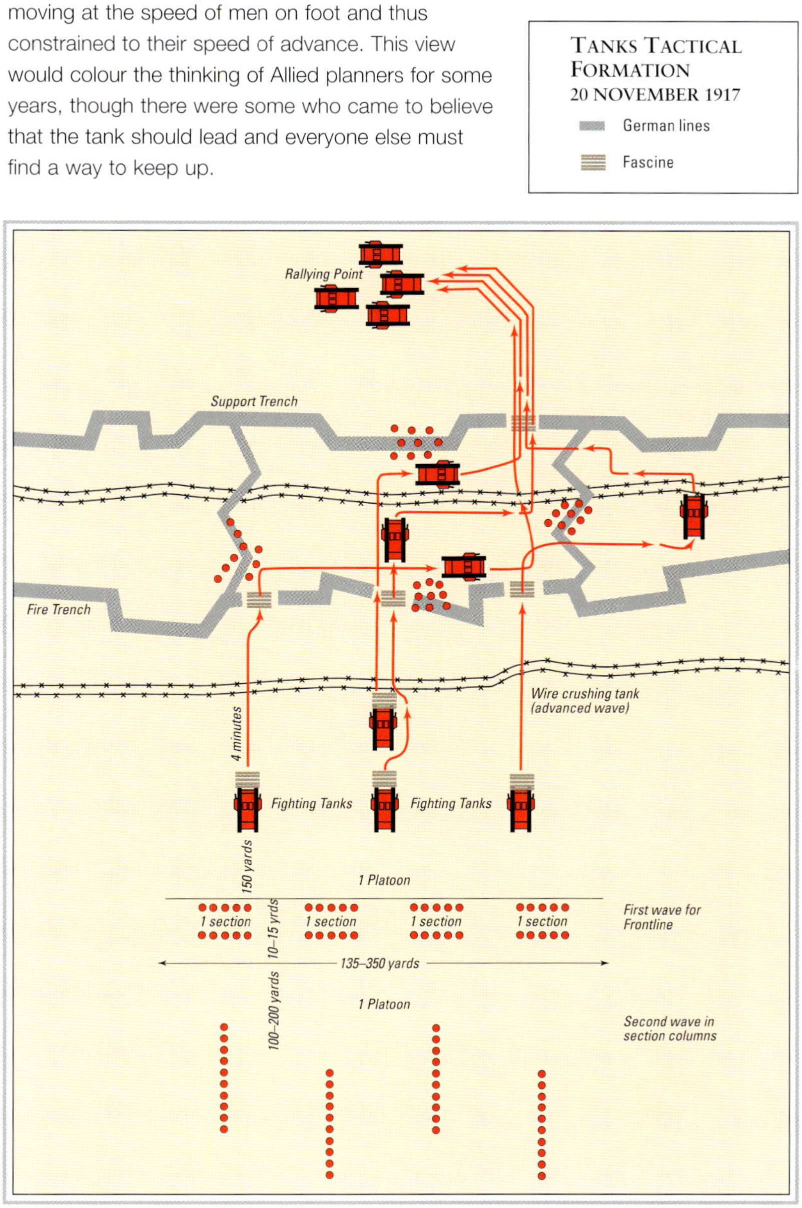

TANK-VS-TANK VILLERS-BRETONNEUX 1918

Germany did not attach the same importance to armoured vehicle development as did Britain and her allies, and so embarked on her own tank programme both late and halfheartedly. Even once the decision was made to acquire an armoured striking force, far greater numbers of captured British and French tanks were in use than German ones. Many of these were captured in more or less working order, having broken down or become ditched during an attack.

The only indigenous German design to see action during the war was the A7V, a badly flawed design that broke down even more frequently than British or French tanks. It was a huge target, with many weak points and shell-traps in its armour making it even more vulnerable to enemy fire. In an era when tanks

WHITE ELEPHANT
The German A7V was certainly a fearsome sight, but as a weapon of war it was a waste of resources. Too few were completed to be any use, and those that saw action achieved virtually nothing.

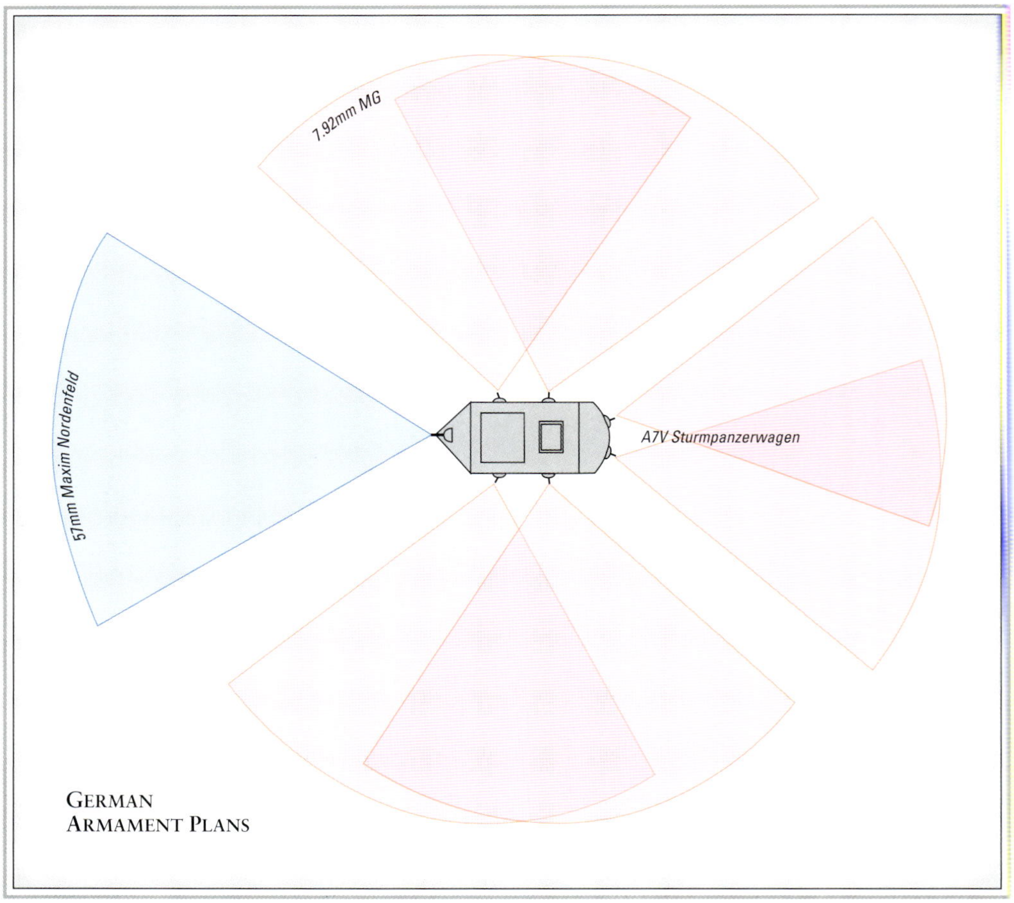

A7V ARMAMENT
The A7V's armament was laid out to be able to fire in several directions at once, but individual weapons suffered from a very limited arc of fire which created several dead zones.

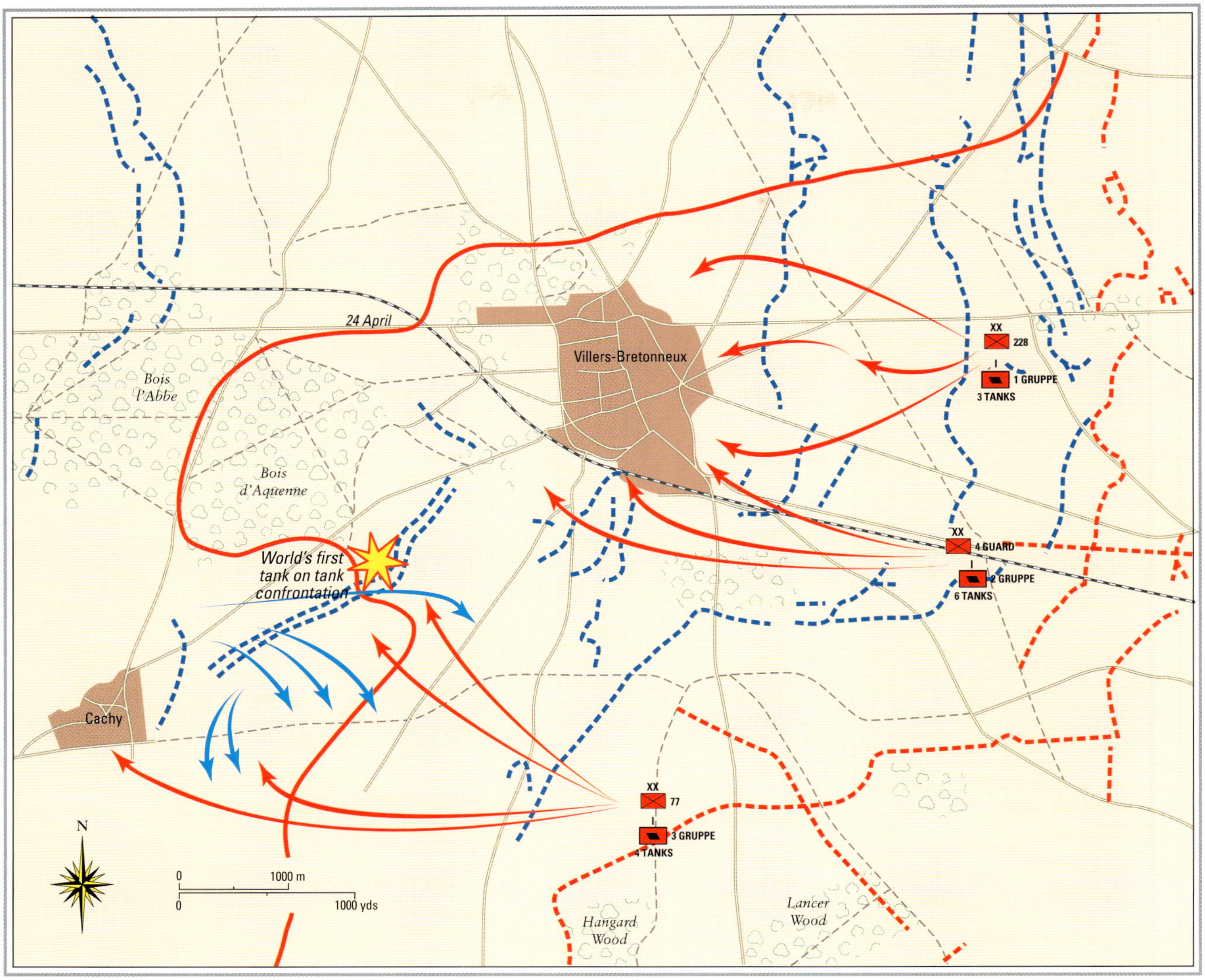

VILLERS-BRETONNEUX
24 APRIL 1918, GERMAN
TANK ATTACK

- German tank attacks
- Allied counter attacks
- German frontlines with dates
- German trenches
- Allied trenches
- Woods/Forests

TANK VS. TANK
The tank engagement at Villiers-Brettonneux was not planned by either side, but a clash was inevitable sooner or later, given the number of tanks on the Allied side.

were regarded as assault vehicles to assist a trench assault, the A7V had an extremely low ground clearance and easily became stuck. It also required a large crew who could not easily communicate with one another.

Other German tank designs were put forward but failed to appear before the end of the war. These ranged from a light vehicle based on an armoured car design and equipped with a turret-mounted machine gun, to a massive 'breakthrough tank' not dissimilar to a huge version of the British Mk IV and similar vehicles. This tank would have mounted four main guns of 77mm (3in) calibre, as well as seven machine guns, and been crewed by no less than 22 personnel. Although even larger than the A7V, this vehicle was in some ways more practical. It was lower, making it a less attractive target, and had better rough-ground performance. However, it was still an impractical design which would probably not have been any more effective than the A7V under combat conditions.

Attack at Villers-Bretonneux

In April 1918, hoping to break through to Amiens, German forces launched an attack assisted by gas shells, and succeeded in capturing Villers-Bretonneux. This opened the way to Amiens, creating the possibility of a decisive breakthrough. To counter this threat, Australian infantry were ordered to counterattack before any gains could be consolidated. The attack was made at night and was successful in forcing a German retreat.

A force of three British Mk IV tanks was sent forward to help hold the area against any further attacks. In keeping with the practice of the time, one of these tanks was a 'male' and two were 'females'. 'Female' tanks were armed with machine guns and intended primarily for anti-personnel work, while

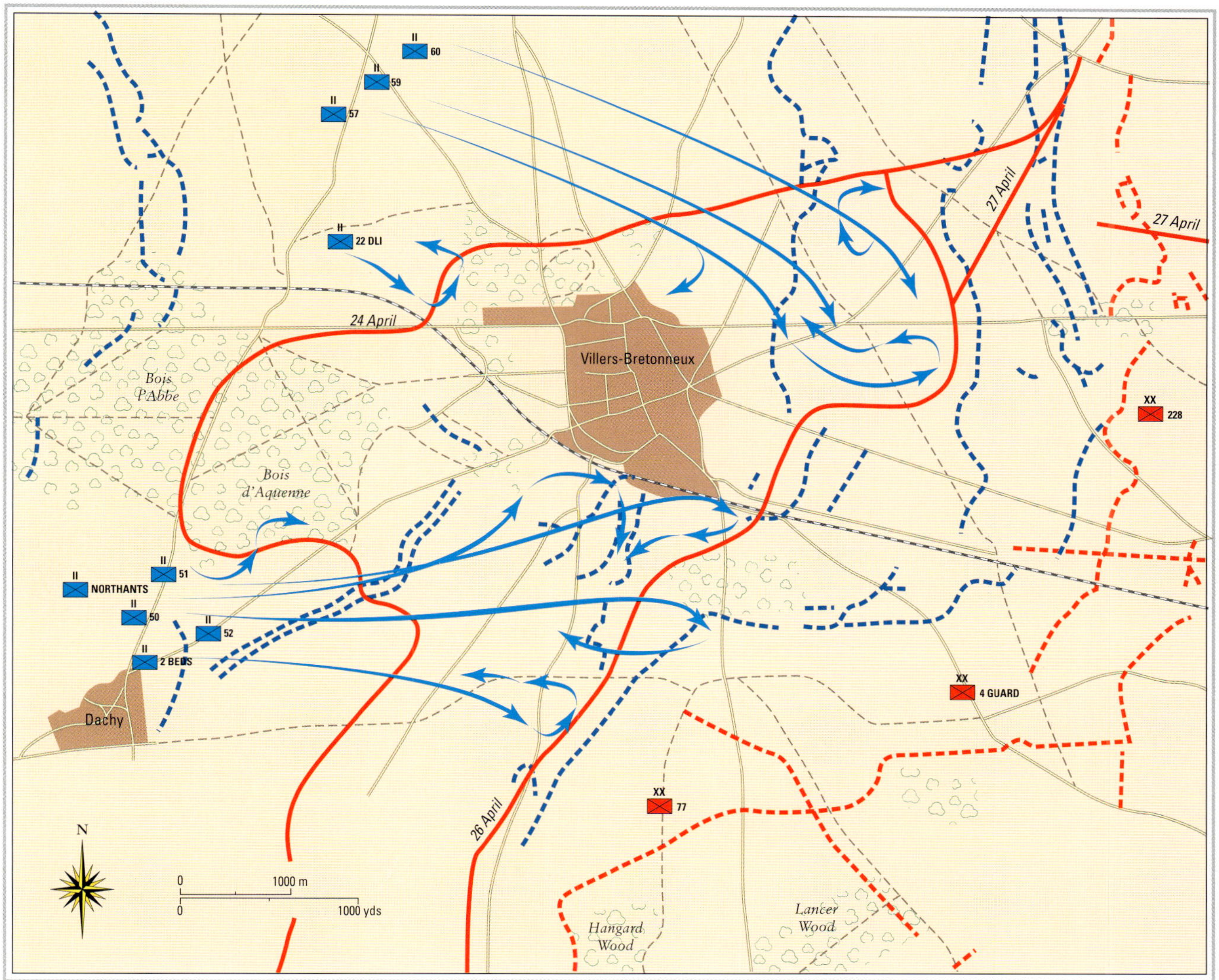

VILLERS-BRETONNEUX TO 27 APRIL 1918, ALLIED COUNTER-ATTACKS

- Allied counter attacks
- German frontlines with dates
- German trenches
- Allied trenches
- Woods/Forests

ARMOURED COUNTERATTACK
Several British Whippet light tanks took part in the counterattack, engaging enemy infantry with their machine guns and at times running them down. Some tanks were disabled by artillery during the action.

'males' carried 6-pounder guns which could be used against emplacements or fortifications as well as their machine gun armament.

Although 'male' tanks carried guns capable of destroying enemy armoured vehicles, there was no thought at the time of anti-tank work; tanks were infantry support vehicles which might have to face anti-tank guns or infantry positions, but were unlikely to encounter enemy vehicles. Up until that point, it had never happened.

German tanks arrive

As the British armoured force advanced, it was sighted and fired upon by a German A7V. In addition to its six machine guns, the A7V mounted a 57mm (2¼in) gun equivalent to the British 6-pounders. The main gun was mounted frontally, with the machine guns in limited-traverse mounts, causing the A7V to resemble a mobile bunker more than an armoured vehicle. Thus although the A7V mounted considerable firepower, it could not be concentrated against a single target.

British tanks by contrast mounted their main armament in sponsons with a wider arc of fire, although this was still limited compared to the all-round firepower offered by a turret. Turreted tanks were at that time under development, but had yet to be fielded. The lack of a turret made the tank simpler to construct, but was a handicap in several ways. A single weapon in a turret could face in any direction, whereas a vehicle using sponsons or internal mounts needed several guns to cover all arcs. This increased the number of crew needed and the size and weight of the vehicle.

In addition, a turret could traverse far more quickly than a tank could pivot to bring its main armament to bear, and allowed the tank to hide 'hull-down' behind cover. Thus once tank-versus-tank warfare

became likely, turreted vehicles were necessary. The ability to engage moving targets in any direction, while keeping the overall size of the vehicle down, is essential to combat effectiveness.

The A7V opened fire against the British 'female' tanks, which replied with their machine guns. The A7V was much better armoured than the British tanks, with a maximum of 30mm (1in) of armour to their 12mm (½in). As a result, their machine guns proved entirely ineffective against the German vehicle. The 'females', whose crews were already depleted by gas casualties, were damaged and forced to withdraw from the engagement.

Multi-tank engagement

The British 'male' tank then came up and engaged the A7V with its 6-pounder guns. Although hits were scored, the German tank's armour was not penetrated. Non-penetrating hits can still injure the crew or damage systems, however, largely by causing spallation – dislodging hot fragments of metal from the inside of the tank hull. Thus some damage was done to the German tank, which manoeuvred as best it could to bring its weapons to bear while avoiding British fire.

The A7V ran up on to a bank where its huge size and high centre of gravity caused it to tip over. Although the first tank-versus-tank 'kill' in history was something of an anti-climax, this did put the German machine out of action. However, two of its companions arrived and engaged the British Mk IV.

The action continued inconclusively for some time, largely due to the difficulty of obtaining a hit on a moving target, with the primitive gun-laying systems of the time. While adequate for attacking static emplacements, the tanks' guns were unable to hit one another much of the time. However, after a while, one of the German crews abandoned their vehicle and fled on foot, which prompted the other to retire.

RECOGNITION MARKS
British tanks often used vertical coloured stripes to identify them from machines captured by the German Army and put into service.

A7V IN ACTION
The A7V offered the enemy a large target with numerous weak points. It could operate only on fairly flat ground, and was therefore of no use in forcing a breakthrough of enemy trenches.

SOISSONS JULY 1918

SCHNEIDER CHAR D'ASSAULT
Built around a Holt tractor chassis, most of the Schneider tanks that were built entered service as unarmed supply vehicles.

With the Allied sea blockade slowly strangling the German war effort, war-weariness sapping morale, and increasing difficulty in providing replacements for casualties, it was clear to German planners that they needed to break the deadlock and win the war before American troops could arrive in great numbers. To this end, they conceived what became Operation Michael.

The Michael offensive was intended to drive between the British and French armies, pushing the British towards the Channel ports, while the French were driven back towards Paris. This was not greatly different to what had been attempted in 1914, and the strategy was sound. However, it had proved to be beyond the capabilities of the impressive German Army of 1914. For the weary divisions of 1918, it was a desperate gamble.

The Michael offensive was initially successful, notably due to the use of light infantry 'stormtrooper' tactics, whereby small units tried to infiltrate the enemy front lines and to bypass strongpoints rather than eliminating them head-on. Where the Allies had come to rely on tanks to lead their assaults, the Central Powers had similarly good success with stormtroopers. A series of assaults pushed the Allied front lines back at several points, though the offensive was eventually fought to a standstill.

Both sides suffered huge numbers of casualties, but while the Allies were receiving American reinforcements, German losses could not be made good. Nevertheless, attempts were made to renew the offensive over the next few months, at times achieving significant gains. By 18 July a large salient had been driven towards Paris, but after the Second Battle of the Marne in August 1918, the German Army was forced to go over to the defensive. Further attacks were beyond its capabilities.

French tank operations

During the German offensives of spring 1918, French tank assets were concentrated for use in counterattacks. These were primarily heavy designs of the St Chamond and Schneider types. The newer and lighter Renault design had been commissioned early in 1917, initially as a command vehicle to be integrated into heavy tank companies. By early 1918 the Renault was in full-scale production and light tank battalions were being formed.

The heavier tanks had not performed well up to this point. Even when assisted by specially trained detachments of infantry whose task was to help the tank across difficult ground, they frequently became stuck, and breakdowns were as common as in any tank design of the day. The St Chamond design was particularly prone to get into difficulties without any intervention from the enemy.

Nevertheless, the tank formations were successful in recapturing terrain objectives lost to German attack. These were for the most part small-scale actions, but early in June a much larger tank force was used to counterattack between Noyon and Montdidier. This was successful, though tank losses meant that Renault light tanks were used to replace heavier machines lost in action, creating mixed battalions along with the light tank battalions equipped exclusively with Renaults.

Allied counteroffensives

The final major offensive launched by the German Army began on 15 June 1918. It was halted within two days, leaving the German salient with two large bulges. One cut strategic rail links near Amiens, disrupting logistics and troop movements; the other threatened Paris. Reducing these was a primary objective of the Allied counteroffensive. The task of reducing them fell on four French armies, assisted by British and American divisions.

Some of the U.S. troops were integrated into the French command structure, but the eight divisions held in reserve to support the offensive were under American command. The American divisions were oversized compared to those fielded by the

European combatants and were manned by fresh troops. Their fighting power was thus significantly greater than that of a French or German division of that period.

The French counteroffensive was, for the first time, based on tank power. Never before had they combined a tank attack with a surprise bombardment, nor committed their full tank strength. The assault was led by all the available heavy tanks. These were followed by 16 infantry divisions, of which two were American. Their opponents, 10 German divisions, were critically undermanned due to losses taken in the earlier offensives. Much of their artillery had been transferred to Flanders, to support a planned offensive there, and they had been in possession of their positions for only a short time. Thus the Allied offensive was launched against hasty defences rather than the well-established trench lines of earlier in the war.

Tenth Army takes the lead

The attack was opened on 18 July by the French 10th Army, with three other armies in support. Six battalions of heavy tanks were allotted to the initial attack, with a brigade of three light tank battalions held in reserve. The attack came as a complete surprise, taking advantage of morning fog, and within three hours the lead elements had advanced some 3-4km (1-2 miles) into the enemy positions.

The Tenth Army fielded 324 tanks on the first day, of which 225 took part in the fighting. Only one of the light tank battalions was engaged, while all of the heavy tank elements took part. Despite the absence of much of the German artillery, 62 tanks were disabled by artillery fire, with another 40 being

GERMAN SPRING OFFENSIVES
The German spring offensives represented a final attempt to win the war before American manpower made it impossible. Although impressive success was achieved, victory was already beyond the power of the exhausted German Army.

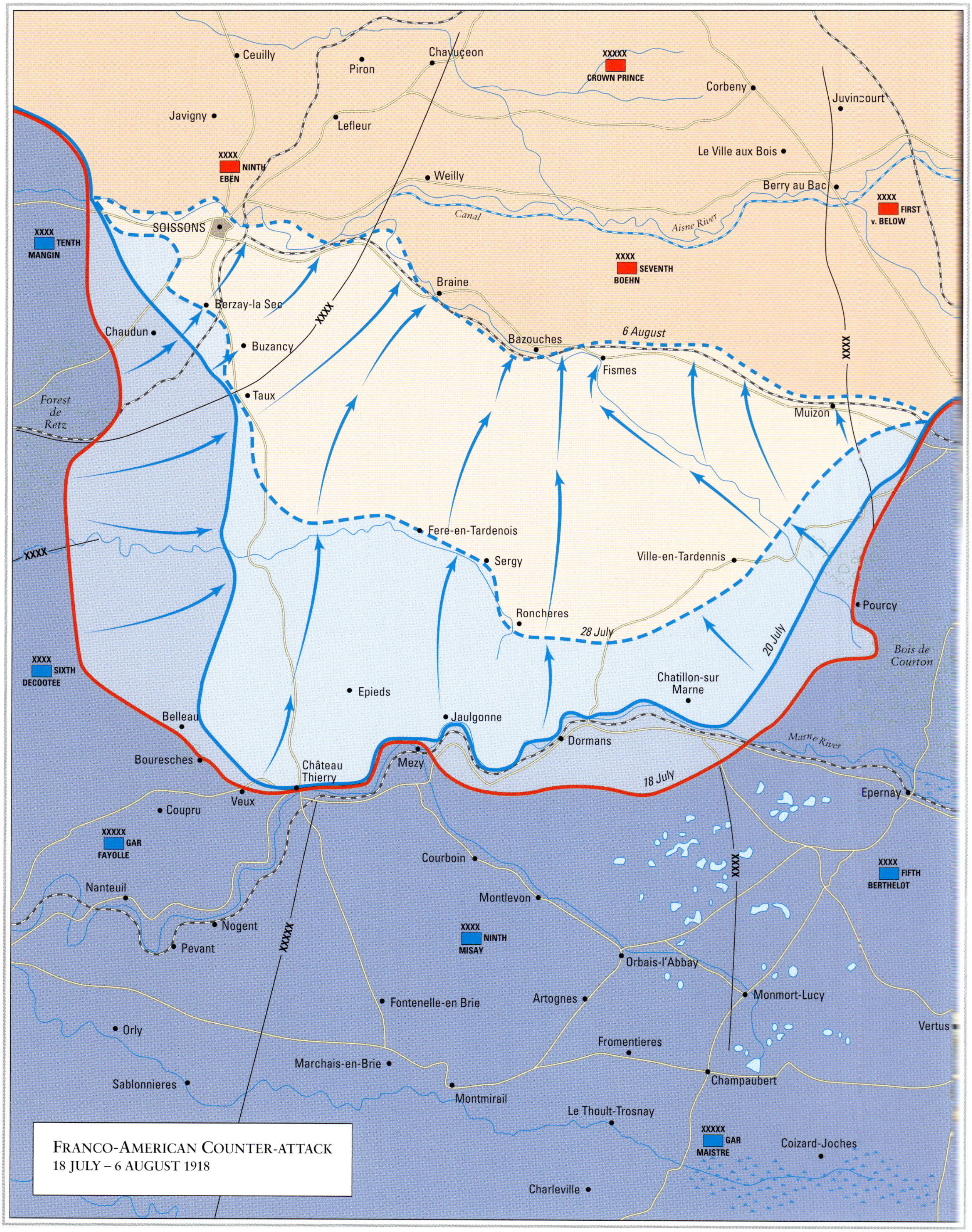
Ceuilly
Piron
Chavuçeon
XXXXX CROWN PRINCE
Corbeny
Juvincourt
Javigny
Lefleur
Le Ville aux Bois
XXXX NINTH EBEN
Weilly
Berry au Bac
XXXX FIRST v. BELOW
Canal
Aisne River
SOISSONS
XXXX TENTH MANGIN
XXXX SEVENTH BOEHN
Braine
Berzay-la Sec
XXXX
Chaudun
Buzancy
Bazouches
6 August
Fismes
Taux
Forest de Retz
Muizon
XXXX
Fere-en-Tardenois
Sergy
Ville-en-Tardennis
Pourcy
Roncheres
28 July
20 July
Bois de Courton
XXXX SIXTH DECOOTEE
Chatillon-sur Marne
Epieds
Belleau
Jaulgonne
Dormans
Marne River
Bouresches
Château Thierry
Mezy
18 July
Veux
Epernay
Coupru
XXXXX GAR FAYOLLE
Courboin
XXXX FIFTH BERTHELOT
Nanteuil
Montlevon
Nogent
XXXX NINTH MISAY
Pevant
XXXXX
Orbais-l'Abbay
Monmort-Lucy
Fontenelle-en Brie
Artognes
Orly
Vertus
Fromentieres
Marchais-en-Brie
Champaubert
Sablonnieres
Montmirail
Le Thoult-Trosnay
XXXXX GAR MAISTRE
Coizard-Joches
Charleville
FRANCO-AMERICAN COUNTER-ATTACK
18 JULY – 6 AUGUST 1918

lost to other causes. Naturally, these losses fell most heavily on the leading heavy tank units.

With so many of the heavy tanks disabled, composite formations were hurriedly put together and the advance resumed on the following day. Some 105 tanks were committed to action on the 19th, again suffering heavy casualties to breakdowns, ditching and enemy action. The 3rd Heavy Battalion, which had started the 18th with 27 tanks, had just two machines operational at the end of 19 July. The 12th Heavy Battalion lost 29 of its 30 tanks – not counting replacements swapped in during the night of the 18/19 July – by the time it reached its objectives.

After a day spent consolidating their gains, in which 17 French tanks were lost out of 35 committed to counterattacks, the French continued their advance, with 100 tanks committed to lead a large-scale attack. It succeeded, although the ground was retaken by German counterattacks. After regrouping, the tank force was able to put in a final attack with 82 machines on the 23rd. Exhausted, the armoured units were then transferred to the reserve.

Tank operations elsewhere

Meanwhile, the 6th Army had made an attack in support of the main drive. Assigned one heavy tank battalion and three light battalions, the 6th Army was generally successful in its sector on the 18th, although tank losses as usual were heavy. These were largely due to breakdowns, some of which were quickly repaired. Nevertheless, armoured support dwindled fast. On the 19th there were typically five tanks per infantry battalion, but by the 25th few were still combat capable.

Meanwhile, other tank formations were transferred to sectors near Rheims, which were threatened by a German attack there. Once this was contained, they were redeployed to take part in operations conducted by the 9th Army. By 23 July, the situation had largely stabilized. Fighting went on longer than this, but French gains on the 23rd were slow and little more was achieved. However, the operation was a resounding success for the French tank corps and for the Allies in general, and contributed to the declining morale of the German armed forces.

Results and aftermath

The French counteroffensive eliminated the threat to Paris and recaptured much of the ground lost to the German spring offensives. The effect on morale in the German Army was considerable; not only were the effects of massed tank assault keenly felt, but the newly arrived combat-ready American forces proved to be formidable opponents.

THE SHAPE OF THINGS TO COME
Although distinctly modest compared to other tank designs of the era, the Renault FT-17 was an effective infantry support platform and the longest-lived of all early tank designs.

Both the Schneider and particularly the St Chamond tanks had under-performed in action, while the new Renault vehicle had established its effectiveness. Mass production by several companies ensured that huge numbers were rapidly made available. At the peak of production, it was possible to outfit a battalion in a week. Thus the Renault eclipsed the earlier models, becoming by far the most important French tank design of the war.

It was Renault tanks that bore the brunt of later actions, both in terms of small-scale counterattacks and the renewed offensives that took place in August and September 1918. Although smaller and far lighter than the St Chamond or Schneider designs, the Renault mounted its armament in a turret. This permitted all-round fire by a single weapon, making a Renault with its crew of just two scarcely less effective than a St Chamond, which carried nine.

The Renault mounted either a machine gun or a 37mm (1½in) gun in its turret; a few received a short 75mm (3in) gun instead. The tank was designed for infantry support and thus had a short range and very low speed. To some extent, its very success became a liability in the interwar years. With so many Renaults available, France did not aggressively pursue tank design and thus was still using these outdated first-generation machines in 1940, many in static positions on the Maginot Line.

One other significant outcome from the Battle of Soissons was some years in making itself felt. The German High Command concluded that the defeat had occurred mainly because of the high concentration of armour fielded by the French. This became a key factor in German military thinking over the next 20 years or so.

ALLIED COUNTEROFFENSIVES
A combination of innovative tactics and American reinforcements allowed the Allies to rapidly reverse the gains made by the German spring offensives.

BLACK DAY OF THE GERMAN ARMY AUGUST 1918

Although unable to make good the losses suffered in the Second Battle of the Marne, the German Army seemed secure in its positions at the beginning of August 1918. The Allies had won an essentially defensive victory at the Marne and had not moved to exploit it. The situation had not, apparently, been greatly altered.

This appreciation was incorrect. At the time of the Marne offensive, the British and French were massing armoured forces for an attack of their own in the Amiens region. The area had been chosen partly because it offered suitable ground for tank operations, and it was upon tanks that the success of the offensive rested. Where the German attack at the Marne had included some 52 infantry divisions, the Allied offensive was launched with a much smaller force, but this was spearheaded by around 600 armoured vehicles.

Operational secrecy was maintained by the Allies, enabling them to mass forces for the assault without alerting their opponents. There was no extended period of bombardment before the attack. Instead, the Allied guns were registered to map co-ordinates using aerial reconnaissance, and opened fire as part of a detailed plan that first attacked the German artillery and defensive positions, then switched to a creeping barrage to support the advance.

Most of the infantry involved in the attack were Australian and Canadian, and had gained

BRITISH MK V TANK
Appearing in 1918, the Mk V tank had a more powerful engine than its predecessor and was simpler to operate, requiring only a single crewman to drive, although two gearsmen were still carried.

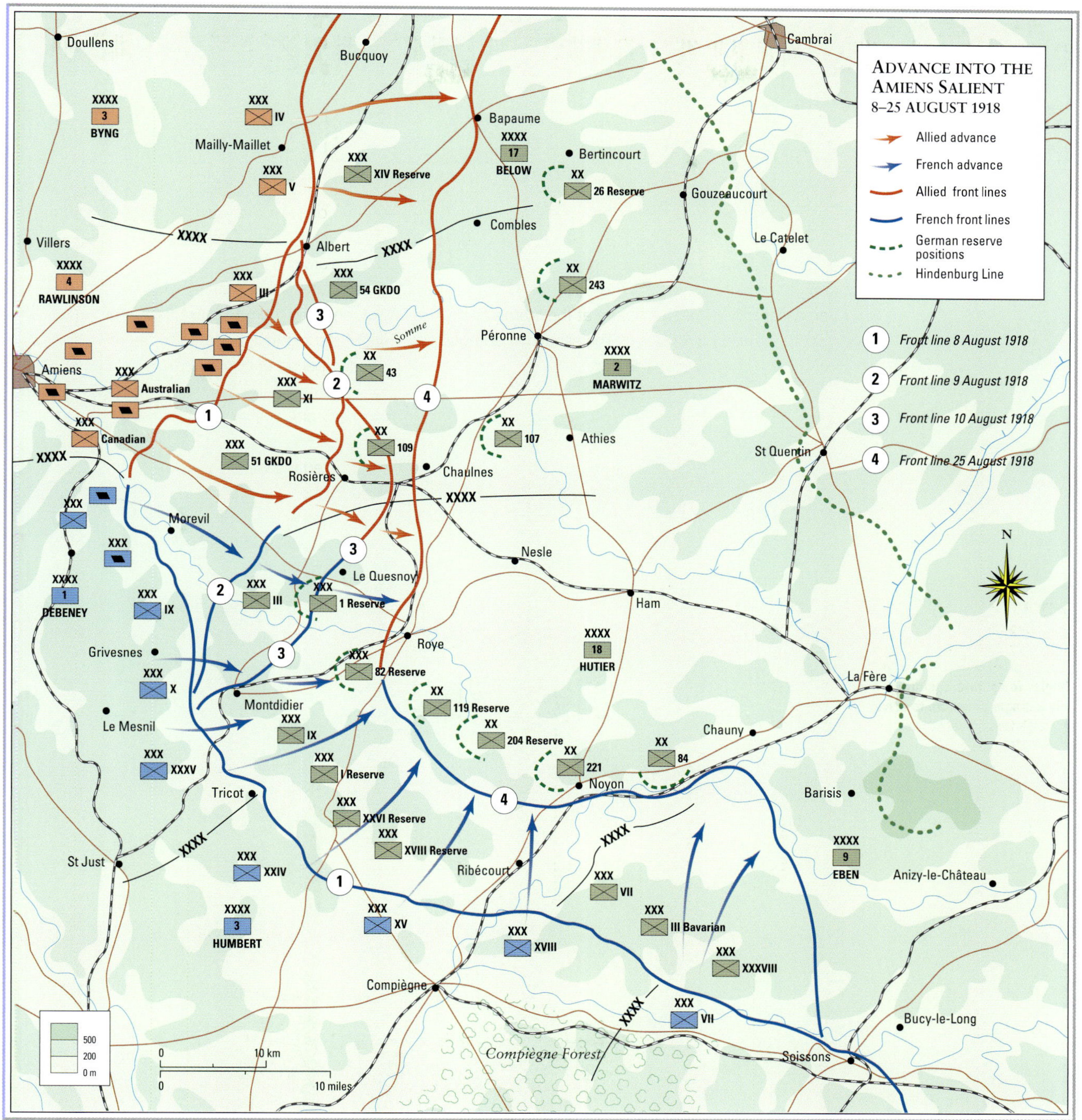

experience in smaller-scale surprise attacks. The artillery, infantry and operational techniques used in the offensive had been successfully employed in the past by German forces, but the Allies had the additional advantage of armoured support.

The German forces were reasonably alert, as there had been a series of small raids and retaliatory attacks in the region over the past few days. However, there was absolutely no expectation of a major offensive and the response was slow. By the time artillery began firing on Allied concentration areas the troops were already well forward.

The French, led by 70 tanks, attacked at Montdidier while the main blow was struck by British forces led by 530 armoured vehicles. These included combat tanks, but also armoured supply carriers and unarmed tanks used to transport infantry squads. Assisted by surprise, supported by

SUCCESSFUL OFFENSIVE
Through a combination of surprise and overwhelming force, the Amiens offensive caused a collapse among the German troops opposing it. Armoured forces overran command posts, paralyzing the response even where troops were willing to fight.

FRENCH/U.S. COOPERATION
An American tank arm was formed for service in World War I, but initially U.S. troops received support from French tank units equipped with Schneider heavy tanks.

ALLIED ADVANCES (OPPOSITE)
During the last months of the war, mobility returned to the Western Front. The exhausted German Army was unable to create a defensive line capable of resisting armoured assault.

NOVEMBER ARMISTICE
By November 1918, Germany was wracked by internal political troubles and in a state of economic collapse. An armistice was the only hope of preventing an Allied advance into Germany itself.

well-coordinated artillery and concealed by morning fog, the assault quickly broke through the first German line and had punched a hole in the second within four hours.

The primary anti-tank weapon at the time was artillery or specialist anti-tank guns, and these were slow to redeploy as they relied on horses for mobility. The fast-moving tanks and their supporting infantry suffered relatively low casualties largely due to the speed and aggression of their advance and continued pressure made it impossible for the Germans to establish a new defensive line.

Utterly exhausted by the fresh Allied advance, many defending German troops simply collapsed on 8 August, earning it the nickname the 'Black Day of the German Army'. Armoured units cut lines of communication and overran command posts, crippling the defending forces in a manner that would be repeated time and again in armoured assaults. Many units broke and fled, some shouting abuse at those trying to form a defensive position.

Relatively small losses enabled the Allies to maintain the momentum of the offensive, though by the end of four days the armoured force was reduced to almost nothing. This was largely due to breakdowns or ditching, however, and many of the lost tanks were recovered and returned to action. During this time the old Somme battlefield was retaken, and continued attacks throughout the rest of August drove the Germans all the way back to the Hindenburg Line. These tank-driven attacks marked the beginning of the end of World War I.

ALLIED ADVANCE
September–November 1918

Area retaken by Allies to 11 Nov. 1918
Still occupied by German forces at Armistice

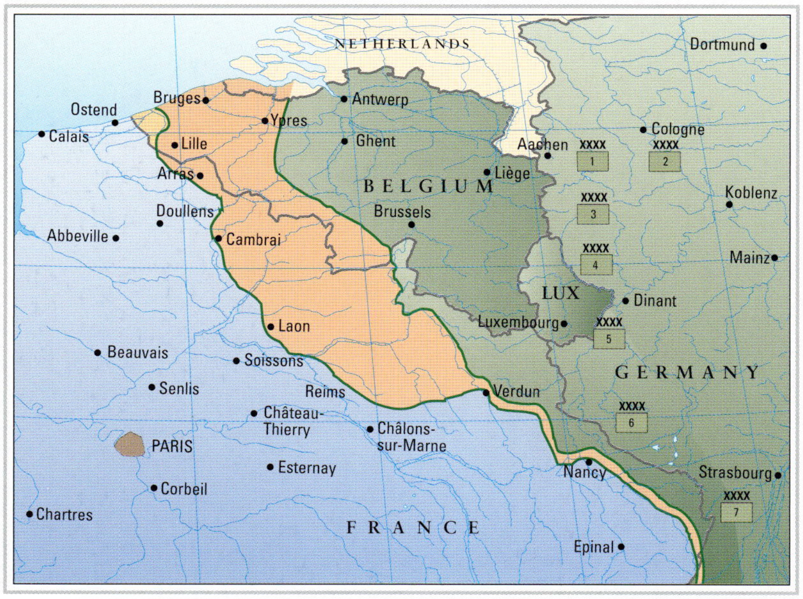

ADVANCE TO VICTORY
5 October – 11 November 1918
British advance
French advance
Belgian advance
British front lines
French front lines
Belgian front lines
German armistice line
1 Allied front line, 26 September 1918
2 Allied front line, 17 October 1918
3 Allied front line, 9 November 1918
4 Allied front line, 10 November 1918
5 German armistice line, 11 November 1918
North Sea
NETHERLANDS
BELGIUM
FRANCE
N
20 km
20 miles
Zeebrugge
Knocke
Zuyenkerke
Ostend
Bruges
Eecloo
St Nicolas
Nieuport
Furnes
Dixmude
Thourcut
Thielt
Ghent
Termonde
Dunkirk
XXXX KING ALBERT
XXXX 4 ARMIN
Roulers
XXXXX RUPPRECHT
Alost
Passchendaele
Brussels
XXXX Poperinge
Ypres
Courtrai
Audenarde
XXXX 2 PLUMER
Menin
Comines
Messines
St Genois
Grammont
Bailleul
Hazebrouck
Armentières
Roubaix
Aire
Merville
Lys
XXXX 5 BIRDWOOD
Lille
Tournai
Ath
Neuve-Chapelle
Béthune
Festubert
XXXX 6 QUAST
Soignies
Lens
St Amand
Mons
Vimy
Aubigny
Douai
Scarpe
Binche
Namur
Charleroi
Meuse
Thuin
XXXX 1 HORNE
Valenciennes
Arras
XXXXX BRITISH EXPEDITIONARY FORCE HAIG
XXXX 17 MUDRA
Bouchain
Le Quesnoy
Maubeuge
Beaumont
Dinant
Bourlon
Escaut
Cambrai
Solesne
Philippeville
Givet
Bapaume
XXXX 3 BYNG
XXXX 2 CARLOWITZ
Avesnes
XXXXX BOEHN
Mailly-Maillet
Le Cateau
Chimay
Fumay
Gouzeaucourt
Albert
Combles
Le Catelet
Bohain
La Capelle
Somme
Péronne
XXXX 4 RAWLINSON
Oise
Hirson
Rocroi
Guise
Athies
Rosières
St Quentin
Vervins
Morevil
Nesle
XXXX 18 HUTIER
Serre
Liart
Le Quesnoy
Ham
XXXX 1 DEBENEY
Mézières
Sedan
Roye
La Fère
Goudelancourt
XXXXX CROWN PRINCE
Montdidier
Chauny
Barisis
Laon
Noyon
Sissone
Tricot
XXXX 7 BOEHN
Ribécourt
Rethel
Le Chesne
St Just
Aisne
Rethondes
Craonne
Compiègne
XXXX 10 MANGIN
Soissons
Vouziers
Buzancy
XXXXX FAYOLLE
XXXX 1 EBERHARDT
Grandpré
Fismes
Monthois
XXXX 3 EINEM
Villers-Cotterêts
Reims
XXXX 5 BERTHELOT
Ourcq
Fère-en-Tardenois
XXXXX MAISTRE
Vesle
XXXX 4 GOURAUD
Vienne
Suippes

AMERICAN OFFENSIVES SEPTEMBER-OCTOBER 1918

STRATEGIC MOVEMENT
Just getting tanks to the battle area posed some serious problems. Where marching infantry could scramble around an obstacle, armoured forces required an unbroken chain of intact bridges, rail lines and roads.

U.S. ARMOURED FORCES
By the last weeks of World War I, the US armoured corps was a significant fighting force largely equipped with French FT-17 tanks. The corps was rapidly run down after the war and had to be virtually recreated two decades later.

American involvement in the war had been discounted by some German commanders early on. The U.S.A. might not be drawn into the war at all, and its troop ships would have to run a gauntlet of submarine attacks, which would surely make it impossible to deploy sufficient troops.

Arriving at first in small numbers, the Americans contributed to British and French operations as their strength increased. At the Second Battle of the Marne, five American divisions were under French command. Each of these formations contained around 28,000 fresh troops. Not only had these men not been tired by years of warfare and suffering, but they were also well-fed and well supplied, in contrast to their German opponents.

By June 1918 a quarter of a million American troops were landing in France and were gaining battle experience alongside their allies. Towards the end of August these forces were regrouped under American command and deployed south of Verdun with the intention of eliminating the St Mihiel Salient. This operation was agreed by the Allies on the condition that U.S. troops would be available to support an attack in the Argonne region on 26 September 1918.

Despite the tight time frame, US commander General John Pershing believed that he could accomplish his objectives, so went ahead with the St Mihiel assault. German forces had held this area since 1914, but were planning to withdraw from it into better defensive terrain. Before this could be accomplished, the American artillery opened fire with 2,900 guns and the U.S. I and IV Corps launched an assault.

The operation was the first action for the newly formed U.S. Tank Corps, and was also the first attack to be made solely by American troops. In some ways it resembled the early offensives of the war – the fresh U.S. divisions were more aggressive than their allies had become, and were willing to take serious losses in order to achieve their aims. Troops of all other nations had become increasingly sensitive to casualties as the war went on, and fought with more attention to survival than victory.

The U.S. tank contingent supporting I Corps was equipped with 144 French-supplied Renault tanks and was supported by French-crewed St Charmond and Schneider models, making 419 armoured vehicles committed to the offensive. The French vehicles proved unreliable and prone to breakdowns, with many becoming stuck in the muddy battlefield. Only three tanks were disabled by enemy action; about 40 broke down or became stuck.

An overwhelming success

Despite an inauspicious beginning for the Tank Corps, the offensive was an overwhelming success. The defenders were forced out of their positions and put to flight, leaving behind more than 450 guns and over 13,000 prisoners. The attack also had a serious effect on morale throughout the German forces. Faced with seemingly endless numbers of American troops, many German soldiers finally lost faith in their ability to win the war. Anything that prolonged the struggle seemed pointless and reduced their chances of survival. Demands for an armistice became increasingly common, though the German Army fought on.

By the 26th, Pershing's force had redeployed to the Argonne valley where they hoped to repeat their achievement. Victory would have far-reaching consequences, as it would allow the capture of Sedan and the subsequent disruption of the German Army's vital railway system. The Tank Corps went with them, quickly improvising measures to deal with the worst of its problems.

U.S. TROOPS IN TRAINING
Lacking significant armoured forces of their own, U.S. troops in France trained with tanks donated by their British and French allies. Familiarity was important in establishing mutual trust between the infantry and armoured forces.

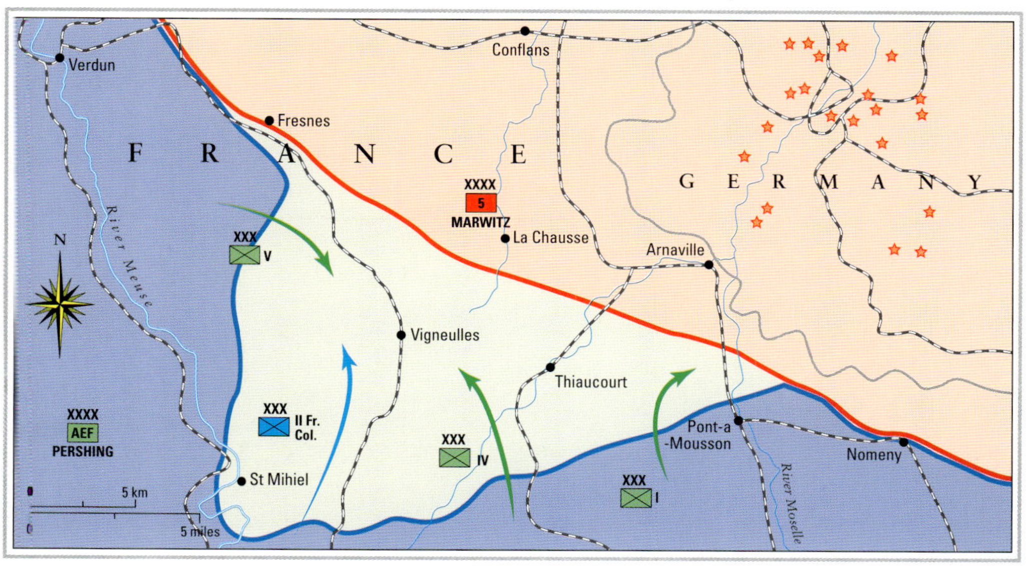

ST MIHIEL SALIENT
Assisted by French artillery and tanks, US troops successfully drove the German defenders from the St Mihiel salient, which they had held since 1914 but were preparing to abandon.

MEUSE-ARGONNE OFFENSIVE
Slowed by logistical difficulties and strong defensive terrain, American forces made slow gains throughout September and October, but were able to advance more rapidly in November.

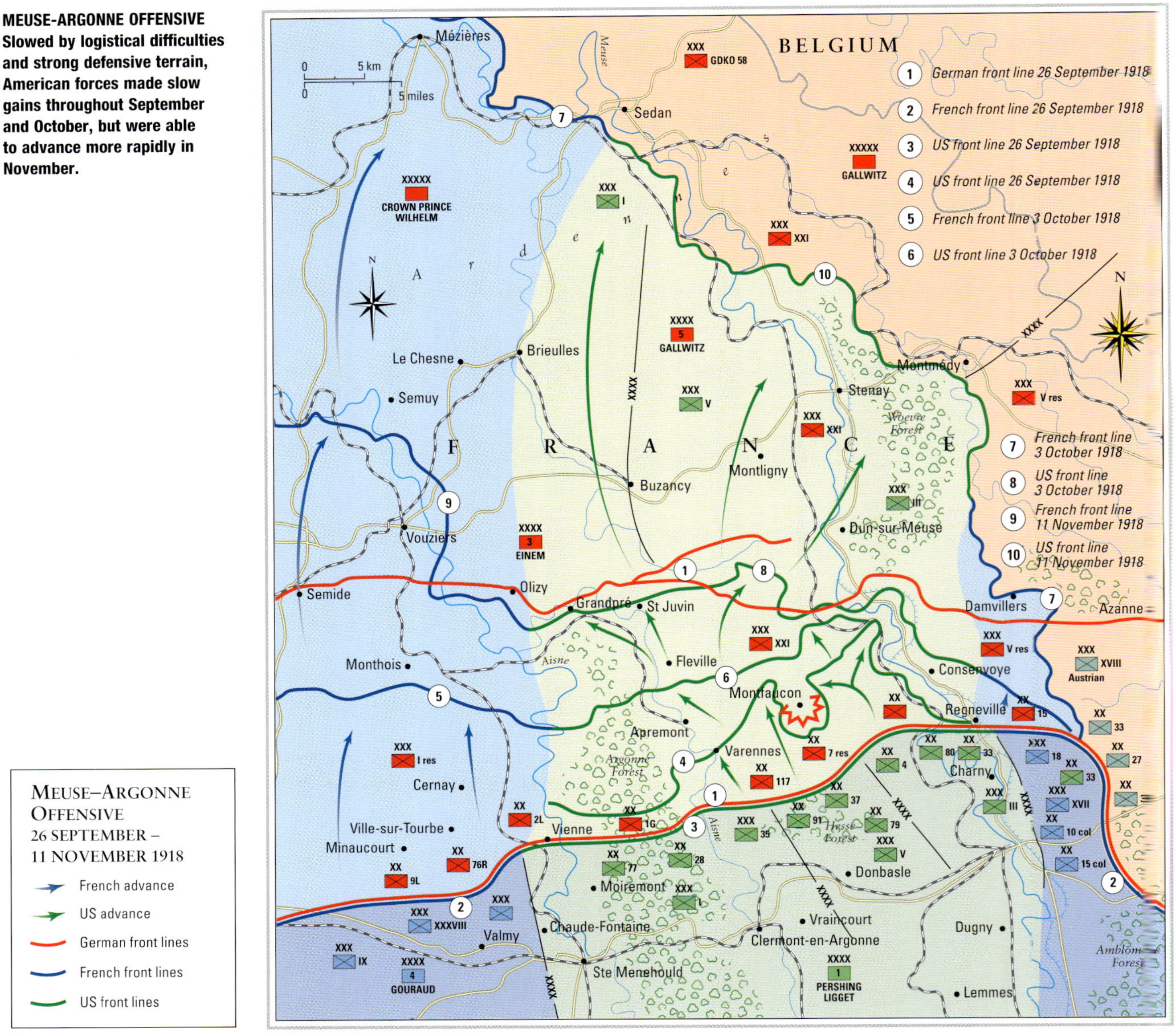

To offset logistical problems, U.S. tanks went into action with drums of fuel strapped to their hull. This was an obvious risk; if a drum were punctured then it would almost certainly catch fire. However, it was hoped that the ropes securing a drum might burn through, causing it to roll off, before the tank was destroyed. In any case, the loss of some vehicles in this manner was considered less serious than the consequences of the tank force running out of fuel.

Another innovation was the conversion of one tank in each company into a mobile workshop loaded with critical spares. Certain components had been noted as being common sources of failure, and in many cases these were quite minor items. A mobile armoured workshop could reach a disabled tank and quickly return it to action during the battle, and could bring up spares afterwards to allow disabled vehicles to self-rescue. Some vehicles were repaired and returned to action several times.

The plan for the Argonne offensive was over-ambitious, especially considering that half of the U.S. troops committed had not previously been in action. The terrain was also more favourable to the defenders than at St Mihiel, with steep slopes and thick woods. Nevertheless, supported by 2,700 guns and led by U.S. and French armoured troops, the assault force began its advance.

The first day of the offensive went well, but the defenders were able to pull back to a new line and bring in reinforcements. The U.S. advance was slowed to a crawl and dogged by logistical difficulties. By 4 October a number of more experienced divisions had been swapped in to replace inexperienced formations, and a renewed offensive began. Still only minimal gains were made, in return for heavy casualties.

Relatively few American tanks were disabled by enemy action, and many of those were indirectly disabled. One lethal trap was a water-filled ditch, which would flood the engine of a tank that entered, and possibly drown the crew if they did not escape in time. Other countermeasures were more direct, such as anti-tank rifles, which were of marginal effectiveness at best. A hit in the engine region might stop a tank, but the best defence remained to use artillery weapons in the direct-fire role.

Despite the difficulties they encountered, U.S. troops continued to push forward and this had an effect on German morale at all levels. General Ludendorff retired to his office on 28 September, in a fury of impotent rage, and later emerged to inform his colleagues that the war could not now be won. Many demoralized soldiers agreed, but all the same, their defence remained extremely tenacious.

Inexperienced troops

It was not until 13-16 October that U.S. troops began to break through the German positions, finally reaching their first-day objectives. This was achieved at the cost of heavy casualties, not least because of the aggression and inexperience of the newly arrived troops. American losses could be made good, however, whereas the Germans had few reserves left, and this permitted a renewed offensive in November which made rapid gains. By 5 November, American troops were across the Meuse. Sedan fell to French forces just as the war ended.

Attention to resupply and field repairs allowed the U.S. Tank Corps, which by this time was receiving British vehicles as well as French ones, to remain effective throughout the last months of the war. By the time of the armistice on 11 November, the Corps was down to about 50 effective vehicles at any given time, and had suffered an average attrition rate of 123 percent, mostly from breakdowns.

FT-17 IN U.S. SERVICE
Although a humble vehicle beside the great monsters fielded by Britain and France, the FT-17 became the backbone of French armoured forces. It also served with distinction in American hands, in this case with 345 Tank Battalion.

VISIBILITY AND VENTILATION
With its armour panels open the FT-17 offered its two-man crew good visibility. Once 'buttoned up' for action, their field of view was limited to what could be seen out of a tiny slit.

PLAN 1919
ATTACK BY PARALYZATION

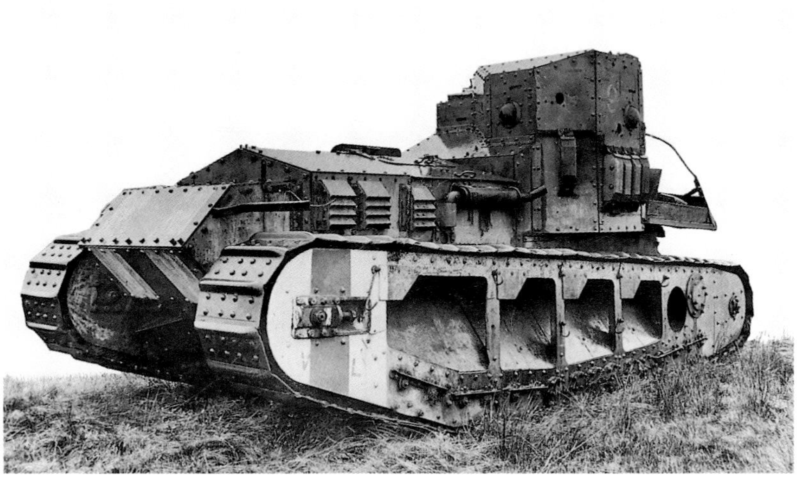

BRITISH MEDIUM MK A 'WHIPPET' TANK
The 'Whippet' was originally intended to have a rotating turret, but engineering problems forced the substitution of a fixed armoured box. The Whippet was the first tank that could be driven by a single crewmember, albeit with some difficulty.

Experience of fighting on the Western Front during World War I only served to reinforce what military thinkers had always known – that decisive results are rarely achieved when only the enemy's combat assets are engaged. For all of history, an army that was defeated but managed to pull back in good order has usually been able to return to action with much of its fighting power intact. Only when a collapse occurred and a vigorous pursuit could be mounted were lasting results likely. Conversely, the loss of bases and logistics capability has defeated many armies that could otherwise have remained in the field.

The enormous advantages enjoyed by the defender for much of World War I ensured that decisive results were impossible to achieve, except perhaps by exhausting the enemy's reserves of manpower or the fighting will of their troops. The French Army mutiny of 1917 was the closest the latter came to occurring, and even then morale was restored in a matter of weeks. However, for much of the war, the means simply did not exist to win by any other methods.

By early 1918, the Central Powers were exhausted, and could not win a protracted war. The blockade of Germany was causing severe shortages at home and in the trenches, and the army was fought out. Yet even after the failure of the last great gamble, the spring offensives of 1918, it seemed that the German Army planned to cling to one defensive line after another, each one costing the Allies massive casualties in their efforts to break them. In the event, internal politics caused the Central Powers to finally admit defeat, but had this not occurred, the Allies intended to bring the war to a swift conclusion by offensive action.

Decisive results

The availability of large numbers of tanks, and to a lesser extent aircraft, made it possible to consider the sort of offensive that might bring about decisive results. Deployed in sufficient numbers, tanks should be capable of breaking through the main and secondary enemy defence lines, and of defeating any counterattacks that attempted to seal the breach in the line. In and of itself, this would be only a tactical success, but it made possible exploitation of the breach.

The world's first armoured exploitation was made in August 1918 by the British Medium Mk A Tank, better known as the 'Whippet'. A group of Whippets were able to break through the enemy defences and overrun enemy artillery, and during this action a tank named 'Musical Box' became detached from the main force. Having pushed deep into enemy territory, 'Musical Box' was unable to return to Allied territory and instead ran amok for a period of nine hours. During this time the tank overran an artillery battery and transport convoy, and also inflicted

ARMOURED WARFARE THEORIST
J.F.C. Fuller was an advocate of rapid armoured warfare, where the tank led and all other arms had to keep up. His theories attracted more attention in interwar Germany than in Britain.

heavy casualties on enemy infantry before finally being disabled.

The results of this one-tank rampage were vastly out of proportion to the force involved, and reminiscent of the days when cavalry might get into an enemy rear area and cause chaos. The implication was obvious – if a force of tanks could get into the enemy rear then it might force the collapse of the entire sector, or at least severely weaken defences.

Key targets

For much of the war, the main targets of an offensive were enemy combat troops and the defensive terrain they occupied, not least because this was the only attainable goal. However, while tank attacks might facilitate victory by a series of captures of strategic ground, a more ambitious strategy was put forward by a British officer named J.F.C. Fuller.

Fuller advocated the viewpoint that attacking enemy troops and positions was only a means to an end. The key targets were those that would cripple the fighting power of many enemy units. Even high-value targets such as artillery batteries were secondary to the elimination of two vital enemy capabilities.

The first was command and control. By eliminating enemy headquarters, or at least disrupting their operation, it would be possible to decapitate enemy units within that headquarter's chain of command. These units might be able to continue fighting under local control, but the enemy force would become disjointed and open to a crushing attack at a point of the Allies' choosing. Some units might be induced to retreat or even surrender out of confusion or simply from lack of orders.

Higher-level headquarters would provide greater returns, of course, but these were located further from the battle lines. Thus a very deep penetration would be necessary to overrun a corps or army headquarters. There were still limits on what could be achieved, many of them technical. Tank assaults were limited by fuel capacity, breakdowns and the ability of infantry to support the tanks adequately.

The other key target was enemy logistics assets. A vast amount of food and ammunition was needed to keep the army functioning on a daily basis, and even more if offensive operations were to be considered. Thus an attack on enemy supply dumps could disrupt the performance of forward units and their supporting artillery. Troops that could not be fed might have to be withdrawn from their positions.

The logistics apparatus could also be attacked. Road and rail transport was necessary to the work of supplying the combat units; the loss of transport might be as significant as the destruction of the supplies they carried. Previous offensives had considered the possibility of cutting enemy rail links, but finally the means to sever them was available.

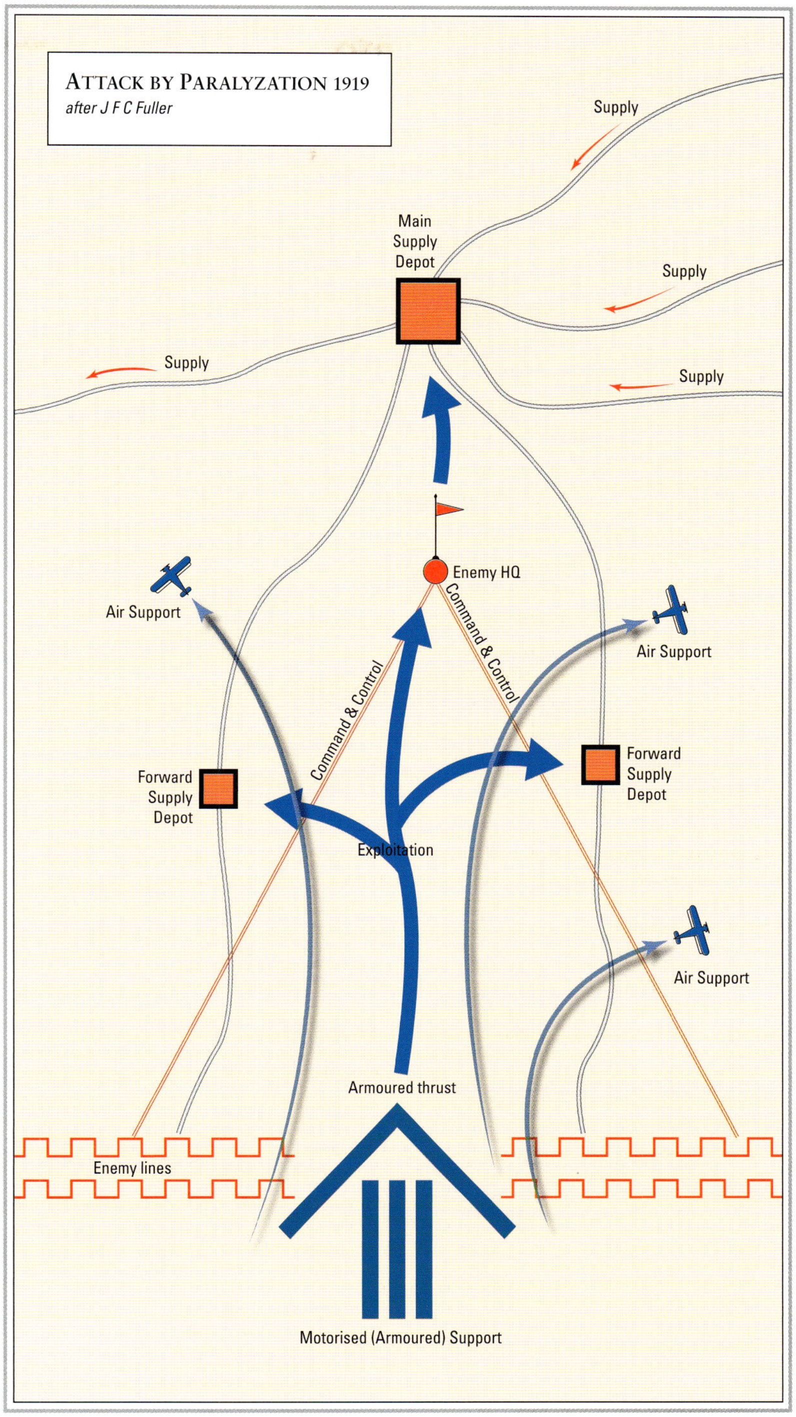

ARMOURED ASSAULT
Breaking the enemy line was a means to an end, not an objective in itself. The aim of Plan 1919 was to cripple the enemy by smashing his command and supply apparatus using a rapid armoured advance.

Armoured assault

In order to strike these targets, it was of course necessary to break through the enemy lines with

FT-17 TANKS IN TRANSIT
The French contribution to Plan 1919 would have been largely made with FT-17 tanks, which were available in large numbers. The short range of these vehicles might have limited the results that were possible.

sufficient force remaining to exploit the breach. It was also necessary to be able to hold open a line of retreat for the armoured force, in case the enemy did not obligingly flee his positions. This created a number of requirements, some numerical and some technical.

The original version of Fuller's 'Plan 1919', as the concept of a massed armoured offensive was known, called for an assault on a 145km (90-mile) front. This would require at least 5,000 tanks, all of them of designs superior to those existing at the end of 1917.

Two types of tank were considered desirable – 'breakthrough' tanks and 'exploitation' tanks. The former were heavy vehicles, able to take punishment and still break through enemy positions. They would not need a large operating range; firepower and protection were the paramount requirements. 'Exploitation', or medium, tanks would be lighter and faster, able to get out into open country and fall on the enemy rear before preparations for defence could be made. Range and speed were critically important, as well as mechanical reliability.

Fuller's plan was for a three-stage attack. Initially medium tanks, supported by aircraft, would attempt to infiltrate through the enemy defences and cause as much disruption and confusion as possible. The breakthrough would then be made by a force of heavy tanks and infantry, again supported by aircraft. Once a suitable hole had been torn in the enemy line, the heavy tanks and infantry would hold it open while a torrent of medium tanks and cavalry would rush into the enemy rear and cause as much mayhem as they possibly could.

New tank designs required

The tanks of 1917-18 were not up to the task before them. The French Renault light tank might be capable of conducting an effective massed exploitation, but its British equivalent the Whippet, was too lightly protected and lacked the range. British heavy tanks were effective, but prone to

MK VIII 'INTERNATIONAL' TANK
The Mk VIII 'International' or 'Liberty' tank was ordered in huge numbers for British and American armoured forces. Construction of a dedicated Mk VIII production plant in France was well advanced at the end of the war.

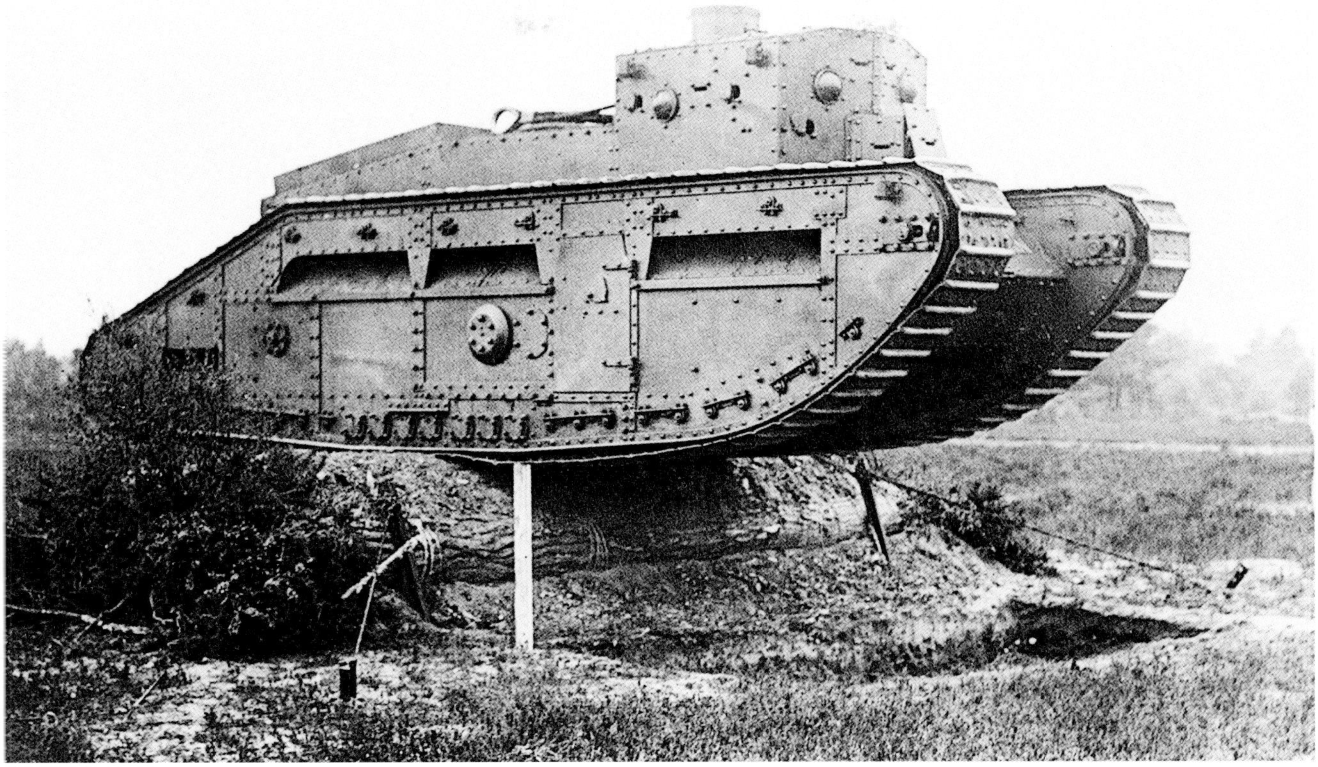

breakdown, while the French heavy tank designs were simply inadequate.

A new generation of tanks was needed, one that incorporated significant improvements. One key area was habitability; early tanks were prone to internal fuel leaks and often allowed exhaust fumes into the crew compartment, slowly poisoning their operators. Noise, and the general difficulty of operating the vehicle, contributed to both inefficiency and crew exhaustion.

Thus the new generation of tanks needed to be easier to operate and less prone to breakdowns, and must provide the crew with good protection while retaining armament levels. Trench-crossing ability was also of great importance, and for this reason the rhomboidal design was retained for both new heavy and medium designs.

The 'breakthrough' or heavy tank that would have carried out Plan 1919 was the Mk VIII 'International' or 'Liberty' tank, built around a Ricardo or Liberty engine, and built jointly by Britain and the U.S.A. at a plant in France. Orders for 4,500 of these vehicles were placed, though few were built before the war ended.

The Mk VIII was significantly larger and heavier than its predecessor, with better armour and similar armament. The two 6-pounder guns (with 57mm/2¼in shells)were carried in sponsons, with seven machine guns giving all-round anti-personnel capability. The crew compartment was separated from the engine by an internal bulkhead, which represented a significant increase in crew safety and habitability.

The Medium Mk C, or 'Hornet' tank, intended for the exploitation role, was similar in many ways, but was smaller and more lightly armoured. Some

MEDIUM MK C TANK
Designed with input from tank crews, the Medium Mk C marked the beginning of a transition from lozenge-shaped tanks to a more modern design with a much lower hull profile.

DRIVER'S CONTROLS
Even after the need for a team of drivers and gearsmen had been done away with, driving an early tank was a complex and physically demanding business even without the added stress of combat.

6,000 were ordered, of which one in three was a 'male' armed with a 6-pounder gun, and the rest 'females' carrying only machine guns. As with the Mk VIII, the crew were grouped together in a single compartment, with the engine in the rear.

French designs were also put forward, although none entered service before the war ended. The French breakthrough tank was designated Char de Fortresse 2C and was based on an enlarged version of the successful Renault light tank. In keeping with its designation as a 'mobile fortress' the Char 2C mounted a 75mm (3in) gun in a turret located forward on the hull, with a smaller turret at the rear mounting a machine gun. Additional machine guns in ball mountings along the hull flanks provided protection from infantry attack.

The Char 2C was the first multi-turreted tank to enter service, and represented a line of thinking that considered heavy tanks to be 'land battleships'. Like a ship, the Char 2C required a 12-man crew, which made command and control a problem, especially since the commander also served the main gun. By contrast the Mk VIII, similarly sized although much less heavily armoured, required a crew of eight.

Multi-turreted designs continued to appear throughout the Interwar years, including some vehicles with up to five turrets. Inefficient and often clumsy, those that saw combat proved much less effective than their vast armament might have implied. The Char 2C was still in service at the outbreak of World War II, but all examples were destroyed while being transported by rail towards the battle front. It is unlikely that they could have achieved much even had they been successfully deployed.

Air-land cooperation

Aircraft were an integral part of the Plan 1919 structure. In addition to conducting reconnaissance aircraft could also provide close support. This was important when the armoured advance outran the capability of artillery to support it. Bomber aircraft could break up enemy counterattacks and neutralize artillery positions, and could also conduct interdiction missions, disrupting the enemy's access to the battle area by bombing roads, railways, bridges and any units that tried to move along them. Lighter aircraft could strafe enemy troops or bomb them, causing confusion and preventing a coherent response.

Aircraft were also an important means of liaison with the rear. Wireless communications were at best unreliable, but an aircraft could obtain up-to-date information on what was occurring in the combat area and relay this to commanders, returning to drop messages to units operating behind enemy

MOBILE LAND FORTRESS
The Char 2C was the first of the multi-turreted tanks. It was immensely heavy, clumsy and a huge target, and required special railway wagons to transport it to the battle front. All existing examples were destroyed in transit in 1940.

lines. Communications with aircraft conducting 'contact patrol' operations, and therefore with the operational commanders behind friendly lines, could be maintained with pre-arranged signals made with flares or strips of cloth laid out on the ground.

This in turn necessitated a particular kind of aircraft optimized for contact patrols. Forced to fly low in a predictable operating area, a standard two-seater was vulnerable to enemy fire. Specialist designs, trading manoeuvrability and speed for armour plate, were put into production to support the armoured assaults of Plan 1919.

Likewise, specialist ground-attack aircraft were designed to support the tank offensives. These, too, carried armour plate to protect the crew and were optimized for low-level operations. Among the designs put forward was one that mounted no less than eight downward-firing machine guns. However, none of these specialist aircraft arrived in time to see action.

Although tank-aircraft co-operation was never implemented on a grand scale before the war ended, the concept was explored during 1917-18, laying the groundwork for German 'Blitzkrieg' tactics two decades later. Another type of aircraft inspired by the planned mass tank assaults was an early form of 'tank-buster'. One German design carried a 20mm (¾in) cannon which was sufficient to punch through the top armour of any tank then existing. This concept was never put into practice, but would have been used to try to contain the allied armoured thrusts had Plan 1919 been implemented.

T-18

Manufactured	1928
No. built	960
Weight	5.9 tonnes
Length	4.38 m
Crew	2
Armour	6–16 mm
Main Armament	37 mm M28
Secondary Armament	7.62 mm MG
Speed	10 mph (16 kph)

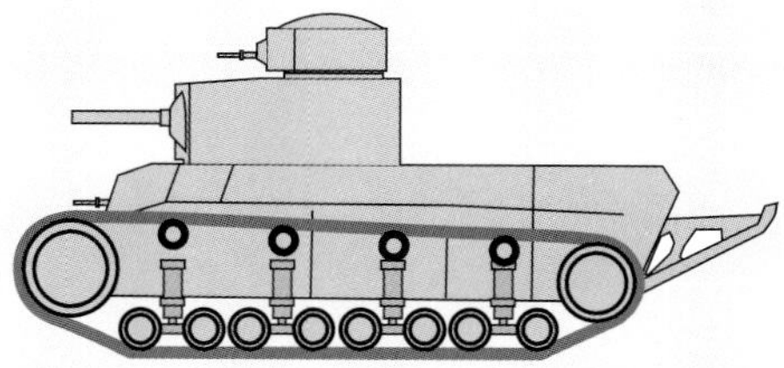

T-24

Manufactured	1931
No. built	c. 25
Weight	18.5 tons
Length	6.5 m
Crew	5
Armour	8–20 mm
Main Armament	45 mm M32
Secondary Armament	3 x 7.62 mm MG
Speed	16 mph (26 kph)

PzKpfw 1

Manufactured	1934
No. built	1,493
Weight	5.4 tons
Length	4.02 m
Crew	2
Armour	7–13 mm
Main Armament	105 mm or 120 mm
Secondary Armament	—
Speed	31 mph (50 kph)

MATURING TANK DESIGN
In the years between the end of World War I and the beginning of World War II, tank design matured. By the mid-1930s tanks began to take on a modern appearance, leaving behind many features of first-generation armoured vehicles.

Implications for the future

Although Plan 1919 was not implemented, the thinking behind it had important implications for the future. Up until this point the tank had been a support system, a device for getting infantry across no-man's land and into the enemy positions. Plan 1919 postulated the tank as the arm of decision, advancing at a rapid pace to achieve decisive results.

The logical extension of this thinking was that the tank would in future dictate the speed of advance, rather than being tied to the pace of the infantry. Armoured forces would advance as fast as the tanks could go, and other arms would have to keep up. Self-propelled guns were many years off, but already aircraft had been put forward as a possible alternative, serving as 'flying artillery'. Similarly, armoured personnel carriers had been experimented with as early as 1916, and would eventually form an essential part of the armoured force.

Ironically, perhaps, the nations that had made the greatest contribution to the theory of armoured warfare did not pursue it vigorously after the war, to the point that exponents such as Fuller were largely ignored in their home country. In Germany, however, the lessons of 1917-18, and the implications of Plan 1919 were much better absorbed. Forbidden to own tanks for many years by the Treaty of Versailles, once Germany decided not to be bound by its terms any longer, a powerful tank arm was created. Even then, there was much debate about whether tanks were to be used as part of an infantry/armour/artillery combined structure or as the main striking force of an army.

The decision in 1940 to give the tank forces their head and allow them to launch a deep-penetration assault into France was a world first. Even in the 1939 invasion of Poland, the armoured forces were used much more cautiously. However, the origins of what became known as Blitzkrieg can be clearly seen in the preparations for Plan 1919.

ANTI-TANK DEVELOPMENTS 1917–19

INTERLOCKING DEFENCES
The most effective tank defences did not rely on one type of weapon. Attacks were funnelled by mines, obstacles and strongpoints, and attacked from all sides by light and heavy anti-tank weapons.

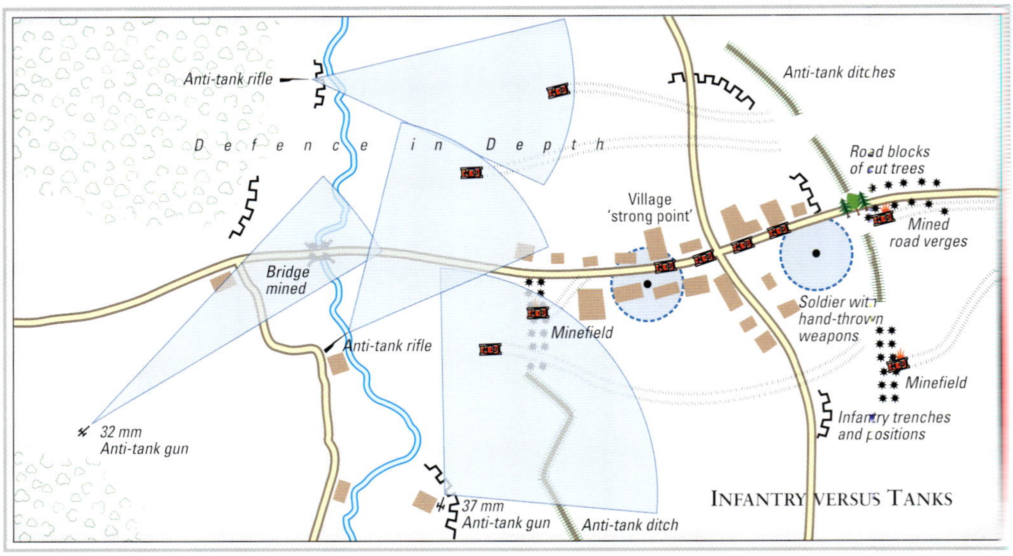

Any new weapon system inevitably stimulates a rapid search for an effective countermeasure. The tank was no exception; indeed, although Germany attempted to create an armoured vehicle development programme, much greater efforts were directed towards anti-tank capabilities. This was a matter of some urgency, as although the initial armoured attack at Flers-Courcelette did not achieve impressive results, the psychological effects of the advancing tanks on defending troops was profound. Effective countermeasures were necessary to prevent panic as much as to actually defeat the tanks.

In 1917, there seemed to be two potentially useful anti-tank defences. One was to prevent the tanks from coming within range of the defenders, or at least to slow them down. Rough terrain might cause tanks to ditch, and the longer they were in operation the greater the likelihood of a breakdown. Delay also granted additional time to call in indirect artillery fire on the avenues of attack, and to engage the tanks with direct fire. This was also more likely to be effective if the tanks were slowed down or forced to seek a way around an obstacle which might send them into a predetermined 'killing zone'. Obstacles would, of course, stop a few tanks, but unless covered by reinforcing fire they were nothing more than a delaying tactic.

Arguably, the best all-round anti-tank weapon is another tank. However, armoured vehicles dedicated to engaging and destroying tanks, such as tank destroyers and tanks equipped with armour-piercing weapons, were not deployed during World War I. Thus tanks had to be attacked and disabled by artillery and infantry. Initially, this was a matter of using weapons that already existed, but soon specialist ant-tank weapons began to appear.

Passive measures

Natural obstacles and shell craters could be an effective anti-tank defence, and troops soon learned to make use of whatever strongpoints presented themselves. Bunkers, buildings or the ruins of structures could provide infantry with concealment and shelter from the tank's weapons, as well as creating an area that could not be entered. Barricades were improvised where possible to strengthen such positions. Thus urban terrain, even on the scale of a small village or ruined farm, allowed infantry to get close to tanks if they sought a way in, and protected them from more distant ones.

The first tanks were designed with trench crossing in mind, but by simply widening a trench it could be made more difficult to cross. There was a limit to how wide a trench could be without reducing the protection it offered from shellfire unacceptably, however. An alternative was to create anti-tank ditches or 'tank traps', which might be filled with water, in front of defended positions. Water-filled obstacles were particularly dangerous as they could flood a tank's engine or drown the crew, or at least force the abandonment of the vehicle. A tank that became stuck but which could still fire was a threat to any enemy forces nearby; at times stuck tanks fought on for hours as a sort of forward-deployed pillbox.

Concrete obstacles known as 'dragon's teeth' were also used to create a barrier that tanks could not penetrate. These used fewer resources and yet were more effective than a simple wall, which could be demolished by gunfire. A row of dragon's teeth would cause a tank to run up on to one and become stuck, with its tracks unable to gain enough purchase to push itself over the obstacle or even to back off it. Dragon's teeth were extremely difficult to remove or destroy by artillery bombardment, and could effectively deny an avenue of attack.

The Allies responded to these measures by using air reconnaissance to identify the worst obstacles and planning a route between them, and by deploying fascines – bundles of sticks which could be rolled off the top of a tank into a ditch – to facilitate crossing wide trenches and anti-tank ditches.

Later British tanks were also longer, requiring a wider ditch to create a suitable obstacle. Ironically, perhaps, one effective countermeasure to anti-tank ditches was to bombard them with heavy artillery. This created very rough terrain, but tanks could cope with that; the steep sides of a ditch were likely to be turned into less difficult slopes by a near miss.

Anti-tank mines were also fielded. Often, obstacles were used to funnel attacking tanks into a minefield, which was created by burying mines where the pressure of a tank passing overhead

ROUGH-GROUND CAPABILITY
Early tanks could cope with quite large slopes as long as they were not too steep and the ground was firm enough for the tracks to grip. Although slow, tanks must have seemed unstoppable to their intended targets.

DITCHING
This British tank, captured and put into service by the German Army, has become ditched as a result of trying to climb a steep trench side. The lozenge shape and all-round tracks helped in situations like this, but there were limits to what a tank could achieve.

HIGH-ANGLE FIRE
Powerful anti-tank rifles lay at the outer end of what a single man could carry and shoot, but provided infantry with a measure of anti-tank defence at relatively low cost.

ANTI-TANK DITCH
Any ditch whose width was about equal to or was longer than the tracks of a tank was a potentially lethal obstacle. Providing the sides were fairly steep it would not be possible for a tank to unditch itself under its own power.

would detonate them. The first anti-tank mines were field expedients created by replacing the fuse of an artillery shell with a board and nail, which was pushed into the nose of the shell by pressure from above. Production of specially designed anti-tank mines began in 1918; sufficient numbers were made that some were still available at the outbreak of World War II.

Infantry weapons

Initially, the only options available to infantry under tank attack were small arms fire and grenades, and until specialized weapons were fielded, troops had to use what they already had available. These were of course the weapons that the tank had been designed to counter, and were consequently not very effective. One way for infantry to attack a tank was to throw grenades on top, though this was only marginally effective at best. Grenades often rolled off the top of a tank, especially one that was moving over rough ground. Wire mesh screens were used to prevent grenades from landing on top of the tank; instead they rolled off the angled mesh and fell to the ground. The typical fragmentation grenade of the time was not very suitable for anti-armour use in any case, and would usually be defeated by the tank's armour. Arguably, grenades were more of a hazard to the attacking infantry than to the tank. However, they could sometimes obtain a result, especially against a tank that was immobilized or caught in close terrain.

A more effective option was to throw a bag of grenades or an explosive charge under the tank in the hope of blowing off a track. Bundles of grenades, arranged around a central stick grenade, were already in use as demolition charges to attack enemy trenches and strongpoints. These were readily adaptable to anti-tank work and offered a reasonable chance to do some damage. An attack of this sort required getting very close and was extremely hazardous, but it was one of the better options initially available, as it used an existing weapon.

Rifle grenades were available during World War I, and could be fired at tanks, but they were not effective. Anti-tank rifle grenades did not appear for two decades; those available to German troops in 1917-18 were anti-personnel fragmentation weapons and not well suited to use against armoured targets. They were launched by inserting the grenade's long tail down a rifle barrel and then fired using a blank cartridge, requiring a switch of ammunition and lengthy preparation time. The high arc and inaccuracy of rifle grenades also made them unlikely to hit a moving target even if the infantryman survived being exposed to fire long enough to aim and launch his weapon.

The infantryman's rifle could also be effective against early tanks, especially if sufficient volume of fire were directed at one. While it was disheartening to see a tank rumbling onwards despite a hail of rifle fire, the effects on the crew were unsettling even if penetration were not achieved. The interior of an early tank was a hellish place already, even without the endless din of rifle rounds striking the armour. Non-penetrating rifle fire contributed to the exhaustion of the tank crew and, in the case of some tank designs, could damage components that

were not properly protected by the armour.

It was also possible to indirectly harm the crew with small arms fire. Hits on the driver's vision blocks could shatter them, sending fragments into his face. 'Bullet splash' or spallation was also a threat to the crew, as small fragments of hot metal were dislodged from the inside of the tank's armour by non-penetrating hits. These were unlikely to disable a crewman, but caused many injuries. Chainmail masks were issued to counter the threat, but were not widely used.

Lucky rifle hits could penetrate a weak spot or enter through a gun port, vision slit or other gap in the armour. While the chances of any given round doing so were low, a tank was a big target that attracted a lot of fire. Statistically, the occasional round would get through, with the chances increasing if a tank was sprayed by several machine guns. The threatening nature of tanks was such that they were likely to be treated as priority targets by machine gunners, whether or not orders had been issued to fire at them.

To enhance the effectiveness of infantry weapons, armour-piercing ammunition was issued to snipers, machine gunners and some riflemen. These 'K' rounds had been available for some time, and had been used by snipers to disable machine guns by shooting out the breech block. Their effectiveness against a metal target was proven, but in any case they were available to be quickly issued, which was of vital importance if morale was to be maintained.

Armour-piercing rifle ammunition could penetrate most early tanks, though there was no guarantee that a penetrating hit would strike anything important within the vehicle. Ironically, although British Mk IV and later tanks were much less vulnerable to K rounds, the average German infantryman did not know this and was able to face tanks with greater (albeit misplaced) confidence than previously. The availability of a countermeasure was an important factor in reducing the effects of 'tank terror'.

Specialist anti-tank rifles were also developed or converted from big game rifles. These were large-calibre rifles firing armour-piercing rounds which could punch through up to 20mm (¾in) of tank armour at ranges of around 100m (109 yards) and remained reasonably effective out to 300m (328 yards) or more. The slab sides of many early tanks were highly vulnerable to such weapons; sloped armour greatly reduced the effective range of anti-tank rifles.

The 13.2mm (½in) anti-tank rifle issued in 1918 used a scaled-up version of the bolt-action infantry rifles then in use, and may have been developed from a big game hunting weapon. Like all anti-tank rifles it was extremely heavy and suffered from fearsome recoil that could injure the operator. As a result, some anti-tank rifles were mounted on a light carriage and used like a miniature artillery weapon. Some heavy machine guns were also redesigned as specialist anti-tank weapons and directed to fire at tanks in preference to attacking infantry.

Other infantry weapons were used against tanks from time to time, such as flamethrowers. Early tanks were very vulnerable to flame weapons and burned relatively easily, but the short range of early flame weapons made using them a highly risky

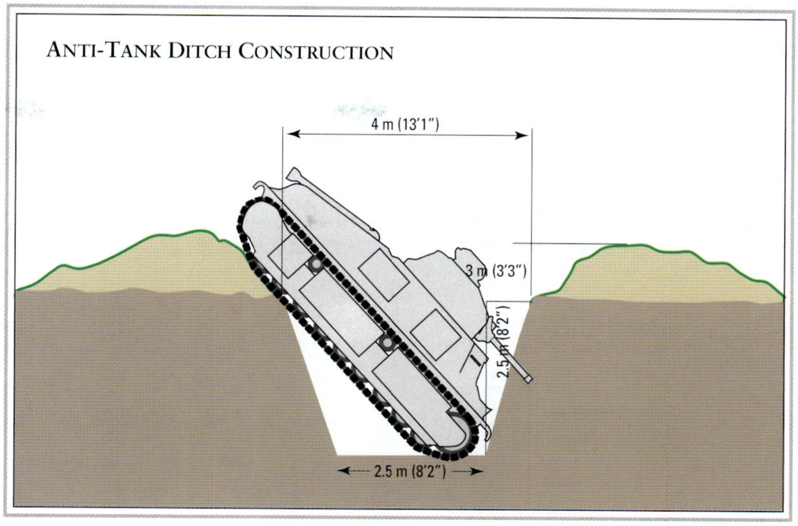

TANK TRAP
Any ditch whose width was about equal to or was longer than the tracks of a tank was a potentially lethal obstacle.

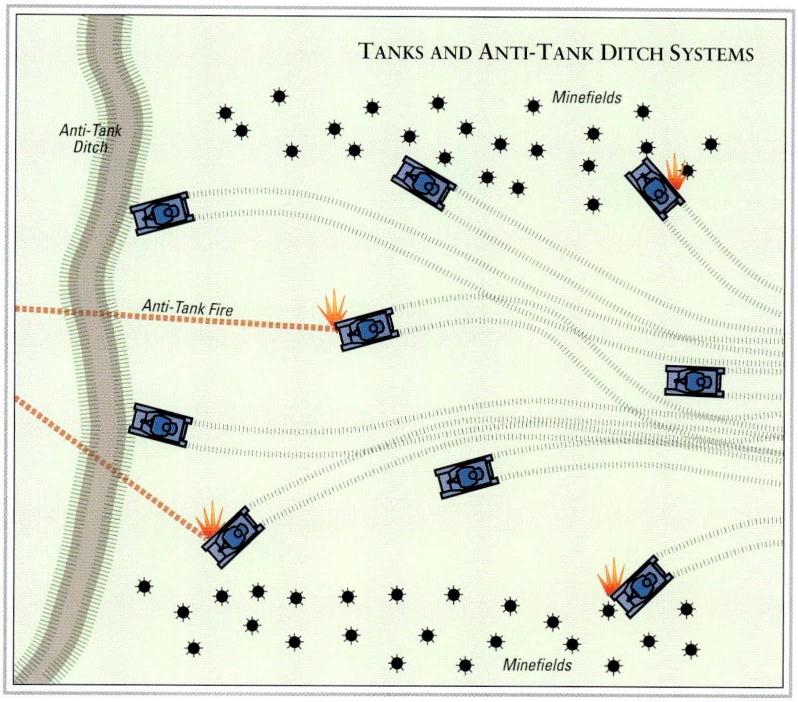

CHANNELLING AN ATTACK
Minefields and ditches were used to channel an assault into a 'killing zone' for anti-tank weapons, and to delay the tanks as they crossed the danger area. A ditch might not stop tanks but it if slowed them then the guns had more time to do their work.

FLAMMENWERFER
Early flamethrowers were effective trench-clearance weapons but were very clumsy. Their short range made it difficult to bring one to bear against enemy tanks under conditions where the user could survive firing his weapon.

prospect, especially as the user would often have to leave cover in order to use his weapon effectively. The large numbers of defensive machine guns carried aboard British 'female' tanks, in particular, made any sort of close-range assault a near-suicidal prospect.

Indirect fire with artillery could disable tanks or create ground too cratered for them to quickly cross, but it was in the direct-fire role that artillery was most effective in stopping a tank attack. This initially occurred unintentionally, as tanks approached artillery positions whose crews chose to stay and fight rather than flee. On those occasions where the gunners did not panic, they were often able to hold up a tank assault or even repel it, although they were using normal anti-personnel high-explosive ammunition instead of armour-piercing anti-tank rounds which, at the time, did not yet exist.

In response to the threat of a tank attack breaking through the forward infantry positions, artillery was considered as an anti-tank system. Almost any artillery shell would disable or at least damage a tank, and lighter guns were generally favoured. This was not least because they were likely to be closer to the front line than heavy artillery, and thus more likely to be available when an attack occurred. The guns would function in their normal bombardment role unless tanks came within range, at which point they would engage over open sights.

Guns used in this role were extremely vulnerable and often suffered heavy losses, as the typical artillery position was not set up to defend its crew from direct machine gun fire. It soon became practice for guns that might be used to repel an assault to be dug in as well as any forward infantry position. Obviously this was only possible where pre-sited guns were attacked, or where guns were moved into the likely path of an attack. Batteries that were flanked or which had to be moved during an action remained extremely vulnerable, but nevertheless this was the best available counter to a tank assault.

Specialist anti-tank guns began to appear at the very end of the war. These were 37mm (1½in) weapons set up to fire in a flat trajectory which limited their range, but made a penetrating hit more likely. Most of these weapons arrived too late to make any difference, though some saw action. In the meantime, existing artillery weapons had to fill the breach. By 1918, a segment of every artillery formation was usually designated for the anti-tank role, and was issued with steel-tipped penetrating shells. Even trench mortars were pressed into the anti-tank role. So seriously was the tank threat taken by the German Army that virtually any weapon with a chance of penetrating a tank's armour was considered a potential anti-tank system, and instructions were at times issued that all artillery and heavy machine guns should consider tanks their primary target. This was something of a redundant order, as the

DESTROYED TANK
Relatively few tanks were completely destroyed by enemy action; most were disabled but could be repaired or salvaged. Those that were comprehensively destroyed were usually the victims of artillery shells.

appearance of tanks was a severe threat which would be unlikely to be ignored by weapon crews.

This desperation to stop the tanks in some ways contributed to their effectiveness in their intended role – to get the infantry into position to make an assault. Guns that were engaging tanks were not firing at the infantry, and even if a tank were disabled, if it drew enough fire it would still have contributed to the success of the mission. Tank losses were in some ways less serious than those inflicted on the infantry – many disabled tanks could be repaired or at least scavenged for spares, but infantry casualties were rather more permanent.

Improved tactics

Soon after the initial attacks, tank defence began to be a major factor in creating a defensive position. Anti-tank weapons were found to be more effective if they were grouped rather than spread out all along the front line, and were generally sited where an attack seemed most likely. Weapons were usually employed in groups of at least two, and positioned so that a tank directly facing one weapon exposed its flank to another.

Ditches, minefields and obstacles were also factored into the creation of anti-tank zones, and were used to force tanks to advance along a predictable route; ideally one that exposed them to fire from the flanks as they turned to negotiate obstacles. Where possible weapons were concealed, to enable them to open fire from an unexpected direction. Initial shots were likely to be far more effective if the anti-tank weapons were not under fire, and of course concealment protected the guns from preparatory measures designed to suppress the anti-tank defences.

German forces also experimented with the creation of anti-tank strongpoints, creating an interlocking defence by positioning lighter weapons such as anti-tank rifles well forward, covered by mortars, artillery and anti-tank guns further back. The approaches to these strongpoints were protected by obstacles, and infantry positions prevented enemy troops from clearing the way for their tanks to advance. In addition to increasing the effectiveness of anti-tank weapons, this system made it more likely that the infantry would remain in their positions and fight it out rather than falling back in the face of a tank assault. The tanks' effect was not all physical, and by creating an effective defence the German Army also protected its men from the psychological effects of an armoured assault.

ABANDONED TANKS
A crew whose tank was disabled had only two choices; to attempt a return to their own lines across no-man's land or to stay with the tank and use whatever weapons still functioned, which would draw fire on the immobile vehicle.

THE INTERWAR YEARS

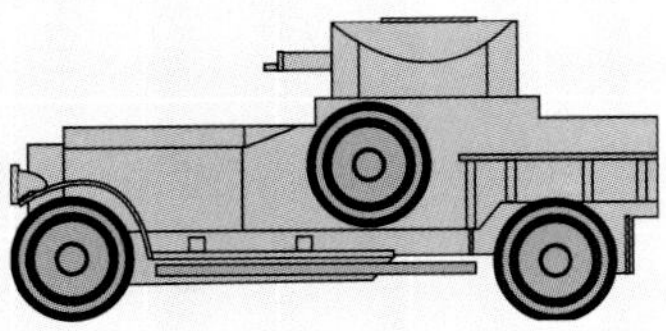

Rolls-Royce Armoured Car

Manufactured	from 1914
No. built	over 250
Weight	4.7 tonnes
Length	4.93 m
Crew	3
Armour	12 mm
Main Armament	1 x 0.303 MG
Speed	45 mph (72 kph)

ROLLS-ROYCE ARMOURED CAR
The Rolls-Royce armoured car saw worldwide service, mainly in colonial policing operations, in the 1920s and 1930s. It made a useful mobile support weapon platform but was badly outdated by the late 1930s.

The manner in which World War I ended had a number of implications for tank design and development during the next two decades. The sum total of worldwide tank experience was compressed into just a few months of rapid development and improvement, and similarly, experience both of combat and of armoured operations in general was restricted to the relatively few actions of the war.

Armoured vehicles had arrived just in time to convince military planners that they were going to be necessary in the future, but too late to establish their precise place in the scheme of things. In some nations, the sheer numbers of early-model tanks built for the war inhibited development in the 1920s.

Most nations found themselves with large numbers of legacy vehicles built for the war, and although these were primitive, using them was far cheaper than developing new designs and building them in suitable numbers. Thus France, in particular, neglected tank design and development to such a degree that 1917-vintage FT-17 tanks made up the majority of the tank forces that unsuccessfully opposed the German armoured assaults of 1940.

Britain, the nation that had at the outset led tank design and development, also neglected the new combat arm. Some thinkers did put forward tank theories including designs and tactical concepts, but they were largely ignored. Germany, the nation that had suffered the most from tank assaults, and had initially chosen not to make development of a tank arm a priority, was perhaps the most keen to experiment with armoured fighting vehicles. However, Germany was banned from owning tanks by the Treaty of Versailles.

This may have worked in Germany's favour, as it meant that Germany had no large stocks of legacy equipment and had to create a tank arm from the ground up. Rather than be influenced by existing designs, German machines were developed to suit the needs perceived by her tank proponents. Overall, this allowed freer and fresher thinking, and required that the German armoured forces, when they were finally created, be bound by fewer compromises than those of other nations.

Tankettes and armoured cars

One concept that emerged in the interwar years was the 'tankette', which was essentially a very small tank-like vehicle with light armour and

TANK-INFANTRY COOPERATION
The Soviet BT-7 combined excellent mobility and good firepower. Its armour was light but was well sloped to make the most of its protection. Subsequent Soviet tanks such as the T-34 used similar features.

weaponry. Most were equipped only with machine guns, and protected by armour that could be penetrated by infantry weapons. Tankettes were capable of performing useful service as fire-support vehicles to assist infantry, and their armour made them somewhat resilient, but they were no substitute for full-scale armoured combat vehicles.

Tankettes did have one great advantage; they were cheap. This permitted cash-strapped governments to acquire a suitably large armoured force and to gain experience of mechanized operations. Tankettes were entirely adequate for colonial policing duties such as those that the British and Italian armies were involved in during the interwar years, and proved effective against ill-armed militia such as the Chinese fielded to oppose the Japanese invasion of the 1930s.

Apart from their limited protection and armament, one main drawback with tankettes was that their crew of just one or two men had to handle a number of jobs all at once. This made already marginally useful vehicles much less efficient than larger tanks. Nevertheless, Britain included a tankette unit in its Experimental Armoured Force in the mid-1920s, while other nations, such as Poland, fielded them as frontline combat assets for lack of anything better.

The tankette ultimately proved to be a flawed concept, but it did pave the way for a number of light armoured vehicle designs. Some of these were developed from tankette designs, some were direct conversions and some were new vehicles created using experience gained with tankettes. These light tracked vehicles included a range of utility, reconnaissance and support vehicles of which the most famous is the Universal Carrier developed in Britain from a tankette designed by John Carden and Vivian Loyd. Better known as the Bren Carrier for its usual armament, the Universal Carrier was an all-purpose, open-topped tracked vehicle which could tow a trailer or an artillery piece, and/or carry troops, supplies and weapons. It was a forerunner of modern 'families' of armoured vehicles.

EXPERIMENTAL VEHICLES
The Morris Martel Tankette used an unusual drive system, with steerable wheels at the rear and most of the vehicle's weight borne by a track system at the front. It was not a success.

M2 MEDIUM TANK
The American M2 Medium Tank, with sponson-mounted hull machine guns, was a transition design which was supplanted by the M3 before it had entered service. Those that were completed served as training tanks.

MK I MEDIUM TANK
The Mk I Medium, the first post-war British tank, was also the first to have a traversing turret mounting an elevating main gun. It also had a spring suspension, greatly improving rough-ground mobility.

It is cheaper and more efficient from a maintenance and logistics point of view to use very similar vehicles for a range of roles rather than creating a custom design for each possible niche. The Universal Carrier thus acted as an early armoured personnel carrier, fire support vehicle and artillery tractor, as well as providing good off-road mobility wherever it was needed. Similar concepts were applied in other nations, and were often successful when they were not forced to replace real tanks in the direct combat role.

Armoured car development also proceeded during the interwar years. These gradually evolved from converted civilian cars into vehicles that, in many cases, resembled wheeled light tanks. Designed from the outset for military applications, the armoured cars of the 1930s possessed much better off-road capabilities than their predecessors and were often incorporated into armoured formations as reconnaissance vehicles.

Although a tracked vehicle is much more capable of crossing rough or boggy terrain, or climbing over obstacles, wheeled vehicles do offer certain advantages over tracks. Top speed over reasonably flat terrain or on roads is generally much higher, and range between refuelling is usually greater as well. Wheeled vehicles do not require as much maintenance as track-layers, and can move over long distances under their own power where tanks require rail transport or road movement via dedicated tank transporters.

Thus armoured cars offered better strategic mobility even if their tactical mobility might be less than that of tanks. Like tankettes, they were also cheap and were not banned by treaties. Germany, for example, was permitted to own and operate armoured cars for security purposes, but could not build tanks until the Treaty of Versailles was abrogated. The creation of a large armoured car force allowed German troops to gain experience of armoured operations and to pave the way for the creation of an effective tank force without incurring an international backlash. Thus German forces in 1940 used techniques that had been developed in the 1930s and even 1920s, long before Germany had any tanks with which to train.

MK II MEDIUM TANKS
An improvement on the Mk I medium, the Mk II was obsolete by the time World War II broke out. Some examples were assigned to a defensive role as dug-in pillboxes around the south of England.

As with tankettes, armoured cars were found useful in small-scale conflicts such as those that beset nations with colonies. Their combination of mobility, light protection and anti-personnel firepower was highly effective against second-rate or 'colonial' forces, which helped create an impression that heavily armed battle tanks might not be necessary. Indeed, in the 1930s Japanese tank designers pushed for a heavier tank design, but were overruled by the army, whose leaders wanted more light vehicles, as these had proven eminently suitable for the conflicts the nation was then involved in.

Multi-turret madness

False impressions about the needs and capabilities of tank design were not restricted to light vehicles. Right from the start the armoured fighting vehicle had been dubbed a 'landship' and this naval connection influenced a number of development decisions. In Britain, particularly, there was a tendency to group armoured fighting vehicles into three types.

The most powerful tanks were considered to be 'land battleships', or 'battle tanks'. Slow and short-ranged, they were to be armed with the heaviest weapons and protected by the thickest armour. Battle tanks were to be used to break through enemy positions, permitting lighter machines to exploit the breach. This role required a longer-range, more lightly equipped vehicle whose relationship to a battle tank was rather like that of a naval cruiser to a battleship. Indeed, many nations referred to these medium tank designs as 'cruiser tanks'.

Cruiser tanks were intended to be able to range all over the countryside, attacking targets of opportunity and causing mayhem once the battle tanks had broken the enemy line. They had to be able to take care of themselves and to beat off any counterattack, even when operating solo. This required a mixed armament of anti-personnel machine guns and heavier artillery, though at this time the tank was still thought of mainly as an infantry support vehicle. Many cruiser designs were thus very well protected, but equipped with wholly inadequate main guns.

Lighter tanks would fulfil the naval role of the frigate, as the 'eyes of the fleet', and some nations actually referred to their light tanks as 'frigate tanks' for a time. Their role was primarily reconnaissance and support of heavier machines, but they were quite capable of overrunning an infantry position on their own. This battle-cruiser-frigate tank model was not adopted wholeheartedly in every nation. Many, notably Germany, preferred a simpler, and arguably more sensible, designation of heavy-medium-light, though within these broad designations there was room for specialist types such as reconnaissance tanks or breakthrough tanks.

The 'landship' concept gave rise to a phenomenon which lasted only a few years: multi-turreted tanks. The concept behind these vehicles was not dissimilar to that of World War I-era vehicles: multiple weapon mounts permitted the tank to fire in several directions at once, bringing the correct weapon to bear on each target. Thus a multi-turreted land battleship was expected to be able to fight its way through a sea of enemies, fending off infantry assaults with its machine guns, while demolishing strongpoints and artillery with its big guns.

The reality of these multi-turreted monsters was that they were inefficient at best. Several turrets meant more space for (the more numerous) crew members, weapons and ammunition, as well as armour to protect them. Each turret needed a traverse and elevation mechanism, and had to be mounted so as not to impede the fire of others excessively. All this made the vehicle larger and heavier, requiring a more bulky transmission and powerplant, and this in turn added to the overall size of the vehicle.

Thus multi-turreted tanks tended to be large and slow, where they were workable at all. The least fanciful mounted a main turret and a defensive turret with a machine gun; the most outrageous carried up to five turrets. None proved more than marginal in combat, and far from battling their way to victory

MS LIGHT TANK
Derived from the French FT-17, the Soviet MS light tank incorporated an improved engine and redesigned hull form. It was an effective vehicle in the 1920s but the surviving examples that faced the German invasion of 1941 were grossly outmatched.

through a sea of enemies, most failed at the prototype stage.

There are sound reasons why a modern tank carries its weapons the way it does, with a main gun and co-axial machine gun for anti-personnel work, and a second machine gun for the commander to use as needed. This gives flexibility, while keeping down the weight, and tank designs of the 1930s gradually moved towards this concept. However, the Main Battle Tank as it exists today did not emerge until after World War II, emerging as the fusion of a medium tank's general mobility and protection combined with the powerful armament of a heavy tank.

In the meantime, although multiple turrets gradually faded from the scene, tanks tended to carry several weapons. Some, but by no means all, designs gained a co-axial machine gun in the turret, but bow guns were also common. These forward-firing machine guns had a limited arc and were offensive rather than defensive weapons. Rather than protecting the tank's flanks from infantry attack, they were aimed at targets ahead of the tank as it advanced towards its objective.

Some tanks did carry a defensive machine gun, sometimes mounted at the rear of the turret. This concept disappeared after World War II, largely because of the inefficiency of the mounting and the difficulty of using such a weapon while the main armament was also in play. Hull-mounted weapons also included sponson-mounted guns, although this was another concept that gradually faded away in time. It is easier to carry a large-calibre gun in a limited-traverse hull mount than to build a suitably large turret to carry it, but a hull-mounted gun is limited in many ways.

Not least, a hull-mounted gun cannot traverse to engage rapidly emerging targets to the flanks, and cannot fire over obstacles or cover. A tank whose main anti-tank gun is low down on the hull can be hit by a conventional tank long before it can fire back, and while this is acceptable for specialist tank destroyers, it is not effective for fluid armoured warfare.

Towards a modern tank design

Throughout the 1930s, tank design became increasingly more 'modern' in outlook. Advances in suspension and transmission permitted tanks to move faster over rough ground, though primitive fire control meant that halting to shoot was absolutely necessary if accuracy were to be achieved. The tank designs of the late 1930s, i.e. those that were in action at the beginning of World War II, were far ahead of the lumbering monsters of the Great War. However, the tank concept had not yet fully developed.

There were still questions about the tank's role to be answered. Most nations fielded a range of different designs for various applications, whereas postwar thinking would concentrate on a single general-purpose design. The main oversight, however, was the fact that most tanks were woefully underarmed for a clash with their own kind.

Many tanks, particularly British designs, used small-calibre guns that simply could not penetrate

M3 (STUART) LIGHT TANK
Developed from the M2, the M3 was given a gyro-stabilised main armament, enabling accurate fire on the move. The sponson-mounted hull machine guns were deleted on later models.

SOVIET TANK INDUSTRIES
The Soviet Union developed a large tank-production industry, often using concepts developed in other countries which were brought together to create quintessentially Russian vehicles.

other nations' cruiser or medium tanks. Many fairly large vehicles carried only machine guns and were expected to engage in anti-personnel work, or at most to attack soft targets, light artillery batteries and transport assets. Large-calibre guns were often short-barrelled, low-velocity weapons of limited utility against an armoured target.

This began to change in the late 1930s, notably after experiences in the Spanish Civil War. There, machine gun-armed Panzer Mk Is could penetrate Russian-supplied enemy tanks only by using armour-piercing ammunition at close range, prompting the creation of an upgunned design with a 20mm (¾in) main armament. Tank designs began to emerge with guns suitable for tank-versus-tank warfare, though these were still often very light. Guns in the 30-40mm (1-1½in) range seemed adequate for anti-tank work in 1939, based on experience to date and the thinking of the interwar period. The rapidity with which tank design evolved after the outbreak of World War II suggests that interwar conclusions were not always valid.

THE ROYAL TANK CORPS EXPERIMENTAL FORCE

EARLY ARMOUR
This British Mk IV exemplifies the ponderous design of World War I-era tanks. Note the sponson-mounted cannon and the forward-firing machine gun. British Mk IVs fought a German A7V in the first documented tank-versus-tank battle.

The earliest tanks in action during World War I were designated as part of the heavy branch of the Machine Gun Corps. Following the deployment of armoured cars on the European continent in 1914, tanks first saw action on the Somme in 1916. By the end of the war, 25 tank battalions had been formed, with the original eight companies upgrading to battalion strength and an additional number coming into service from the autumn of 1916 through December 1918.

In the midst of War War I, the heavy branch was separated from the rest of the Machine Gun Corps to form the Tank Corps under the command of General Hugh J. Elles. The first recorded tank-versus-tank action took place near the French village of Cachy on 24 April 1918, as the German A7V tank named 'Nixe' engaged three British Mark IV tanks. Two of the British tanks were damaged before the German vehicle was put out of action.

Tenuous future

With the end of World War I, the status of the British armed forces, particularly the Tank Corps, came into question. Maintaining a large wartime force was costly, and sceptics remained unconvinced as to the relative value of the tank on the battlefield. Reduced in strength to a mere five battalions, the Tank Corps deployed units to Russia during that country's civil war, while some armoured cars and tanks had already been sent to the Middle East.

Despite the lack of military spending and direct emphasis on the organizational structure of the armoured forces of the British Army, in 1923 King George V officially redesignated the unit as the Royal Tank Corps. On the eve of World War II in 1939, the word corps was changed to regiment, and in the preceding months, 14 additional battalions had been organized. Through a series of redesignations the entire armoured force became known as a 'regiment', while its components were grouped into units with the same name. One battalion was formed in Egypt in 1933. By the end of the war, several battalions had been redesignated as regiments and a total of 24 regiments were engaged in combat across the globe.

Cold War consolidation

In 1945, the Royal Tank Regiment was again reduced in strength. Its eight regiments remained on extensive deployments in troubled regions such as the Middle East, Korea, Southeast Asia, and Northern Ireland. In 1969, regiments were disbanded or combined to form single units, and the 1st and 2nd Royal Tank Regiments remain the operational forces of the Royal Tank Regiment today. Elements of the Royal Tank Regiment participated in peacekeeping efforts in Kosovo, in the 1991 Gulf War, Afghanistan, and the 2003 invasion of Iraq.

ARMOURED PIONEER
General Hugh J. Elles was placed in command of the British Army's Tank Corps following its organization as a separate unit. Originally, British tanks had been formed as the heavy section of the Machine Gun Corps.

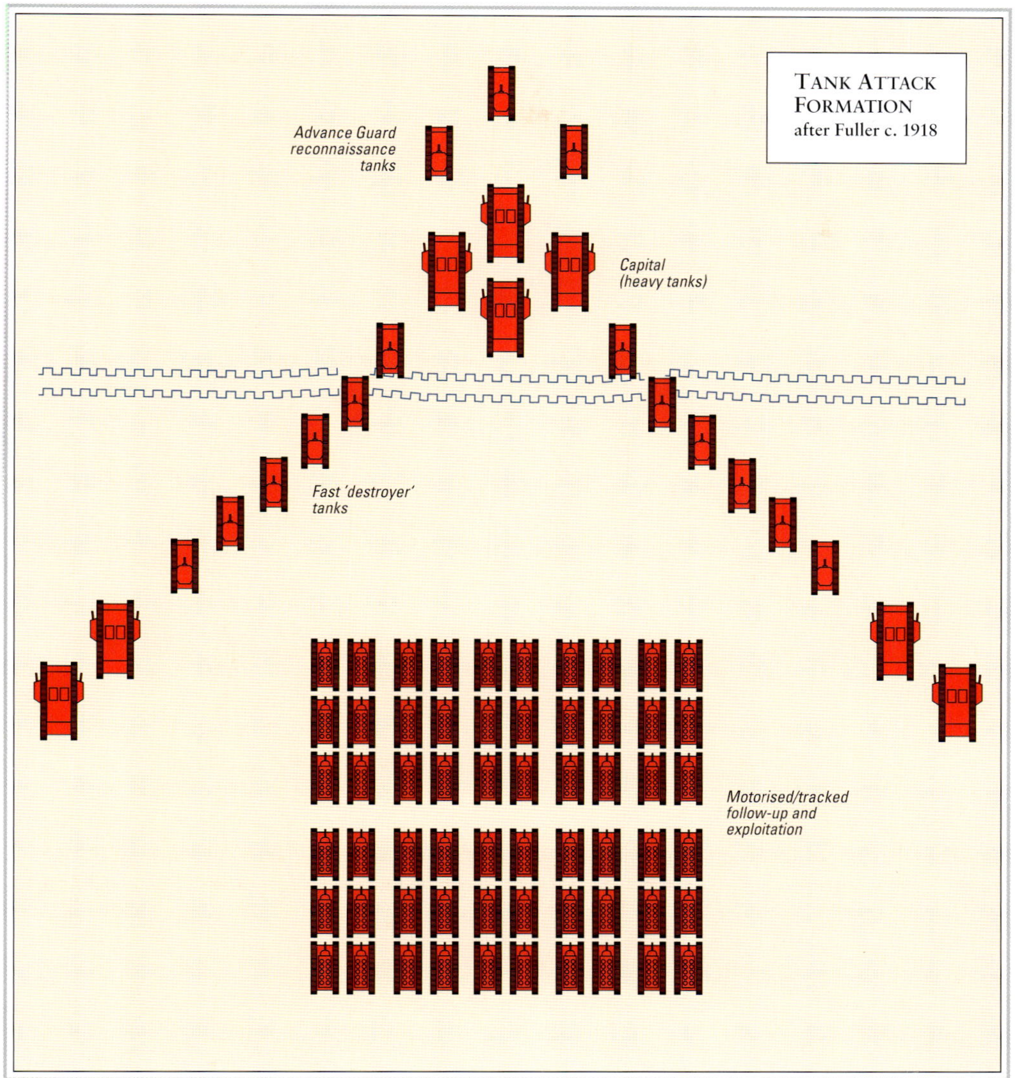

FULLER PHALANX
British armoured warfare theorist J.F.C. Fuller conceived a combined arms approach to tank deployment on the battlefield. Although his 1918 plan of attack was never enacted, his innovative doctrine influenced later generations of armoured tacticians.

Tank tactics

Throughout the development of the armoured vehicles fielded by the Royal Tank Regiment and the successive conflicts in which it has been engaged, the tactics of tank warfare have been discussed and refined, adapting to the conditions of the modern battlefield. Technological developments have yielded more potent tanks and anti-tank weapons.

The individual who was perhaps most responsible for the concept of tank warfare had actually doubted the probability of the tank being deployed successfully on the Western Front during World War I. Following two years as an infantry officer, J.F.C. Fuller was appointed as Chief General Staff Officer of the Machine Gun Corps Heavy Branch in December 1916. Although he was not familiar with tanks, Fuller rapidly began to grasp the potential for the tank to break the stalemate of trench warfare. However, he asserted that the new weapons had to be deployed on favourable ground that was suitable for the heavy, tracked vehicles to manoeuvre. Early engagements that resulted in several tanks bogging down in mud and falling into shell holes vindicated his belief.

Fuller and Cambrai

After somewhat disappointing performances at Ypres and the Somme in 1916, the future of the tank in the British Army was more in doubt than ever. Good ground for the British Mark IV tank was identified at Cambrai, east of the River Somme where costly fighting had taken place in 1916.

Fuller seized the opportunity to formulate a plan of armoured attack that would take full advantage of the favourable terrain around Cambrai. His plan included substantial artillery bombardment preceding

BRITISH WHIPPET
The British deployed the Whippet tank as a fast, maneuverable armoured vehicle capable of exploiting a breakthrough in German lines. British forces introduced the Whippet during action around Amiens in the summer of 1918.

and during the attack, including a creeping barrage moving forward ahead of the tanks as they advanced. Fuller also intended to commit more than 300 tanks to the battle and to employ them in strength rather than piecemeal. The tanks would be supported by ground-attack aircraft, advance rapidly and retire after their infantry occupied the territory that had been gained.

As the battle developed, good progress was initially made. However, an error on the part of one divisional commander resulted in supporting infantry following the tanks into battle at a 200m (218-yard) interval rather than 100m (109 yards) as prescribed, leaving the tanks without their all-important screen. In response, the Germans lay in wait with their clusters of hand grenades bound together as improvised anti-tank bombs and 77mm (3in) field guns poised to fire as the ponderous Mk IVs topped a ridgeline. In minutes, nearly 30 British tanks were knocked out in a small sector of the battlefield.

Within a week, the offensive at Cambrai lost its momentum, scores of tanks had been lost to enemy fire, mishap, or mechanical failure, and the British had retired.

Forward-thinking Fuller

The bitter lessons of Cambrai led Fuller to conclude that tanks had to be more mechanically reliable, that it was crucial for their supporting elements to be within reasonable distance and coordinated with one another. Finally, he realized that armoured vehicles needed speed and freedom of movement.

With the introduction of better protection for low-flying aircraft, ground support from the Royal Flying Corps became more effective, while Fuller reiterated his doctrine of the effective armoured spearhead. At Hamel in July 1918, a force of 60 Mk IV tanks performed well beneath an umbrella of aircraft protection and the fire of nearly 600 artillery pieces.

The following month at Amiens, the British introduced the fast, manoeuvrable Whippet tank, a light vehicle capable of rapid exploitation of a breakthrough in German lines. The Whippet, also known as the Medium A tank, was armed with four 7.7mm (1/3in) machine guns and capable of speeds up to 13km/h (8 mph) with a range of 64km (40 miles). The Whippet was served by a crew of three.

CAMBRAI BATTLEFIELD
A disabled British Mk IV tank lies amid barbed wire entanglements on the battlefield of Cambrai. The performance of the British tanks at Cambrai was disappointing due to poor command and the rapid reaction of German artillery.

In sharp contrast, 18 men were aboard the slow, unwieldy German A7V tank.

While the Germans continued effective anti-tank tactics, the British made deep penetrations at Amiens and consolidated their gains as senior German commanders recognized the changing dynamic on the Western Front. The German High Command was taken aback by the rapid British advance and remained off balance for some time as an air of defeatism began to permeate the ranks of the German Army.

Tactical refinement

Even as his Tank Corps was at long last realizing some of its potential on the battlefield, Fuller was continuing to refine his tactical armoured doctrine. Anticipating a continuation of the war beyond 1918, he saw the growing capacity within the British military for co-ordinated offensive action and understood that the key to future success for the tank in warfare was contingent upon support from the air and the artillery, while infantry protected the vulnerable blind spots of the armoured force during an advance into enemy territory.

Fuller envisioned a comprehensive battle plan that influenced the conduct of future ground wars, but this was never put forward by the British Army in World War I once peace was declared on 11 November 1918. His plan, however, served as a template for wars yet to come.

Fuller's plan envisaged that as the offensive opened, British tanks would attack in a formation resembling a gigantic wedge. While an artillery barrage softened enemy defences and maintained close fire support for advancing formations, a large number of aircraft would strike ground targets, disrupting communications and preventing reinforcements. A few reconnaissance tanks were to perform scouting duties and probe enemy lines for weak points. These were followed by the powerful sledgehammer blows of the heavier Mk V tanks that would achieve a breach in German lines.

Once the decisive breakthrough was achieved, the light Whippet and newer models of fast, manoeuvrable tanks that were still in development when the war ended, would perform the function of shock cavalry, pouring through any substantial breach and creating havoc in the enemy rear. The latest light tanks were capable of speeds of up to 30km/h (20 mph), and Fuller considered these essential for the attack. They would destroy German command and control, render ineffective their attempts to counterattack, and, ideally, capture forward defensive positions from the rear.

Fuller's plan also gave birth to the concept of motorized infantry, keeping pace with the rapidly moving tanks. The infantrymen were to mount trucks and roll forward through the breach, eventually dismounting to attack fixed fortifications, reduce pockets of resistance, and occupy territory gained by the tanks during their rapid, slashing advance.

Practical application

While it is a moot point whether the German Army of the 1930s embraced Fuller's tactical vision as a basis for their future Blitzkrieg across Europe, it must be acknowledged that Fuller influenced the thinking of battlefield tacticians. Although Fuller had his detractors and was confronted by those who doubted the worth of his theories, his ideas were vindicated on the field of battle.

J.F.C. Fuller was fortunate in that the opportunity to prove his tactical armoured doctrine presented itself, at least partially, during the latter stages of World War I. The campaigns fought on the Western Front served as the laboratory for his innovative experimentation and allowed for adjustment and a measure of trial and error in the process. Otherwise, his ideas would have existed only in the realm of the training manual and the war game. As it turned out, these early operations of the experimental Royal Tank Corps continue to influence the deployment of armoured vehicles and battlefield coordination to this day.

BRITISH MK V
The Mk V tank entered service with the British Army late in World War I and saw limited action. Its firepower included two six-pounder sponson-mounted guns and four 7.7mm (⅓in) machine guns.

SOVIET DEEP PENETRATION THEORY

The Soviet Red Army doctrine of 'Deep Penetration' was born out of the frustration and military decline experienced by the armed forces of Imperial Russia from the Crimean War (1854–56) to the negotiated peace with the Germans that took the nation out of World War I in 1917. Other stinging defeats of the Tsarist forces had taken place at the hands of the Japanese during the 1904–05 Russo-Japanese War and in the conflict with Poland during the early 1920s.

Following the Bolshevik Revolution, several visionary senior commanders of the Red Army sought to remedy the decline in Russian military performance and to establish both tactical and strategic cooperation in combined arms. The proponents of Deep Penetration further introduced a third element into military planning and execution that of operations. Among the Red Army officers who crafted the innovative doctrine were Mikhail Tukhachevsky and Vladimir Triandafillov. Tukhachevsky was a veteran of World War I and the Russo-Polish conflict, who was eventually executed by Soviet Premier Josef Stalin during the great purge of the Red Army officer corps in the mid-1930s. Triandafillov had also served in World War I. He rose to the position of deputy chief of the general staff, wrote extensively on modern military operations, and was killed in a plane crash in 1931

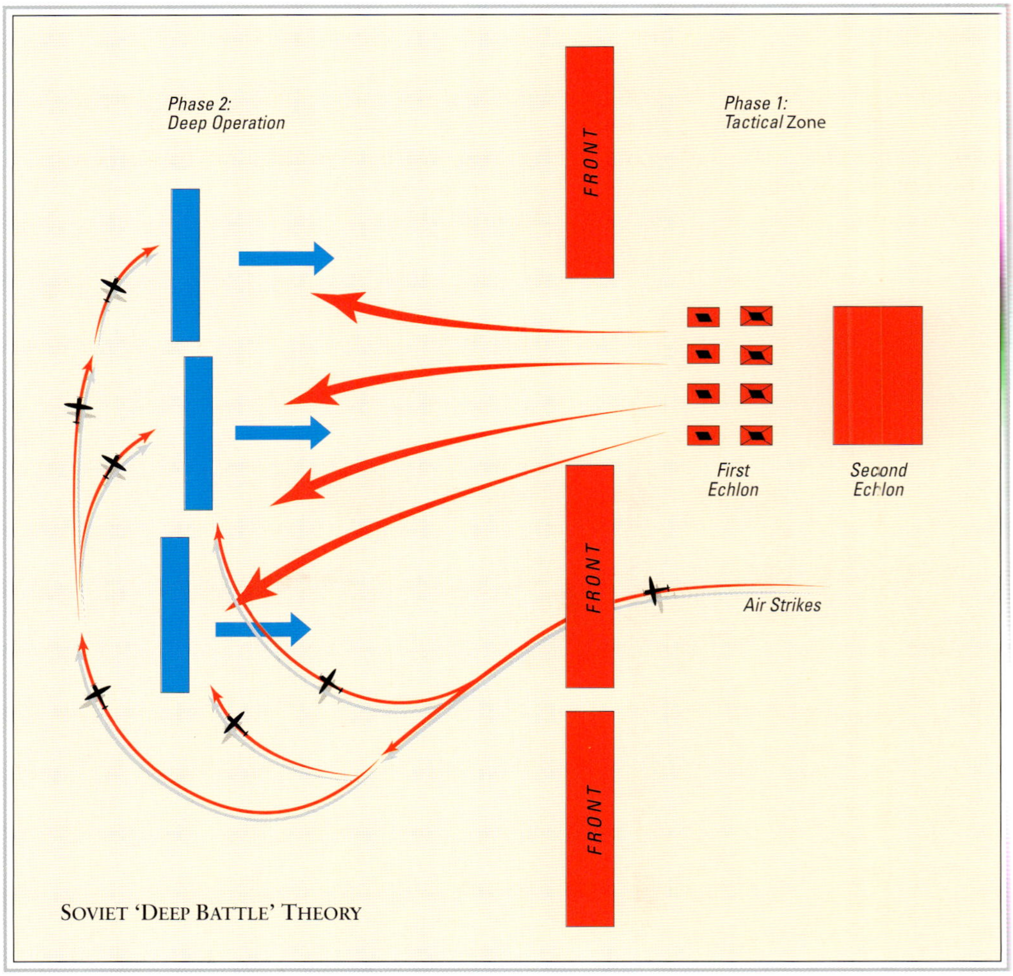

DEEP PENETRATION
The Soviet concept of Deep Penetration involved keeping the enemy off balance and achieving a breakthrough for armoured exploitation. Soviet commanders were further tasked with envisioning the culminating fight of a campaign as they prosecuted the initial battle.

A new perspective

The essence of Deep Penetration theory was to attack the enemy at multiple points along his tactical front simultaneously, keeping senior commanders off balance and unable to commit reserves to an area clearly identified as the focus of a breakthrough effort. At the same time, the enemy was to be attacked in depth throughout his area of ground operations. Such thorough offensive action would serve to erode combat effectiveness rapidly and eventually result in a breakthrough of the enemy tactical perimeter.

Following the achievement of the breakthrough, mobile forces would strike swiftly and deeply into the enemy rear, destroying command and control centres, wreaking havoc with supply and reinforcement efforts, and creating general confusion. Ultimately, Deep Penetration would conclude with the capture of the Red Army's strategic objectives.

As Triandafillov conceived of Deep Penetration, Tukhachevsky advocated that Soviet commanders alter their perspective of the battlefield. The battle was not simply a linear confrontation between two opposing forces. The battle and the ensuing campaign must be considered as a whole and continually managed. The victorious commander was the one that was capable of envisioning the final battle of the campaign, even as he was preparing to fight the first.

MARTYRED MARSHAL
Remembered on a postage stamp, Soviet Marshal Mikhail Tukhachevsky was an architect and early advocate of the Deep Penetration doctrine. However, he fell victim to Stalin's paranoia and was executed during the great purge of the Soviet officer corps.

Such overarching coordination of military forces required the unprecedented cooperation of combined arms. The role of the tactical unit, typically up to divisional strength, was to fix the enemy along the front. Once a breakthrough was achieved, the mobile forces, primarily the armoured corps and the shock armies, would exploit the advantage to its fullest, rapidly slashing into the enemy rear and causing resistance to collapse within the operational area as enemy troops realized they were being assaulted from behind. Heavy artillery bombardment and tactical air strikes were essential elements of the concerted offensive effort.

SOVIET KV-1
The thickly armoured Soviet KV-1 tank mounted a 76.2mm (3in) gun. Its thick armour was impervious to German 37mm (1½in) shells; however, it was costly to build and was supplanted by the more economical T-34.

STALIN STEEL
The heavy Soviet IS-2 tank entered service in the autumn of 1943 and saw action during the Battle of Kursk. Mounting a 122mm (4¾in) main gun, the IS-2 was also heavily armoured to counter German 88mm (3½in) shells.

The successful Deep Penetration would bring about profound shock among the enemy and an inability to respond effectively.

Tukhachevsky the teacher

It was Tukhachevsky who primarily advanced the concept of an offensive Deep Penetration among the Red Army leadership after the death of Triandafillov. The Tukhachevsky theory consisted of five major components. Initially, it was to be understood that tactical units functioned in support of operational manoeuvres. Continuing offensive operations along a broad front further inhibited the enemy's ability to respond to any penetration that might occur at one or more points along an extended perimeter. The depth of penetration and the speed with which that dagger thrust was executed would increase the shock and confusion among enemy forces, both forward and to the rear. Advancing technology, Tukhachevsky further reasoned, would be instrumental in augmenting firepower and manoeuvre to increase the effectiveness of the doctrine. Lastly, the Red Army commander must be capable of strategic vision as well as tactical skill, viewing the current and future battles as a continuous operation, planning for the battles yet to come while in the midst of the demands of an ongoing fight.

Formal Adoption

By 1933, a manual titled *Provisional Instructions for Organizing the Deep Battle* had been produced by the Red Army. In the execution of the Deep Penetration, multiple Red Army Fronts, essentially army groups consisting of several shock armies, would function in both simultaneous and successive operations to bring about the collapse of enemy defences and an inability to respond to the Soviet pressure effectively. Continuing pace and progress would facilitate the attainment of strategic goals by supporting fronts, although the Soviet High Command would identify the main strategic objective of the entire offensive.

Planning for the launch of a Deep Penetration offensive began with the identification of the primary strategic objectives. From there, the tactical aspect was refined. Typically, the rifle corps was the principal element of the tactical fighting, and the formation was reinforced with tanks and artillery. Attacks were prosecuted in three echelons with the first initiating combat and breaking through in one or more of the tactical zones. The second echelon would solidify the gains made at the point of breakthrough and immediately beyond. Finally, the third echelon would move rapidly through the breach and begin the swift exploitation of it. Meanwhile, holding elements, primarily consisting of rifle corps troops, would defend the flanks of the advance, fending off counterattacks or conducting limited offensive operations of their own.

Although the focus of Deep Penetration was offensive, it also accounted for defensive preparation, identifying strategic geographic locations that might be attractive to an enemy and setting up stout defences involving tactical zones of responsibility that would be fortified and held by infantry and artillery formations. Beyond the immediate defensive perimeter, extensive minefields were to be sown to impede the progress of any advancing enemy force. Located some distance from the strategic position, the tactical zones were intended as a critical defensive element to wear down the combat effectiveness of an attacking force, causing it to lose momentum and absorb casualties as its will to wage an offensive campaign inevitably dwindled, leaving it open to a decisive counterstroke.

Pragmatic application

Deep Penetration theory seems well suited for the Red Army of the 1930s, which included a large number of troops that were, for the most part, inadequately trained. Red Army planners might utilize the masses of poorly trained troops and the vastness of Russia itself to execute Deep Penetration tactics, operations and strategy along a front that extended many miles. Multiple breakthroughs might be achieved, heightening the effect of the doctrine as the enemy was overwhelmed in attempting to contend with large numbers of Soviet troops and armoured vehicles which rapidly advanced into rear areas.

In contrast to Deep Penetration, the German theory of Blitzkrieg relied on the tactical concentration of armed might at a single location, the *Schwerpunkt*. Tactical combined arms cooperation would achieve the single decisive breakthrough for the Germans. Insomuch as the two combat theories were dependent upon combined arms, they were alike. However, from there the similarity faded.

Numerical superiority

Wherever possible, it was assumed that the Red Army would maintain numerical superiority during the execution of Deep Penetration. Although the forces allocated to specific functions varied from time to time, in 1943 the rifle army included three corps and as many as 12 rifle divisions, four artillery regiments specifically designated as field, anti-tank, anti-aircraft and mortar, along with communications elements to coordinate with air assets and other formations. The rifle corps was comprised of three divisions, an artillery regiment, signals and engineer troops. Each rifle division included three rifle regiments, an artillery regiment, an anti-tank battalion and supporting formations. Altogether, a single rifle division included nearly 9,400 soldiers, 44 field artillery pieces, nearly 50 anti-tank guns, and 160 or more mortars. Elite guards rifle divisions numbered more than 10,500 troops.

Carrying the fight

Obviously, momentum had to be maintained for the operation to succeed. Merely breaking through the enemy lines was insufficient to be deemed successful, and recent experience on the battlefields of World War I had revealed that initial tactical successes often withered on the vine as the enemy

ABANDONED HULK
Its turret completely blown off the chassis, a knocked out German PzKpfw IV tank lies on a desolate battlefield in the winter of 1943. The workhorse of the Panzers, the PzKpfw IV was produced in great numbers.

TAMED TIGER
Captured in the Ukraine in 1943, this German Tiger tank appears to have been damaged and left on the battlefield by its crew. The Tiger's 88mm (3/12in) main gun and hull mounted machine gun are visible in this photo.

WINTER WASTELAND
Abandoned in the wake of the retreating German Army, armoured vehicles and tanks litter a frozen winter landscape. In early 1944, the resurgent Red Army initiated a series of winter offensive operations that left the Germans reeling.

recovered and responded in force. Speed and depth were key elements of the Soviet doctrine, and their improving weaponry would help make the Deep Penetration theory plausible on the modern battlefield. Tanks with greater firepower and range could exploit a breakthrough rapidly, while the increasing mechanization of the Red Army would facilitate the movement of motorized infantry and follow-on elements. A new generation of Soviet aircraft was becoming operational as well.

The primary component of the penetrating thrust was the shock army, initially comprised of up to 18 rifle divisions, 20 artillery regiments and 12 tank battalions. The powerful force included up to 300,000 men, more than 3,200 artillery pieces, at least 700 aircraft, and more than 2,800 tanks.

Numerous shock armies were expected to be operating in concert along the broad perimeter, and these were grouped in the Red Army Fronts. Following the death of Triandafillov, the composition of the shock armies in the attack was formalized, with each deploying in two echelons. The first echelon consisted of multiple rifle corps, and the second included reserves and reinforcements that were intended to sustain the momentum of the offensive drive. As these echelons moved to moderate depth behind the enemy's crumbling tactical front, operational troops were unleashed to carry out the devastating and demoralizing penetration that would carry the Red Army to victory.

As the Deep Penetration offensive took shape, concerns were raised that an attack in echelon could prevent sufficient forces from being brought to bear in the event of a counterattack by a superior enemy force. However, this issue was resolved as echelon troops would be directed to strike the flanks of any enemy troop concentrations of consequence until the main Red Army forces reached the scene. Therefore, a major engagement was to be avoided until the strength of the advancing Red Army had achieved critical mass in the area of operations.

Sidelined by purge

Ironically, a number of the Red Army senior commanders who had developed the Deep Penetration theory were executed, imprisoned or dismissed from the service during the purge of the Soviet officer corps in the late 1930s. As a result, Deep Penetration was shelved due to its association with those who were 'guilty' of conspiring to oust Stalin. When Hitler turned on the Soviets and invaded Russia in June 1941, the Red Army had already been crippled by the loss of a generation of its high-level commanders. Further, its operations were impaired by the absence of the Deep Penetration doctrine in its defensive form.

Once the Red Army had managed to stabilize its situation and prevent the Nazis from capturing Moscow in the winter of 1941, the doctrine of Deep Penetration began to reemerge. Apparently, Stalin had begun to realize that its merits transcended any perceived political infighting that had taken place prior to the opening of hostilities with Nazi Germany. Although the Red Army had sustained appalling losses, the reservoir of manpower and impressive Soviet industrial capacity fuelled a reversal of fortune.

In late 1941 and early 1942, the Red Army attempted a Deep Penetration operation to push the Germans back from the Soviet capital. Although there were some initial successes, the offensive failed primarily due to the unsatisfactory performance of Soviet mechanized units intended to advance rapidly during the operational phase. A later offensive failed near the city of Kharkov.

MOBILE ARTILLERY
A Soviet ISU-122 self-propelled gun advances through the Ukraine during the summer of 1944. Mounting a 122mm (4.8in) howitzer gun, the ISU-122 performed various roles, including mobile artillery, close infantry support and tank destroyer.

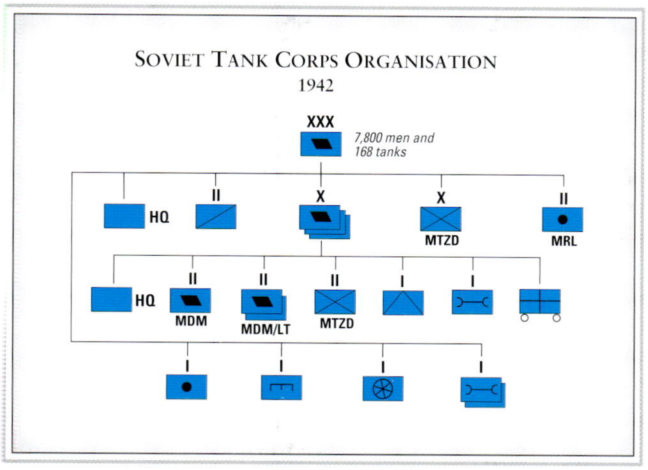

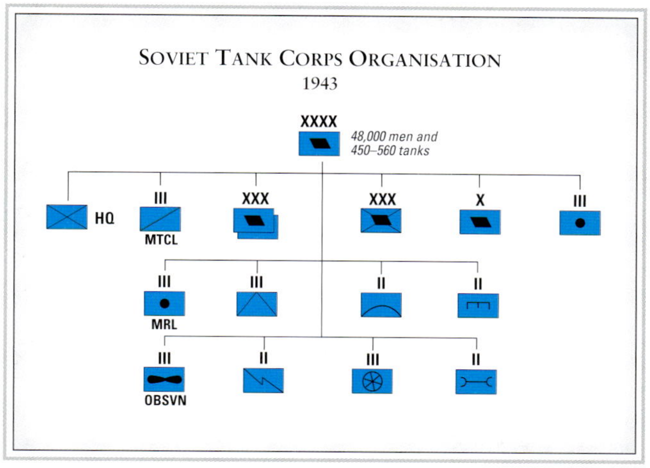

In response to the German offensive against the Kursk salient in the summer of 1943, the Soviets were able to execute their counterattacks based on classic Deep Penetration doctrine. The Germans had neglected their flanks in order to support their forward elements in the offensive, and in the north the Soviets made significant gains, threatening the rear areas of German Army Group Centre before concentrated Luftwaffe air strikes brought the advance to a halt. On the south shoulder of the salient, German forces penetrated Soviet defensive rings and broke through at Prokhorovka, southwest of Kursk. The largest tank battle in the history of warfare ensued, and losses were extremely heavy on both sides. Reinforcements intended for other purposes were reassigned, and the German tide was reversed.

Soviet surge

Following their victory at Kursk, Soviet forces seized the initiative on the Eastern Front. A string of impressive victories throughout 1944 resulted in the validation of the Deep Penetration theory. On several occasions, large numbers of German troops were encircled, cut off from resupply or relief, and forced to surrender. Wide swaths of territory were recaptured by the Red Army, which advanced as a seemingly unstoppable force. By the spring of 1945, the Soviets had advanced to the outskirts of the Nazi capital of Berlin.

Giant Soviet pincers closed around the city, trapping German formations attempting to escape to the west. Intense urban fighting ensued, with Soviet troops forced to root Germans from the rubble of blasted buildings and even from the city's sewer system. At the end of April, the red banner of the hammer and sickle flew from the former bastions of Nazi power.

Continuing influence

Some Western sources indicate that the Deep Penetration theory remained a component of Soviet military doctrine during the Cold War. It has been said that the theory even influenced the NATO doctrine of Follow-on Force Attack, which involved the location and fixing of a Warsaw Pact force on the ground while its following support forces were attacked and neutralized before they could come to the aid of the otherwise occupied frontline troops.

The long shadow of the Deep Penetration theory remains evident in the study of World War II on the Eastern Front, while its influence has been acknowledged in modern warfare for nearly a century. The prospect of an even more devastating Deep Penetration, involving tactical or strategic nuclear weapons, has remained only in the realm of theory and conjecture.

CORPS STRENGTH
From 1942 to 1943, the strength of the Soviet tank corps was increased significantly in personnel and tanks as manpower and equipment became more plentiful. The tank corps formed the backbone of Soviet armoured operations.

CLOSE SUPPORT ASSAULT GUN
This photo taken during the final weeks of the war in May 1945 shows a Soviet ISU-122 assault gun in a Berlin suburb. Scrawled across the wall to the right is the hollow slogan, 'Berlin remains German'.

ACHTUNG PANZER!

PANZERKAMPFWAGEN III
The German PzKpfw III proved versatile in the North African desert and operations in France, the Low Countries and the Soviet Union. A significant improvement to earlier designs, it was upgunned with 37mm, 50mm and 75mm (1½in, 2in and 3in) main armament.

Ironically, it was the German Army that most powerfully demonstrated the offensive capability of armoured forces on the modern battlefield during World War II. The Germans had been the first to feel the weight of enemy armour during World War I, and the immediate lesson was not lost on a few visionary officers. Chief among these was General Heinz Guderian, who rose to prominence during the years between the world wars and is credited with the practical application of a theoretical combat tactic that came to be known as Blitzkrieg.

While some historians have discounted the idea of Blitzkrieg as a real, tangible concept within the German military, there is no doubt that by 1939 the German armed forces were indeed employing coordinated armoured, infantry, artillery and air power to rapidly demoralize and defeat their enemies. The foremost weapon of the German armed forces that ravaged Poland, France and the Low Countries through the spring of 1940 was the tank, concentrated for offensive operations and delivering mobile firepower.

PANZERKAMPFWAGEN I
The PzKpfw I was heavily involved in the Polish campaign after its operational debut in the Spanish Civil War. Although weak in combat, it formed a large portion of Germany's tank strength in numbers and was used in all major campaigns between September 1939 and December 1941.

Covert development

At the end of World War I, the German military was seriously restricted by the Treaty of Versailles. The standing army was limited to only 100,000 men, while Germany was prohibited from producing or placing in the field tanks, armoured cars or any other transport vehicles of offensive capability. Nevertheless, the Reichswehr High Command embarked on a campaign of covert training and development by the mid-1920s.

The embryonic Panzerwaffe, or armoured forces, actually took shape at a secret training facility at Kazan in the Soviet Union where German troops learned to operate British, French and Soviet tanks, while Allied inspectors were oblivious to the treaty violation. At the same time, the Reichswehr maintained a high command structure that was also prohibited by the Versailles Treaty. By the end of the 1920s, the first tanks produced in Germany had been organized into seven euphemistically named 'transport battalions'. Ostensibly formed to carry supplies and move equipment, these formations were intended for future combat operations and trained regularly for that purpose. Opposition to the development of tanks persisted, and General Otto von Stulpnagel, Inspector of Motorized Troops, once informed Guderian, 'You are quite too impetuous. Believe me, in our lifetime neither of us will ever see German tanks in operation'.

Panzer power

Despite political infighting and stubborn opposition, Guderian benefited from one high-ranking supporter, Major General Oswald Lutz, who succeeded Stulpnagel as Inspector of Motorized Troops. Lutz recognized the potential for the tank in modern warfare and promoted Guderian to Chief of Staff of Motorized Troops. The two worked closely together, successfully promoting their agenda and silencing critics. While training exercises involving infantry, dummy tanks, artillery and the dwindling number of operational cavalry units in the German Army continued, the progenitor of the powerful, mobile German armoured vehicles of World War II took shape.

The Panzerkampfwagen I Ausf. A, designed and manufactured by the legendary Krupp firm, entered service with the Panzerwaffe in 1934, following more than a year of design, production and field testing. Armed with a pair of 7.92mm (⅓in) MG13 machine guns, the PzKpfw I weighed 5.5 tons and carried a crew of two. That winter, Hitler attended a demonstration of armoured vehicles at the army proving ground at Kummersdorf. He was reportedly

ARMOURED AUGMENTATION
The speed, range, armament and numbers of armoured vehicles manufactured by the major European powers increased steadily during the interwar years, and a pronounced division of labor among tanks emerged as battlefield tactics were considered.

TANK TABLE
1916–1933

TANK DESIGNATION	COUNTRY	CREW	ARMAMENT Gun in mm	MG	AMMUNITION	ARMOUR in mm	SPEED in km per hr	RANGE in km on one tank	CLIMB in °	WADE DEPTH in m	WEIGHT (mt)	HORSE-POWER	LENGTH (m)	WIDTH (m)	HEIGHT (m)	GROUND CLEARANCE in m
Heavy Mk. I, 1916	GB	8	2 x 57	4	—	5–11	5,2	24	22	1,00	31	105	8,6	3,9	2,61	0.45
Heavy Mk. V, 1918	GB	8	2 x 57	4	2,000 gun & 7,800 m.g.	6–15	7,5	64	to 35	1,00	37	105	9,88	3,95	2,65	0,43
Heavy Schnelder, 1917	FR	6	75	2	96 gun & 4,000 m.g.	5,4–24	6	75	30	0,80	13,5	60	6	2	2,40	0,40
Light Renault FT, 1917	FR	2	37	or 1	240 gun & 4,800 m.g.	6–22	8	60	45	0,70	6,7	40	4,04	1,74	2,14	0,50
Heavy St Chamond, 1917	FR	9	75	4	106 gun & 7,488 m.g.	5–17	8,5	60	35	0,80	23	90	7,91	2,67	2,36	0,41
Medium Mk. A Whippet, 1918	GB	3	—	3	5,400 m.g.	6–14	12,5	100	40	0,90	14	90	6,08	2,61	2,75	0,56
Medium Vickers Mk. II, 1929	GB	5	47	6	95 gun & 5,000 m.g.	8–15	26	220	45	1,20	13,4	90	5,31	2,74	3,00	0,45
Medium A 7 V, 1918	GER	18	57	6	300 gun & 18,000 m.g.	15–30	12	80	25	0,80	30	—	7,30	3,05	3,04	0,50
Light L K II, 1918	GER	4	—	1	3,000 m.g.	bis 14	18	—	45	1,00	9,5	60	5,70	2,05	2,52	0,27
Heavy Vickers Independent, 1926	GB	10	47	4	—	20–25	32	320	40	1,22	30	350	9,30	3,20	2,75	0,60
Heavy Char 3 C, 1928	FR	13	1 x 155 1 x 75	6	—	30–50	13	150	45	2,00	74	1980	12	2,92	4,04	0,45
Light Renault N C 2, 1932	FR	2	—	2	—	20–30	19	120	46	0,60	9,5	75	4,41	1,83	2,13	0,45
Medium T 2, 1931	US	4	47	1 x 12 1 x 7.6	75 gun, 2,000 & 18,000 m.g.	6,35 bis 22	40	145	35	1,20	13,6	323	4,88	2,44	2,77	0,44
Light Mk. II, 1932	GB	2	—	1	4,000 m.g.	8–13	56	210	45	0,75	3,6	75	3,96	1,83	1,68	0,26
Light Renault U E	FR	2	—	1	—	4–7	30	180	38	0,70	2,86	35	2,70	1,70	1,17	0,26
Light Carden-Lloyd (Russkii)	GB	2	—	1	2,500 m.g.	bis 9	9.7 in water otherwise 64	260	30	amp.	3,1	56	3,96	2,08	1,83	0,26
Fast Christie	RUS	3	47	1	—	6,35 bis 16	110 on wheels 62 on tracks	400	40	1,00	10,2	343	5,76	2,15	2,31	0,38
Light Fiat Ansaldo, 1933	ITY	2	—	1	4,800 m.g.	5–13	42	110	45	0,90	3,3	40	3,03	1,40	1,20	0,29
Airmobile Carden Lloyd (Russkii)	RUS	2	—	1	—	6–9	40	160	45	0,66	1,7	220	2,46	1,70	1,22	0,29
Armed Recce. Vickers Guy	GB	6	—	2	6,000 m.g.	6–11	50	220	—	—	9,25	75	6,58	2,35	2,86	0,25
Armoured Recce. Panhard-Kégresse-Hinstin 1929	FR	3	37	1	100 gun & 3,000 m.g.	5–11,5	55	200	35	1,20	6	66	4,75	1,78	2,46	0,25

Motor Vehicle Production
(National Precentage of World Production)

1935		1936	
United States	74.1%	United States	77.2%
Britain	9.1%	Britain	7.8%
France	5.3%	Germany	4.8%
Germany	4.7%	France	3.5%
Canada	3.1%	Canada	3.4%
Italy	1.2%	Italy	0.9%
Others	2.5%	Others	2.4%

TANK PRODUCTION
During the mid-1930s, the United States dwarfed the rest of the West in the production of armoured vehicles. By then, the U.S. was beginning to increase the strength of its armed forces in preparation for national defence.

PANZER DIVISION, 1935
The strength of the early German Panzer division lay in its complement of PzKpfw I tanks. Although the PzKpfw I was armed only with machine guns, its speed facilitated rapid exploitation of a breakthrough in enemy lines.

so impressed with the roar of the engines and the concentrated mobile firepower that he declared, 'This is what I want, and this is what I shall have'.

Out of the shadows

In March 1935, Hitler publicly repudiated the Treaty of Versailles, announced the return of conscription in the German military and accelerated a programme of general rearmament. In October, three self-contained Panzer divisions were created. While the organisations of the Wehrmacht and later SS Panzer divisions were to be revised numerous times during the following decade, these divisions served as the blueprint for the hard-hitting armoured formations that devastated Western Europe and drove to the gates of Moscow.

The 1st Panzer Division initially included two regiments of light tanks, each with two battalions, a motorized rifle brigade, an artillery regiment of two light field howitzer battalions, an anti-tank battalion, a reconnaissance battalion, a signals battalion, and a light engineer company. Theoretically, the prototype Panzer division was capable of operating independently, at least for a time.

As German armoured doctrine evolved, the composition of the prewar Panzer division changed as well. By 1938, its complement included up to 400 tanks in three battalions grouped in two Panzer regiments. Within the battalions were three companies of three platoons, each of which typically included five tanks. With the outbreak of war, other additions included an anti-aircraft battalion, additional combat engineers, and a stronger reconnaissance contingent. The combat experience in Poland led directly to several of these changes.

German Panzer Division
1935

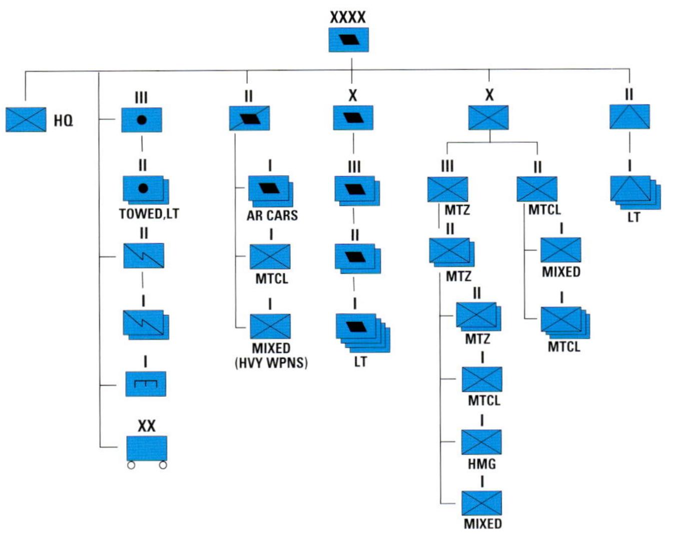

Guderian's armoured fist

While Guderian developed his own theories of armoured warfare and offered such observations in his 1937 book, *Achtung Panzer!*, he was no doubt influenced by the writings of such notable armour advocates as British officers J.F.C. Fuller and Basil Liddell Hart, Soviet General Mikhail Tukhachevsky and French theorist Charles de Gaulle. While he firmly believed that the tank could dominate the battlefield, Guderian realized that this would not be possible without the cooperation and modernization of other weapons systems and tactics.

Therein lay the true genius of the Guderian perspective. The flying artillery of Luftwaffe dive bombers, the sledgehammer of artillery, and the rapid accompaniment of tanks by motorized infantry to consolidate gains and protect the armour from infantry counterattack would produce an irresistible force.

The Blitzkrieg was born of a keen understanding of combined arms. As tactical air strikes disrupted enemy communications and troop movement, artillery paved the way for probing Panzer attacks. When a weak spot in the enemy line was detected, the German armour would strike a decisive, concentrated blow and rapidly exploit any breakthrough, slashing into the enemy rear and creating chaos. Motorized infantry and anti-tank units would protect the Panzers' flanks, but the rapidly moving tanks were not to waiting for follow-on support. Guderian had long advocated the installation of radio equipment on all tanks, and he

believed effective communication would enhance the coordination of each combat element.

Thus, Blitzkrieg and combined arms tactics were logical modern battlefield tools that incorporated long-standing German offensive doctrine, particularly the Napoleonic notion of deep penetration into the enemy's rear and the encirclement of large bodies of enemy troops that would then either be annihilated or compelled to surrender.

Poised to strike

By the time the German Army had conquered Poland and massed its forces in the West to strike at France and the Low Countries in the spring of 1940, the tank strength of the Panzerwaffe totalled nearly 3465 machines in 10 Panzer divisions. A new generation of more heavily armed and armoured tanks was emerging, including the PzKpfw II which had a 20mm (¾in) cannon and could serve as a light reconnaissance and infantry support vehicle; the PzKpfw III, believed capable of engaging in tank-versus-tank action with a 37mm (1½in) gun; and the PzKpfw IV which had a 75mm (3in) gun capable of blasting strongpoints and destroying opposing armour, while its great range of 160km (100 miles) facilitated deep penetrations and encirclement of enemy forces.

During World War II, the German Army formed more than 30 numbered Panzer divisions, along with others that were named in honour of prominent individuals or areas in which the formations trained or their troops were raised. These were augmented by 15 independent Panzer brigades and more than 40 motorized infantry, or Panzergrenadier, divisions. The development of German tanks and self-propelled assault guns reached its zenith in 1943-44 with the introduction of the PzKpfw V Panther medium tank and the PzKpfw VI Tiger and Tiger II, each of these mounting devastating high-velocity 75mm and 88mm (3in and 3½in) cannon respectively.

In the end, however, the German armoured force was defeated. Despite Guderian's sound doctrine, a Panzerwaffe forced on to the defensive was overwhelmed by the sheer number of Allied tanks that were eventually produced and deployed, and devastated by enemy tactical air power.

WEHRMACHT WORKHORSE
Large numbers of the PzKpfw IV were produced for the German Army before and during World War II. The PzKpfw IV was armed with a 75mm (3in) cannon and engaged Allied tanks and fixed fortifications effectively.

BLITZKRIEG ENCIRCLEMENT SCHEMATIC

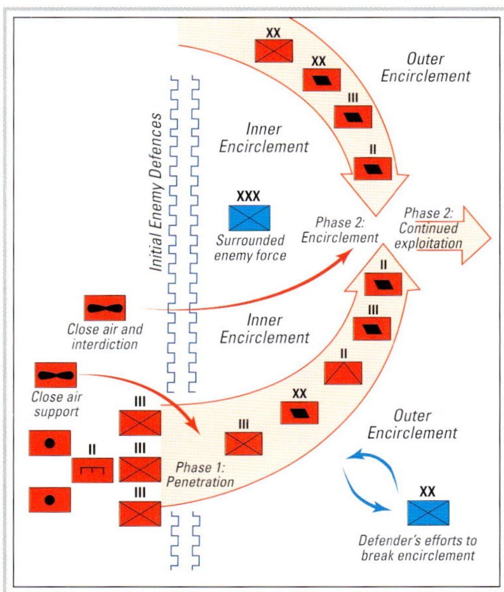

CLASSIC BLITZKRIEG
The German concept of Blitzkrieg involved the coordination of air and ground assets to penetrate deep behind enemy lines and encircle large numbers of troops. Such tactics were in keeping with traditional German battlefield doctrine.

THE SPANISH CIVIL WAR

REPUBLICAN RETREAT
An armoured car of the Spanish Republican forces arrives in the town of Le Perthus as the Nationalists under Generalissimo Francisco Franco press them across the border into neighboring France.

FOREIGN FIREPOWER
During the Spanish Civil War, several foreign nations supplied tanks and armoured vehicles to the warring Nationalist and Republican forces. In the cases of Nazi Germany, Fascist Italy, and the Soviet Union, troops were also committed.

As tank technology advanced and the rudimentary strategy and tactics of armour on the battlefield were explored, the bloody Spanish Civil War of 1936–39 offered not only an ideological clash between the extremes of the political right and left, but also a proving ground for the latest weapons of war, and combat experience for a new generation of soldiers.

Both the Republican forces, and the Nationalists under Generalissimo Francisco Franco received military aid and even troops from outside Spain. Among the military hardware delivered was a number of foreign tanks. The tanks deployed to Spain were most often used in small formations rather than massed for assault, and their crews had little or no training, particularly under combat conditions. The majority of these tanks were lightly armed and armoured, most of them mounting twin machine guns and serving as mobile firing platforms.

One notable exception to the light armament on the Spanish battlefield was the Soviet T-26, armed with a 45mm (1¾in) cannon. Based on the British Vickers 6-ton Mk E and license built in the Soviet Union, the T-26 proved superior to the German PzKpfw I and the Italian L3/33 and L3/35 tankettes it encountered. Of the 331 tanks the Soviets sent to Spain, 281 were the T-26 model, while the others were of the light BT series.

The T-26s in Spain were crewed by both Soviet and Spanish troops, and during a raid against Nationalist forces near the town of Sesena, 15 Soviet-built tanks advanced through the village, firing on enemy positions. Three of the tanks were disabled by mines early in the operation, and three others were lost to artillery fire or Molotov cocktails. When three counterattacking L3/33 tankettes were encountered on a dirt road, one was blasted by the 45mm (1¾in) cannon of a T-26, while another was simply pushed into a ditch and turned on its side by a larger Soviet tank. Although the raiders claimed to have inflicted significant damage on the enemy, it became apparent that the tank would remain vulnerable, particularly in the confines of narrow streets.

German adventure

By the end of 1938, Nazi Germany had deployed two tank battalions, totalling about 120 of the machine gun-armed PzKpfw I, which had never been intended to fight other tanks. When the Soviet T-26 was encountered, it became readily apparent that future German tanks would require heavier armament, armour protection, and greater range. The armour-piercing 7.92mm (⅓in) bullets of the PzKpfw I machine guns could actually penetrate Soviet armour at close range, typically fewer than 150m (165 yards); however, the Soviets adapted quickly and engaged from a safe distance. Soviet-built BA-10 armoured cars were reported to have engaged the PzKpfw I in the autumn of 1936, destroying several of the German tanks with their 45mm (1¾in) cannon at distances in excess of 500m (550 yards), while the German machine guns were ineffective in response.

The Germans soon acknowledged that the

TANKS SUPPLIED BY FOREIGN POWERS
1936–1939

TANK	NATION OF ORIGIN	NUMBER SUPPLIED	SUPPLIED TO
BT-5	Soviet Union	50	Republican
FIAT 3000	Italy	1	Republican
RENAULT FT-17	France Poland	32 64	French Republic Poles to Nationalist
L3/33 and L3/35	Italy	155	Nationalist
PANZER I	Germany	122	Nationalist
T-26	Soviet Union	281	Republican
VICKERS 6-Ton	Bolivia	1	Republican

PzKpfw I was functionally obsolescent, and the PzKpfw II, armed with a 20mm (¾in) cannon, was already in production by the late 1930s. The Germans also realized the value of combined arms operations, with the support of infantry, artillery and air assets. The combat experience in Spain caused German designers to move away from the concept of large numbers of lightly armed and armoured tanks to heavier models that could potentially dominate the battlefield.

Among the other tanks deployed to the battlefields of Spain was the French FT-17, mounting a 37mm (1½in) gun. These were supplied both by France and Poland and used by both the Nationalist and Republican forces. The FT-17 entered service in 1917 and was the first tank to mount its main weapon in a rotating turret, while its layout became a standard for modern tank designers, with the driver forward and the engine to the rear. Its lower track profile was markedly different from other World War I designs. Although obsolete by the late 1930s, it remained in service until the end of World War II.

In Spain, the first notable use of the anti-tank gun revealed yet another threat to the light armour and armament of the majority of tanks developed during the interwar years. The German 37mm (1½in) towed gun and the Soviet 45mm (1¾in) weapon were highly effective against armoured targets, and the stage was set for the continuing augmentation of firepower among anti-tank weapons and the tanks they were intended to defeat.

Following their experiences in Spain, Nazi Germany and the Soviet Union chose divergent paths in relation to their armoured forces. Incredibly, the Soviets even chose to disband their armoured corps for some time, following an assessment that the follow-on motorized troops and artillery simply could not keep pace with the tanks during breakthrough operations. The Germans, however, embraced the concept of the armoured division to devastating effect early in World War II. By the time the Soviets sufficiently rectified their error, they had paid the butcher's bill.

SPANISH CIVIL WAR 1936–39

- Extent of Nationalist control 1938
- Republican areas
- Nationalist offensives
- Republican counter-offensives

SPAIN IN FLAMES
Major European powers rushed to aid the warring factions during the Spanish Civil War. While the Soviet Union supported the Republican forces, eventually Franco's Nationalists emerged victorious with the help of Nazi Germany and Fascist Italy.

SOVIET ARMOUR
The tanks of the Soviet BT series were among the best in the world when deployed during the Spanish Civil War. The superior firepower of this BT-5 provided a distinct advantage against contemporary German and Italian tanks.

ANGLO-FRENCH REARMAMENT 1938–39

RENAULT UE
The Renault UE was considered a tankette, while also serving as a transporter of supplies and ammunition and a prime mover. More than 5,000 of the light tracked vehicle were produced from 1932 to 1941.

The horror of World War I had devastated a generation of British and French men, not to mention its tremendous financial cost. Both Great Britain and France had expended enormous funds to prosecute the Great War, driving their countries to the brink of bankruptcy. During the interwar years, there was much hesitation as to whether the economic condition of either country could sustain an ongoing program of rearmament. Therefore, a policy of arms austerity and even reduction followed the end of World War I, and this persisted into the 1930s. Coupled with the doctrine of appeasement that the British and French governments put forth in response to growing adventurism on the part of Nazi Germany, the dangerous situation became clear to some members of parliament in London and the National Assembly in Paris.

Perilous slumber

By the late 1930s, the two nations had woken up somewhat to the growing threat posed by Hitler's territorial demands. Even while debate continued in the halls of government, the French military assumed a leading role in the development and production of tanks during the 1930s. Clearly seeing a division of labour among their armoured forces, the French considered the exploitation of a breakthrough of enemy lines to be a function of the cavalry tank, while achieving the breakthrough itself was an infantry tank responsibility.

Through the early 1930s, the French Renault FT-17 light tank remained the most numerous of its kind in the world, while French tank designers recognized the need for heavier armour protection and armament. In this regard, the French were pioneers in tank development during the interwar years. The Char B1 bis infantry tank had entered service as early as 1921 and reflected the perspective of forward-thinking French designers. It mounted a heavy 75mm (3in) cannon in the chassis, along with a 37mm (1½in) gun in a rotating turret. Weighing up to 30 tons, it was clearly superior in firepower to other contemporary tanks.

Although the French assumed a leading role in the development of the tank, they failed to formulate an effective methodology for its tactical deployment on the battlefield. Tanks were largely regarded as support weapons and were allocated to infantry and cavalry divisions to be fed into combat piecemeal. Not until December 1938 did the French military establishment authorize the formation of armoured divisions, and this was under way when the Nazis struck in the spring of 1940.

Despite its reliance on the fixed fortifications of the Maginot Line to protect the frontier with Germany, the French Army possessed more than 2,350 modern tanks on the eve of the Battle of France, which erupted on 10 May 1940. Many of these were superior in firepower and mobility to their German

FRENCH AMC 35
A light cavalry tank that served with the French Army during the interwar years, the Renault AMC 35 was armed with a 47mm (1¾in) gun and produced until 1940. It was well protected by up to 25mm (1in) of armour.

counterparts, and more than 250 of them were the new Somua S35 that had entered service in 1936. Its 47mm (1¾in) main gun was more than a match for the machine guns of the German PzKpfw I and the 20mm (¾in) cannon of the PzKpfw II. The blame for the overall dismal performance of French forces in defence of their homeland lay more with the commanders than the machines.

Britain's bold statesman

The British statesman most responsible for raising the alarm that war clouds loomed on the horizon in the 1930s was future prime minister Winston Churchill. Loudly opposing the policy of appeasement, Churchill recognized the necessity of rearmament as Hitler continued to threaten the uneasy peace. In 1933, largely due to his efforts, the British government created the Defence Requirements Committee to study the needs of the military, and designated £71 million over the next five years to improve the quality of the various branches of service.

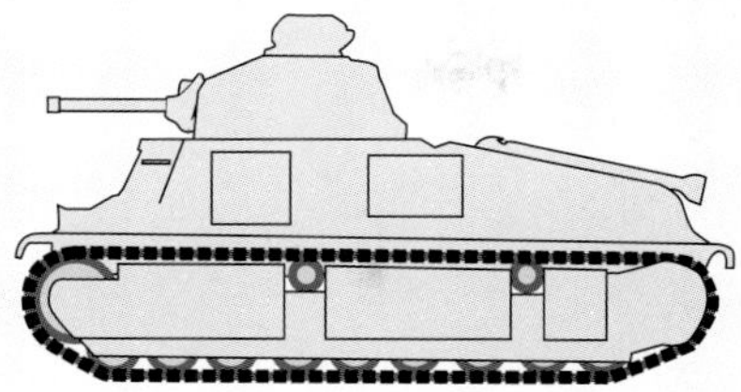

Somua S35

Manufactured	1936
No. built	430
Weight	19.5 tonnes
Length	5.38 m
Crew	3
Armour	47 mm
Main Armament	47 mm SA35
Secondary Armament	1 x 75 mm M1931
Speed	25.3 mph (40.7 kph)

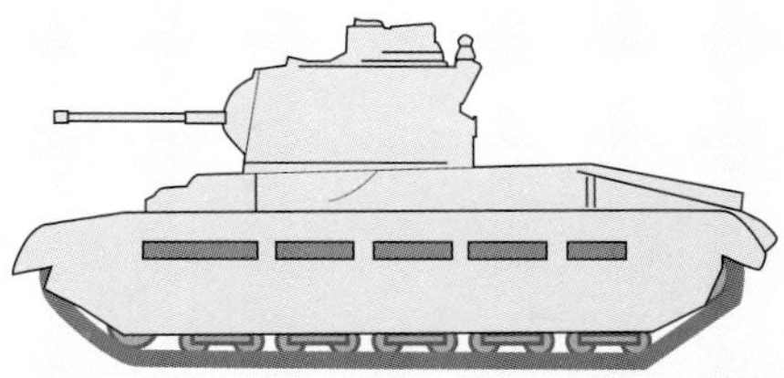

Matilda II

Manufactured	1937
No. built	2,987
Weight	25 tons
Length	6 m
Crew	4
Armour	20–78 mm
Main Armament	1 x 40 mm
Secondary Armament	1 x 7.92 mm MG
Speed	16 mph (26 kph)

ALLIED ADVANCES
The French Somua S35 and British Matilda II tanks set the standard for firepower and innovative design during the interwar years. With the outbreak of World War II, German tanks were only comparable at best.

Fighting Matilda

Despite financial constraints, Britain had taken a lead role in the technology and mobility of the tank. For more than a decade, the armoured corps of the British Army relied on the Vickers medium tank, initially armed with a 3-pdr (47mm/1¾in) gun. These were relatively few in number, as the British armed forces had been substantially reduced in the years immediately following World War I.

By the late 1930s, the British had begun to deploy the Cruiser series of light tanks, and in 1937 design specifications for the Matilda II infantry tank had been completed. Armed with a 2-pdr (40mm/1½in) gun, the Matilda was heavily armoured for its time and gained lasting fame in North Africa. It was an ideal infantry support tank, serving throughout the war and even alongside the Valentine, its intended replacement. However, the Matilda was ponderously slow, with a top speed of only 14km/h (9mph) cross-country.

From an output of 969 tanks in 1939, British factories reached a peak of 7,466 armoured vehicles of various types in 1943. From 1937 to 1943, nearly 3,000 Matildas were manufactured.

FRENCH ARMOUR
The Hotchkiss H35 light tank entered service with the French Army in 1936 and was used as an infantry and cavalry support tank with its 37mm (1½in) main gun. Captured H35s were later pressed into service with the German Army.

NOMONHAN 1939

Japanese expansion on the Asian mainland during the 1930s posed a direct threat to Soviet interests in the Far East, particularly in the northern province of Manchuria, which the Japanese occupied, installing a puppet government and renaming the territory Manchukuo. As the Japanese continued to exert influence in the region, Red Army forces advanced into Mongolia in 1936 and established the Mongolian People's Republic, raising Japanese concerns over Soviet intentions.

While Japanese politicians argued over the course of political and military events in China, the Soviets increased their own military presence substantially. By 1936, more than 100 Red Army tanks were supporting 20 infantry divisions, a substantial escalation of its military presence from the mere six divisions stationed in the East in 1931. As tensions grew along the 4,800km (3,000-mile) Manchurian border, fighting became commonplace. In one year alone, more than 30 armed clashes took place.

The road to conflict

With the incident at the Marco Polo Bridge near Beijing in 1937, the Japanese took a calculated risk. Although seizing further Chinese territory would be an obvious provocation to the Soviets, militarists within the Tokyo government believed that the question of pre-eminence on the Asian mainland would have to be settled eventually anyway.

The Soviets responded to the Japanese aggression with the commitment of the 57th Special Rifle Corps, including four armoured brigades and a motorized infantry division, to Mongolia. Fighting raged at Changkufeng Hill, high ground along the border between Soviet Siberia and northern Korea. Apparently, the commander of the Japanese 19th Army decided to seize the hill without orders from the higher echelons of command. After three weeks the Japanese withdrew. However, the struggle for Changkufeng Hill was simply a precursor to a more intense battle.

Decision at Nomonhan

Aware of Soviet Premier Josef Stalin's purges of the Red Army officer corps during the mid-1930s, sceptical of the fighting prowess of the Soviet soldier, and convinced that their adversaries were preoccupied with the growing threat of Nazi Germany, the Japanese moved toward a showdown with the Soviets in the Far East in the spring of 1939. The result of the five-month struggle at Nomonhan, also known as Khalkhin Gol, was certainly unexpected, as the Japanese were rebuffed.

In May 1939, a series of skirmishes near the Manchurian village of Nomonhan escalated steadily, punctuated in July by an attack by the Japanese 23rd Infantry Division supported by two

TOUGH TERRAIN
A Soviet T-26 tank slogs its way through difficult, marshy ground during Red Army attacks in Manchuria. Favorable terrain was critical to the successful operation of armoured vehicles in all theaters, often dictating the movement of troops and tanks.

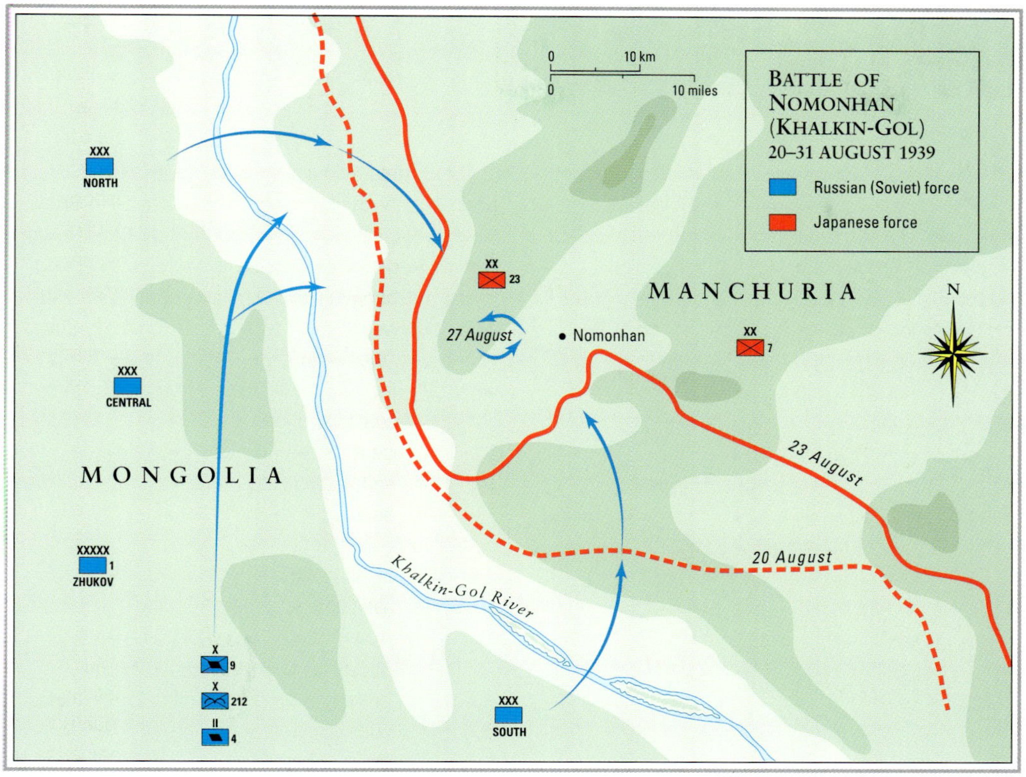

NOMONHAN NEMESIS
At the Battle of Nomonhan, Soviet General Georgi Zhukov skillfully deployed his armoured units and achieved a decisive victory over the Japanese. As a result, Zhukov was elevated to higher echelons of Red Army command.

tank regiments that forced the Soviets to retire to the Khalka River. Confident that Soviet tactics, particularly their piecemeal use of armour that was evident at Changkufeng Hill, would continue, the Japanese were surprised when Red Army forces in the region came under the command of General Georgi Zhukov.

As the summer wore on, Zhukov successfully repulsed the Japanese, inflicting a stinging defeat in July as his armour proved clearly superior to that of the Japanese. While both sides planned further offensive actions for the late summer, Zhukov was continually reinforced and struck the first blow on 20 August, sending three rifle divisions, two motorized infantry divisions, two armoured divisions augmented by two more armoured brigades, and a pair of Mongolian cavalry divisions against the enemy.

With nearly 500 tanks at his disposal, Zhukov fixed the Japanese position with a major attack in the centre and despatched armoured columns rapidly around the enemy flanks, trapping the 23rd Infantry Division in a double envelopment. A Japanese counterattack to relieve the 23rd Infantry Division was thwarted by Soviet tanks that proved impervious to Japanese anti-tank fire and superior in firepower with 45mm (1¾in) main guns. The Japanese lost more than 40,000 killed or wounded at Nomonhan, with the 23rd Division taking 73 percent casualties.

The primary tanks deployed by the Red Army at Nomonhan were the BT-5 and BT-7, forerunners of the famed T-34 medium tank that entered production in 1940 and began to appear in substantial numbers sometime after the German invasion of the Soviet Union on 22 June 1941. A few of the more advanced T-26 design tanks also saw action at Nomonhan. Ironically, one of the foremost attributes of Soviet tank design was the Christie suspension, developed by American engineer Walter Christie, which improved the tank's speed and cross-country performance and allowed the armoured vehicle to exhibit a lower and therefore less vulnerable profile.

TANK TRAINING
A T-26 command tank of the Red Army, radio antennae located near the turret, advances during summer maneuvers in the Kiev Military District in 1936. By the outbreak of World War II, the T-26 was functionally obsolescent.

THE WORLD'S ARMOURED FORCES 1939

BRITISH TANKETTE
The Vickers Carden Loyd Mk VI tankette was the most successful in a series of tankettes developed for the British Army in the late 1920 land was subsequently adopted by several other countries.

On the eve of World War II, the armed forces of the world had embraced the concepts of armoured warfare to varying degrees. Although Great Britain and France had maintained the technological advantage in tank development in the early interwar years, by 1939 the Soviet Red Army dwarfed its contemporaries in sheer numbers with nearly 20,000 tanks in the field.

In contrast, the United States, notwithstanding its tremendous industrial capacity, counted an inventory of fewer than 100 tanks. The U.S. military establishment chose instead to concentrate development efforts on components for armoured vehicles. Although a low number of tanks was produced in the United States, as war approached such research proved invaluable. Among the most prominent US interwar tank designs were the M1 and M2 series, armed with machine guns.

Armoured innovation

Although senior commanders continued to spar over the proper employment of tanks on the battlefield, some even asserting that the armoured vehicle would never substantially influence the outcome of a conflict, revisions to basic designs continued. Early tanks were often unreliable, prone to mechanical breakdowns, and plagued by components with relatively low wear resistance. Therefore, detractors reasoned, the capability of the tank was limited and the presence of armoured formations could in fact create delays and even cause an offensive to lose momentum.

Nevertheless, with an expanding discussion related to armoured doctrine, a division of labour emerged within the armed forces of former belligerents. The role of the tank in future wars was envisioned by such pioneer tacticians as J.F.C. Fuller and Basil Liddell Hart, who foresaw the coordinated and mechanized might of a modern army utilizing the tank for breakthrough, rapid exploitation, firepower, and infantry support. Practical experience during the Spanish Civil War indicated that fast, lightly armoured tanks could indeed perform the function of reconnaissance and mobility, and to a lesser degree, infantry support. However, the introduction of towed anti-tank guns laid bare the weakness of thin armour and the obvious conclusion that the tanks of the future would require thicker armour protection and heavier weapons.

Allied capability

British development of the Vickers Carden-Lloyd light tank, essentially a mobile machine gun platform, influenced tank designs among other nations during the 1920s and early 1930s. The British armoured force included three types of tank: the light tank for reconnaissance, the cruiser tank for exploitation of a breakthrough and pursuit, and the infantry tank, designed to provide fire support for infantry.

Immediately following World War I, the number of serviceable tanks dwindled, ebbing to only five battalions and a handful of scout cars by 1921. Six years later, the Experimental Mechanized Force was organized for testing purposes and became the forerunner of the World War II armoured division. The Royal Tank Corps was formed in the mid-1920s, later to become known as the Royal Tank Regiment. By the late 1930s, the British Army fielded two armoured divisions, and during the course of World War II the divisional structure was altered no fewer than nine times. In the spring of 1939, an armoured division consisted of 349 light, cruiser and infantry support tanks, along with supporting artillery, engineer and reconnaissance troops. These were grouped into two brigades.

In France, the Renault FT-17 tank broke new design ground, incorporating elements that would become universal, such as turrets for command and weapons, lower profiles and compact hulls. France was the first major power to produce welded rather than riveted bodies for its tanks, which improved crew survivability by eliminating the shearing of rivets that created deadly shrapnel in the event of a hit.

French armoured vehicles of the 1930s encompassed two distinct categories. Among these were cavalry tanks, such as the Somua S35, an armoured combat tank intended for infantry support and to engage enemy tanks; and armoured reconnaissance tanks, essentially light tanks such as the Hotchkiss H35, deployed as a screening force. Infantry tanks were classified as light, medium and heavy, with light tanks, such as the Renault R35, supporting infantry and scouting, while the Char B1 bis was meant to operate as an offensive weapon, exploiting any breakthrough. Enthusiasm for the heavy tank waned during the 1930s, and few of them were active by World War II.

Although the French were pioneers in the development of tanks during the interwar years and had produced nearly 2,000 by 1939, tacticians failed to grasp the dominating role that the armoured division was destined to play in the looming European land war. French tanks were assigned either to infantry or cavalry divisions, and, therefore, their collective firepower was diminished. Not until late 1938 did the French military establishment authorize the formation of armoured divisions, and when the war in the West erupted on 10 May 1940, little progress had been made.

Soviet vanguard

The Soviet Union was by far the leading producer of tanks during the 1920s and 30s, and military tacticians advocated the employment of tanks in combination with infantry and air support. However, internal conflict and Premier Josef Stalin's great purge of the Red Army officer corps contributed to a decline in emphasis on armoured warfare, particularly after combat experience in Spain tended to support the assertions of those who wanted to disband the mechanized corps and disperse the tanks to infantry formations.

During the 1920s, the Soviets had imported tanks from Britain and France, as well as prototypes designed by the American engineer Walter Christie. The Christie tanks served as the basis for the BT series of armoured vehicles and later the famed T-34 medium tank that became a symbol of Soviet victory in World War II. In the autumn of 1939, the mechanized corps had been re-established largely due to the performance of Soviet armour in Manchuria. It consisted of three divisions, two armoured and one of motorized infantry. Each armoured division included two tank regiments, a motorized infantry regiment, and a motorized artillery regiment. The motorized infantry division consisted of two motorized infantry regiments, a tank regiment, and a motorized artillery regiment. By 1940, nine mechanized corps had been organized.

German genesis

The development of armoured divisions within the German Army was successful despite the restrictions of the Versailles Treaty and was in large part due to the cooperation of the Soviet Union, particularly following the establishment of a joint armoured training school at Kazan in eastern Russia. Although early German tanks were based on foreign designs, the Panzerkampfwagen I entered service in the mid-1930s, and German tank development accelerated rapidly from then on. Following the occupation of Czechoslovakia in 1938, tanks of Czech design were incorporated into German formations. In 1939, the Germans fielded nearly 3,500 tanks. Under the early leadership of General Heinz Guderian, the Panzer forces became the élite unit of the German Army.

World Tank Forces 1939	
Soviet Union	19,768
Germany	3,400
France	1,900
Britain	1,146
Poland	1,140
Italy	1,000
Japan	c. 400
United States	90

WORLD TANK STRENGTH
On the eve of World War II, the Soviet Union possessed the largest armoured force in the world. Although the Red Army fielded nearly 20,000 tanks, many were lost in the opening weeks of Operation Barbarossa.

EARLY LIGHT
The U.S. M1/M2 light tank was the forerunner of the M3 Stuart design and was initially armed with machine guns. Later variants were upgunned with a 37mm (1½in) weapon. The tank saw action on Guadalcanal early in World War II.

BLITZKRIEG POLAND 1939

PANZERKAMPFWAGEN III AUSF. D
(Opposite) The PzKpfw III combined speed and firepower for German armoured spearheads during the Polish campaign of 1939. Although it was available in fewer numbers than other tanks, the PzKpfw III performed well.

When the German Army launched Case White, the invasion of Poland, on 1 September 1939, General Heinz Guderian and the armoured force that had been formed largely by his efforts during the previous decade would reveal to the world the concept of Blitzkrieg. The Panzers advanced with tremendous speed and swiftly conquered the country in concert with the Soviet Red Army that invaded from the east two weeks later.

While the Polish operation was not a classic Blitzkrieg effort as Guderian had envisioned, and was rather a giant double pincer movement that centred on the Vistula and Bug rivers, encircling Polish forces and capturing the capital of Warsaw, the role of the tank in modern warfare was demonstrated with dazzling success. Guderian was acknowledged as the architect of a formidable armoured fighting force, even though it did not achieve the independence of movement he had sought for the invasion of Poland. Instead, the armoured corps had been assigned to the command structures of two field armies that were primarily made up of infantry formations.

The Panzer force expands

Although the German armoured force that assailed Poland was formidable, it was, even by 1939 standards, somewhat under-equipped. By the spring of that year, five Panzer divisions had been raised, fielding 10 Panzer regiments of five brigades, with four battalions each. A sixth Panzer division was nearing operational readiness and did take part in the Polish campaign. Another four independent Panzer regiments were separated, either to form the nucleus of additional Panzer divisions by the spring of 1940 or to augment other formations. In addition, the Germans assembled four motorized infantry divisions, a Panzer division comprised of both army and Waffen SS troops, and four light divisions that were essentially small Panzer divisions fielding one regiment of armoured vehicles and a motorized infantry force.

Despite their designation as mechanized, these 15 divisions were still dependent on the horse for transportation to varying degrees. Their armoured strength was based primarily on the PzKpfw I and II, both of which were rapidly becoming obsolete. The PzKpfw I, which had entered service five years earlier, was lightly armoured and mounted only a pair of machine guns. While its performance had been adequate during the Spanish Civil War of the mid-1930s, its thin armour offered only limited protection against small arms fire, and the proliferation of the towed anti-tank gun and heavier weaponry made the vehicle quite vulnerable. The PzKpfw II, equipped with a 20mm (¾in) cannon, was a slight improvement in infantry support capability, but provided little in the way of firepower augmentation, particularly against the likes of heavier tanks such as the French Char B1 bis, mounting a 47mm (1¾in) turreted gun and a 75mm (3in) howitzer in its hull.

When World War II erupted, German engineers were already completing the designs for heavier tanks. The PzKpfw III mounted a 37mm (1½in) cannon and was intended to fight enemy tanks. The heavy PzKpfw IV, with a 75mm (3in) main gun and extended range of 160km (100 miles), was conceived to penetrate deeply behind enemy lines, destroy fixed fortifications, and neutralize enemy armoured strength. Both of these were actually being deployed in numbers by the autumn of 1939.

Polish armour

To defend themselves against the marauding Germans and Soviets, the Poles could muster about 800 reconnaissance vehicles and light tanks grouped into two brigades, a total of 11 armoured battalions. Many of the Polish tanks were license-built versions of the light British Vickers Armstrong tank and Carden-Loyd tankettes. Other models

POLAND ABLAZE
The buildings of a Polish village burn fiercely in the background as the commander of a German tank surveys the terrain ahead of his armoured vehicle. The speed of German armoured spearheads surrounded large numbers of Polish troops.

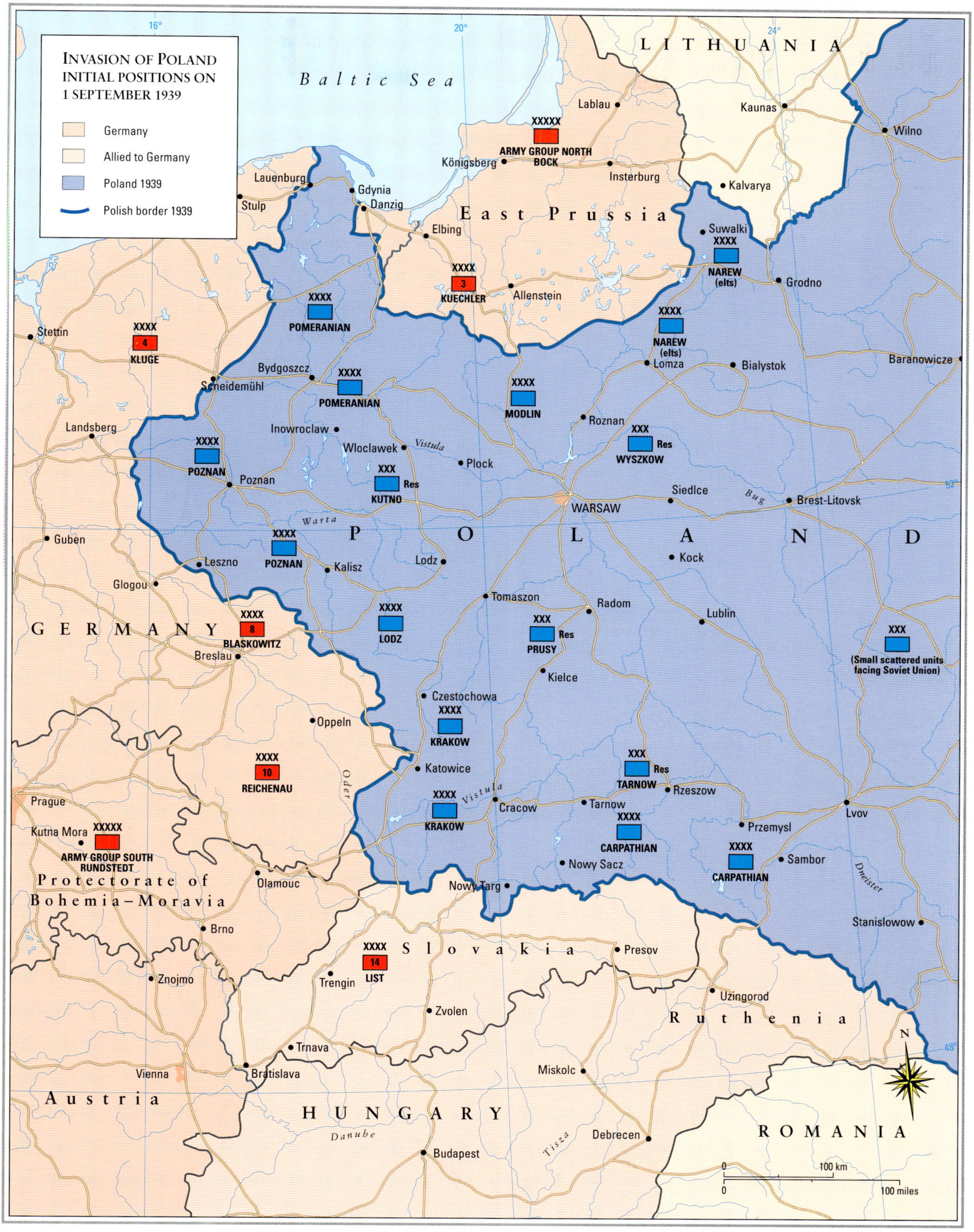

INVASION OF POLAND INITIAL POSITIONS ON 1 SEPTEMBER 1939
Germany
Allied to Germany
Poland 1939
Polish border 1939
Baltic Sea
LITHUANIA
East Prussia
POLAND
GERMANY
Protectorate of Bohemia–Moravia
Slovakia
Ruthenia
Austria
HUNGARY
ROMANIA
XXXXX
ARMY GROUP NORTH
BOCK
XXX
3
KUECHLER
XXXX
4
KLUGE
XXXX
8
BLASKOWITZ
XXXX
10
REICHENAU
XXXXX
ARMY GROUP SOUTH
RUNDSTEDT
XXXX
14
LIST
XXXX
POMERANIAN
XXXX
POMERANIAN
XXXX
NAREW
(elts)
XXXX
NAREW
(elts)
XXXX
MODLIN
XXX
Res
WYSZKOW
XXXX
POZNAN
XXX
Res
KUTNO
XXXX
POZNAN
XXXX
LODZ
XXX
Res
PRUSY
XXX
(Small scattered units facing Soviet Union)
XXXX
KRAKOW
XXX
Res
TARNOW
XXXX
KRAKOW
XXXX
CARPATHIAN
XXXX
CARPATHIAN
Lablau
Kaunas
Wilno
Königsberg
Insterburg
Kalvarya
Lauenburg
Gdynia
Danzig
Stulp
Elbing
Suwalki
Grodno
Allenstein
Stettin
Bydgoszcz
Scneidemühl
Lomza
Bialystok
Baranowicze
Landsberg
Inowroclaw
Roznan
Wloclawek
Vistula
Plock
Poznan
Siedlce
Bug
Brest-Litovsk
WARSAW
Warta
Guben
Leszno
Kalisz
Lodz
Kock
Glogou
Tomaszon
Radom
Lublin
Breslau
Kielce
Czestochowa
Oppeln
Katowice
Oder
Prague
Rzeszow
Vistula
Cracow
Tarnow
Lvov
Przemysl
Kutna Mora
Sambor
Nowy Sacz
Dneister
Olamouc
Nowy Targ
Stanislowow
Brno
Presov
Trengin
Znojmo
Uzingorod
Zvolen
Trnava
Vienna
Bratislava
Miskolc
Danube
Tisza
Debrecen
Budapest
16°
20°
24°
52°
48°
N
0
100 km
0
100 miles

PANZERS FORWARD
A German motorcycle trooper pauses along a dirt road in western Poland, while PzKpfw I tanks roll eastward toward their objective of Warsaw, the Polish capital. Most Polish tanks were quickly destroyed or bypassed during the rapid advance.

included the antiquated French Renault FT-17, R35 and Hotchkiss H35. The Poles had taken some Vickers 6-ton light tanks and modified them to carry a turret-mounted Bofors wz 37mm (1½in) cannon, more than a match for the German PzKpfw I and II. However, fewer than 150 of these tanks were produced. While these 7TP light tanks operated with some success, they were virtually defenceless against German air power and were unable to coordinate effective operations on a scale grand enough to counter rapid German ground movement.

Light armoured vehicles were also included in some Polish cavalry brigades. The Kraków Cavalry Brigade, for example, was comprised of 13 TK tankettes of Polish manufacture, each mounting 7.92mm (⅓in) or 7.7mm (⅓in) machine guns, and eight Samochod pancerny wz.34 armoured cars mounting machine guns or 37mm (1½in) cannon.

Lightning strike

In the predawn hours of 1 September, the Germans unleashed five field armies, consisting of 60 divisions, against Poland. Although the Poles counted an army of a million men, only about half of its strength was mobilized at the time. Coordinating with artillery and air attacks by Junkers Ju 87 Stuka dive bombers, Guderian's XIX Panzer Corps roared eastward and reached its objective at Brest-Litovsk in 15 days. As Guderian had expected, the deep, rapid penetration of Polish lines threw the enemy into chaos. Each time the Poles attempted to take a stand, their positions were overrun or outflanked by the dash of the German tanks.

The Poles had been unprepared for such a sledgehammer blow, and a counterattack planned for 6 September could not be coordinated, mainly due to the fact that the reserve Prusy Army had not been mobilized in time. Within hours of the opening of hostilities, German forces had taken Kraków, capturing large numbers of prisoners while forcing the Poles to retreat to avoid encirclement. The most promising Polish counterattack occurred west of Warsaw near the town of Bzura. For 10 days, the Poles made some headway, but their offensive ground to a halt due to a lack of supplies and the constant threat of attack from the flank and rear. In a week's time, German tanks had rolled to the suburbs of Warsaw, covering an astonishing 225km (140 miles).

Guderian's XIX Corps spearhead consisted of the 1st, 2nd, and 10th Panzer Divisions, and the Motorized Infantry Regiment Grossdeutschland. The XXII Panzer Corps, commanded by General Ewald von Kleist, raced to link up with Guderian east of Warsaw by mid-month.

Although the fate of the Polish nation was swiftly sealed, the Germans encountered stiff resistance in the capital itself. The 4th Panzer Division found its mobility to be severely restricted during street-to-street and house-to-house fighting with Polish troops. Initial attempts to take the city by storm were repulsed on the outskirts at Ochota and Wola. As the remnants of several defeated Polish armies came together to defend Warsaw, only about 40 tanks were initially available. However the Polish anti-tank guns proved their worth in the fighting as the 4th Panzer Division lost 80 tanks in a few days, nearly 37 percent of its total strength. For two weeks, the Poles held out against repeated German attacks, as the city was pounded from the air by the Luftwaffe. As the Soviets closed on Warsaw from the east, German forces finally captured the city on 1 October 1939.

Polish postscript

Following the successes in Poland, a lengthy lull ensued as German forces prepared to invade France and the Low Countries in the spring of 1940. Four of the German light divisions were bolstered with additional tanks and troops and reconstituted as the 6th, 7th, 8th, and 9th Panzer Divisions. The 10th Panzer Division, which had entered the Polish campaign under-strength, was raised to its full complement of tanks and infantry, while the 4th Panzer Division, seriously depleted during the fight for Warsaw, received an additional regiment of motorized infantry.

POLISH ONSLAUGHT
In the early morning hours of 1 September 1939, German columns struck rapidly across the Polish frontier. The Germans had been prepositioning troops and tanks for weeks prior to unleashing their Blitzkrieg.

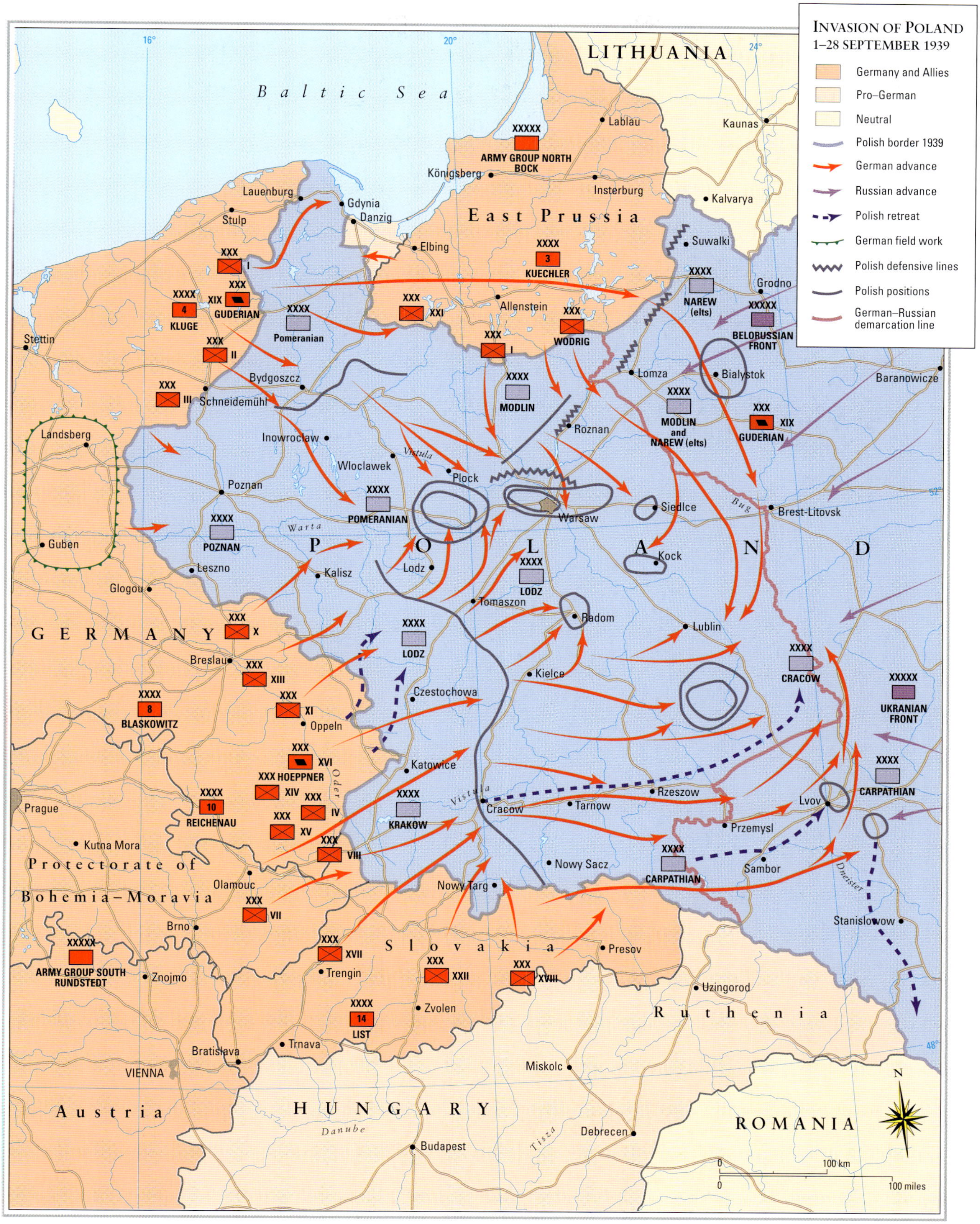
INVASION OF POLAND
1–28 SEPTEMBER 1939
Germany and Allies
Pro-German
Neutral
Polish border 1939
German advance
Russian advance
Polish retreat
German field work
Polish defensive lines
Polish positions
German–Russian demarcation line
Baltic Sea
LITHUANIA
East Prussia
GERMANY
POLAND
Protectorate of Bohemia–Moravia
Slovakia
Ruthenia
Austria
HUNGARY
ROMANIA
XXXXX
ARMY GROUP NORTH
BOCK
XXXX
3
KUECHLER
XXXX
4
KLUGE
XXX
XIX
GUDERIAN
XXX
I
XXX
II
XXX
III
XXX
XXI
XXX
I
XXX
WODRIG
XXXX
Pomeranian
XXXX
MODLIN
XXXX
NAREW
(elts)
XXXX
MODLIN
and
NAREW (elts)
XXXXX
BELORUSSIAN
FRONT
XXX
XIX
GUDERIAN
XXXX
POMERANIAN
XXXX
POZNAN
XXXX
LODZ
XXXX
LODZ
XXXX
KRAKOW
XXXX
CRACOW
XXXX
CARPATHIAN
XXXX
CARPATHIAN
XXXXX
UKRANIAN
FRONT
XXX
X
XXX
XIII
XXX
XI
XXXX
8
BLASKOWITZ
XXX
XVI
XXX HOEPPNER
XXXX
10
REICHENAU
XXX
XIV
XXX
IV
XXX
XV
XXX
VIII
XXX
VII
XXXXX
ARMY GROUP SOUTH
RUNDSTEDT
XXX
XVII
XXX
XXII
XXX
XVIII
XXXX
14
LIST
Lablau
Kaunas
Königsberg
Insterburg
Kalvarya
Lauenburg
Gdynia
Danzig
Stulp
Elbing
Suwalki
Grodno
Allenstein
Stettin
Bydgoszcz
Schneidemühl
Lomza
Bialystok
Baranowicze
Landsberg
Inowroclaw
Roznan
Wloclawek
Plock
Poznan
Warsaw
Siedlce
Brest-Litovsk
Guben
Leszno
Kalisz
Lodz
Kock
Glogou
Tomaszon
Radom
Lublin
Breslau
Kielce
Czestochowa
Oppeln
Katowice
Rzeszow
Prague
Cracow
Tarnow
Lvov
Przemysl
Kutna Mora
Olamouc
Nowy Sacz
Sambor
Nowy Targ
Brno
Stanislowow
Presov
Znojmo
Trengin
Zvolen
Uzingorod
Trnava
Bratislava
VIENNA
Miskolc
Budapest
Debrecen
Vistula
Warta
Oder
Bug
Dniester
Danube
Tisza
16°
20°
24°
52°
48°
N
0
100 km
0
100 miles

SEDAN THE INVASION OF THE WEST MAY 1940

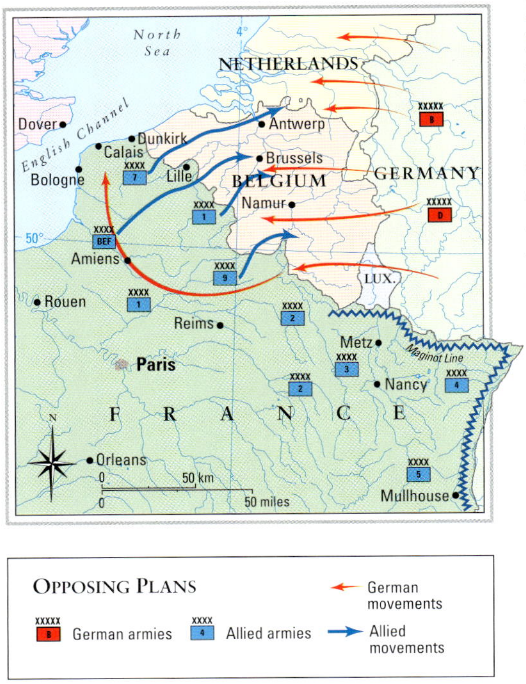

CASE YELLOW
The German plan for the conquest of France involved direct attacks across the Belgian frontier and a rapid armoured sweep through the Ardennes Forest, pivoting northwest to trap Allied forces fighting in Belgium.

Although theoretically crippled by the Treaty of Versailles that ended World War I in 1918, the German Army remained a viable force throughout the 1920s and 1930s. Reduced to an official count of no more than 100,000 troops, the Army was nevertheless preserved and even augmented through clandestine activity, secret treaties with neighbouring countries, and, most ominously, the rise of the Nazis to power in 1933.

General Hans von Seeckt had led the effort to bolster the Army's strength, maintaining a shadow structure that eventually came into the full view with Hitler's repudiation of the Versailles treaty and orders in 1935 to gird for war. By 1939, the German Army was, in the opinion of many senior commanders, still not totally prepared, but Hitler unleashed the Blitzkrieg in Poland on 1 September. Within three weeks, the Poles had been defeated by the Germans from the West and the Soviets invading from the East. Britain and France had guaranteed the sovereignty of Poland and pledged to come to its aid if attacked. Both declared war on Nazi Germany but were powerless to act militarily.

Phony War

Following the rapid success in Poland, Hitler and his generals acknowledged the threat posed by their traditional enemy on the European continent, France. Great Britain had also dispatched an expeditionary force to assist the French against a German attack that was certain to come. Months of relative inactivity followed in the West after the stunning German victory in Poland, and the period was popularly known as the 'Phony War' or 'Sitzkrieg'.

Senior French commanders believed their positions along the German frontier were secure within the formidable string of fixed fortifications known as the Maginot Line. The product of interwar strategic design, the Maginot Line had been constructed as a deterrent to German aggression and with the memory of the horrendous casualties suffered during the trench warfare in World War I. One glaring weakness in the French defensive philosophy was the simple fact that the Maginot Line extended only from the Swiss border to Belgium. Enemy forces, swiftly moving, could conceivably outflank the Maginot Line by striking through Belgium and Luxembourg.

Case Yellow

By the outbreak of hostilities, the German general staff had been planning for war with France for months. Early proposals were similar to the preliminary stages of World War I, including a wide sweep through Belgium, with a massive pivot southward toward Paris, the French capital. Plans formulated by General Franz Halder, chief of the general staff and General Erich von Manstein, chief of staff of Army Group A, were alternately considered and rejected either by Hitler or the general staff as a whole.

Codenamed *Fall Gelb*, or Case Yellow, the plan for the defeat of France was not finalized until General Heinz Guderian, commander of the XIX Panzer Corps, proposed a radical change. Guderian was recognized as the foremost authority in the German Army on the deployment of armour on the battlefield and had authored the book *Achtung Panzer!* in 1937, detailing his philosophy of concentrated

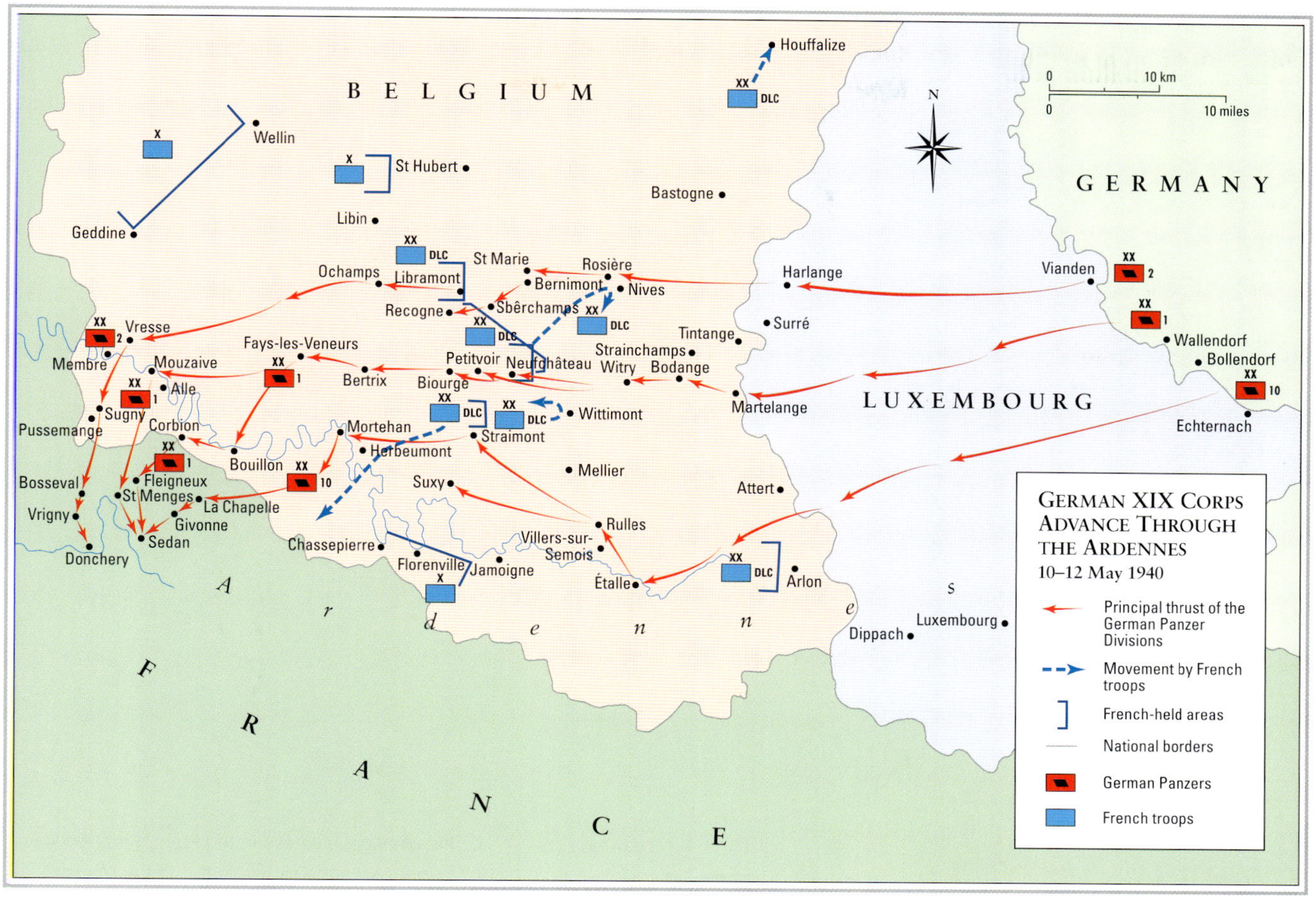

armoured strength smashing through enemy lines and swiftly wreaking havoc in rear areas.

Guderian suggested that the majority of available German armour should concentrate in the area of Sedan, a French town situated on the Belgian border on the banks of the River Meuse and in the heavily wooded Ardennes Forest. The French had long believed that the difficult terrain, old-growth forests and few roads in the Ardennes made the region virtually impenetrable. Guderian further advocated a breakthrough at Sedan and a rapid dash to the northwest by German Panzer formations. Rather than moving to directly confront French troops that had moved into Belgium to meet an invasion threat or turning south toward Paris, a rapid drive to the northwest could indeed precipitate a collapse of the entire French Army and potentially trap great numbers of Allied troops. Amid a storm of protest, this daring version of Case Yellow was finally approved in early 1940.

GUDERIAN'S RUSH

The German XIX Corps, under the command of General Heinz Guderian, swiftly advanced through the Ardennes and broke through French defences near the town of Sedan. Within days, Guderian's Panzers had reached the English Channel.

CHAR LEGER H35

A column of French Hotchkiss H35 light tanks idles along a road through the countryside of northern France. Armed with a 37mm (1½in) main gun, the H35 was a potent reconnaissance and infantry support tank.

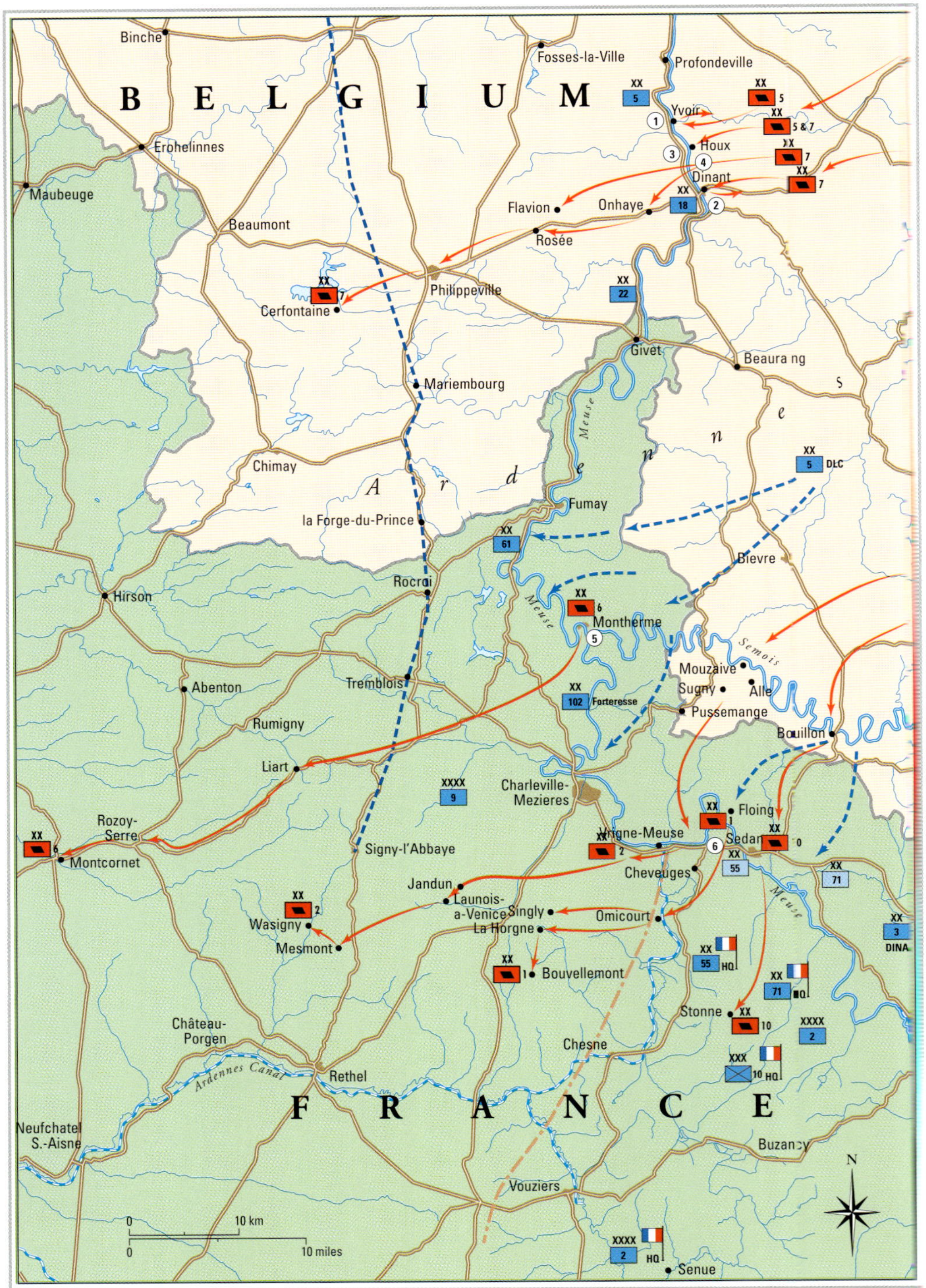

FIGHTING FRENCH
French armoured and infantry units contested the Germans crossing of the River Meuse and inflicted serious losses on some invading formations. The German 7th Panzer Division crossed the Meuse at Dinant and raced for the Channel coast.

Dyle Plan

French commanders, most notably senior General Maurice Gamelin, conceived the Dyle Plan in response to an expected German invasion through Belgium and the Low Countries. Working in co-operation with the Belgian and Dutch armies, French and British forces would advance into Belgium and the Netherlands. Anchored by the Maginot Line in the south, this multinational co-operative effort would maintain a line along the River Dyle, preventing German forces from advancing into France. The French were convinced that an advance to the north and east would contain the Germans offensive on a narrow front.

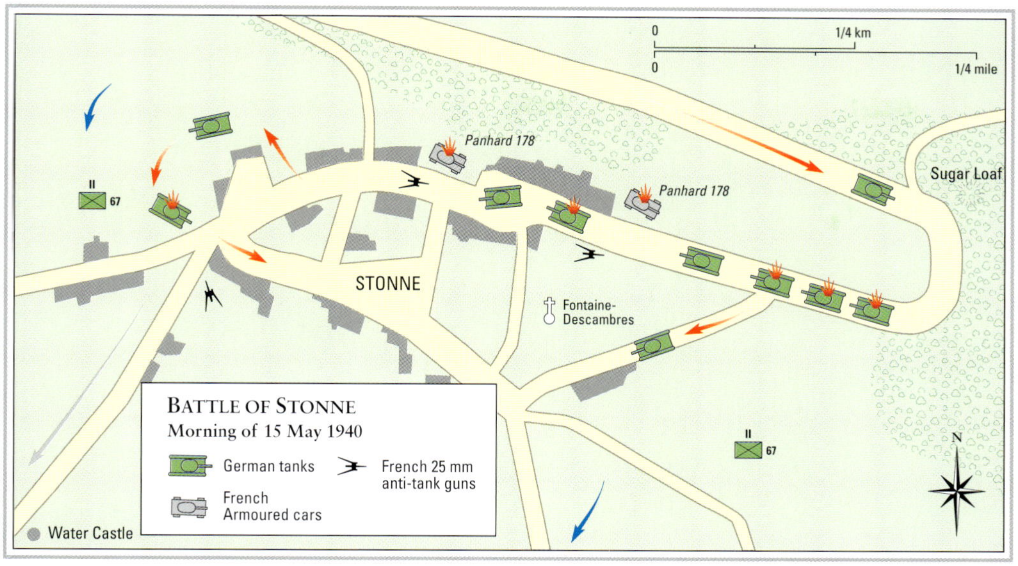

STAND AT STONNE
During the fighting at the village of Stonne, French Char B1 bis heavy tanks and 25mm (1in) anti-tank guns destroyed at least a dozen German tanks; however, the French were unable to hold the town against the overwhelming German tide.

Officially, there was little consideration given to the logical possibility of a German strike through the Ardennes.

French forces

On paper, the French Army appeared vastly superior to its German counterpart, fielding 117 divisions and more than 3,200 tanks. However, the French relied significantly on reserve troop formations, some of which were of questionable combat efficiency, while others were obliged to man the fixed fortifications of the Maginot Line, unable to move northward effectively in the event of a German breakthrough.

The French heavy Suoma S35 and Char B1 bis tanks, the former with a 47mm) (1¾in) gun, the latter with both a 47mm and 75mm (3in) weapon, and each with significant armour protection, were equal in many respects and even superior to the Panzerkampfwagen I, II, III, and IV tanks fielded by the Germans. However, French armoured doctrine negated any superiority in equipment. Despite the outcries of such armoured warfare advocates as Lieutenant Colonel Charles de Gaulle, the man destined to lead the Free French movement after his country's capitulation, French planners relegated tanks to a supporting role for infantry formations and ordered them deployed piecemeal rather than as concentrated units.

In December 1938, the French government authorized the formation of armoured divisions; however, progress was painfully slow, and the execution of the order was not under way until early 1940. Just days before Great Britain and France declared war on Nazi Germany, de Gaulle was in command of the tanks of the Fifth Army, supporting the static Maginot Line defences. These tank formations were still of battalion size and were not intended to assume an offensive role.

German juggernaut

In preparation for renewing the war in the West, the Germans divided their forces into three army groups. In the north, Army Group B under General Fedor von Bock fielded more than 29 divisions and was responsible for luring the Allies into Belgium

TAKING ROUEN
During the German conquest of France in the spring of 1940, a PzKpfw I sits in the center of a main thoroughfare amid heavily damaged buildings in the French city of Rouen. A French automobile lies abandoned nearby.

ARRAS ACTION (ABOVE)
British armour counterattacked the advancing German 7th Panzer Division near the town of Arras and threatened to throw the invaders back before General Rommel ordered 88mm (3½in) guns employed in an anti-tank role to devastating effect.

to confront what they perceived as the main German threat. In the south, the 18 divisions of Army Group C, led by General Wilhelm Ritter von Leeb, were tasked with occupying the French troops along the Maginot Line and preventing an Allied flanking movement. The main German thrust was to come from more than 45 divisions of Army Group A, commanded by General Gerd von Rundstedt, with a heavy concentration of 10 armoured divisions in three corps. Concentrating on the town of Sedan on the east bank of the River Meuse and its nearby bridges, Army Group A was to advance rapidly through the forests of the Ardennes, then dash across open country, cutting off thousands of Allied troops that had advanced into Belgium to confront Army Group B.

Breakthrough at Sedan

The spearhead of Army Group A was the XIX Panzer Corps, which consisted of the 1st, 2nd and 10th Panzer Divisions and the Motorized Infantry Regiment Grossdeutschland. Under Guderian's command, the Panzers slashed through Luxembourg and Belgium, and reached the Meuse at Sedan on the afternoon of 12 May, capturing the town virtually without firing a shot.

Although the French had constructed numerous pillboxes and strongpoints in the area south of the Meuse, these were in various stages of completion, some of them merely concrete shells. The French Army's 55th Division was assigned the primary defensive role in the vicinity of Sedan. The 55th and later the 71st Divisions were both classified as Category B formations, lacking in both training and modern weapons.

STEADY STREAM
German tanks and troops poured into France in May 1940. Infantrymen march toward the front lines in this photo, while PzKpfw III tanks pause along the roadside. Tank crewmen, in their distinctive black headgear, observe the foot soldiers.

The Germans controlled the skies above Sedan, and Luftwaffe Junkers Ju 87 Stuka dive bombers harried the French continually. Although they inflicted relatively few casualties, the effect of the bombing was demoralizing to such a degree that some French units lost cohesion and fled the battlefield.

German infantrymen crossed the Meuse in small boats and captured bridges across the river, at times under intense fire, establishing bridgeheads at three primary locations on 13 and 14 May. Pioneers constructed temporary bridges as well, to enable the armoured vehicles to advance. Guderian's tanks rolled forward and engaged the French in pitched battles south and west of Sedan, particularly at Bulson Ridge and near the town of Stonne. Although at several points, the Germans were slowed by stiff French resistance, by the evening of 14 May, more than 500 German tanks had crossed the Meuse. During the costly French stand at Bulson Ridge, the 7th Tank Battalion lost 30 tanks, a full 75 percent of its strength, while the 213th Infantry Regiment was so badly mauled that it ceased to be an effective fighting force.

At Stonne, a French Char B1 bis heavy tank was credited with destroying more than a dozen German tanks. The struggle there lasted for several days, but, the French were unable to hold the town. The losses were comparatively heavy on both sides, and just under a week of fighting, the Motorized Infantry Regiment Grossdeutschland had suffered nearly 600 casualties.

By May 16, the issue was no longer in doubt. The Germans had penetrated the Ardennes, captured Sedan, and crossed the Meuse much more swiftly than the French High Command had anticipated. Once Guderian had broken out of the Meuse bridgeheads, the flank of the Allied armies in the north lay open. German tanks reached the English Channel within a week.

CHANNEL CHASE (FAR LEFT)
(opposite) German armoured spearheads knifed through France, advancing northwest toward the English Channel while leaving slower moving infantry behind. Hitler grew concerned due to open flanks that were vulnerable to counterattack, but his orders to halt were largely ignored.

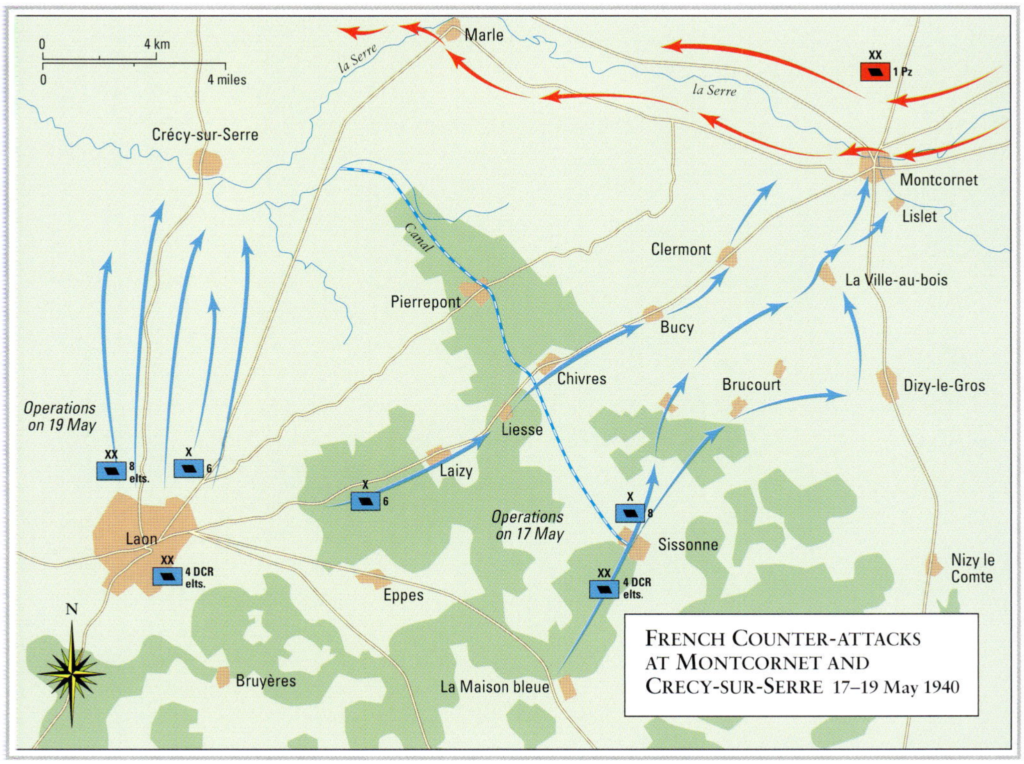

FRENCH FUTILITY (LEFT)
French forces mounted heroic counterattacks against the invading Germans; however, these proved wholly ineffective. Notable among the few French heroes of the period was General Charles de Gaulle, whose 4th Armoured Division mounted two counterattacks.

THE FALL OF FRANCE JUNE 1940

ARMOURED ADVANCE
A PzKpfw I (right) and a PzKpfw 38t (left) in France in May 1940. Superior tactics and radio communications ensured victory of these light tanks over more heavily armoured foes.

As Guderian fought to expand the Meuse bridgehead, taking substantial casualties and with some crossing efforts being repulsed, further north, the 7th Panzer Division, commanded by General Erwin Rommel, exploited a gap between the French 5th Motorized Division and the 18th Infantry Division of the Ninth Army. He reached the Meuse and crossed the river at Dinant on 13 May. By the first week of June, 7th Panzer had reached the English Channel coast at Abbeville, brushing aside resistance and slashing directly through an unsuspecting French armoured unit that had halted as night fell on 14 May. During its rapid advance, Rommel's command covered distances as great as 48km (30 miles) in a 24-hour period.

Once Guderian was clear of the Meuse, his progress was steady. However, senior German commanders became concerned that the pace of the Panzer thrusts was out-distancing the slower-moving infantry support that was considered essential for the protection of the armoured vehicles. As the Panzer penetration widened and the German flanks became more exposed, Hitler himself issued an order for Guderian to halt temporarily. However, even this order from the Führer was largely ignored. By 15 May, German tanks were advancing on an 80km (50-mile) front.

DEFEATED ARMOUR
German troops inspect the wreckage of a French Char B1 bis heavy tank destroyed during the fighting of June 1940. Following a brief pause after their successes in May, the Germans completed the conquest of France swiftly.

Inadequate counterattack

In response to the collapse of the Meuse front, General Gamelin ordered elements of the Seventh Army southward, and seven divisions defending the Maginot Line to the north to counterattack the exposed German flanks that extended more than 64km (40 miles). Just three days prior to the German attack, de Gaulle had been given command of the new 4th Armoured Division, which was actually assembling during its advance toward the Germans and had never operated as a cohesive unit. On 17 May, de Gaulle hit the German flank at Montcornet, blocking the road to Paris with 80 tanks. However, Luftwaffe dive bombers and German reinforcements halted the attack and forced de Gaulle to withdraw to his original positions. A second attack met the same fate, although Guderian's headquarters had been threatened and the German commander spent some anxious moments. De Gaulle also realized that the objective of the German armoured spearheads was not Paris but the Channel coast.

In the north, elements of Army Group B captured the Belgian port of Antwerp on 18 May and had reached the vicinity of Lille by 21 May. German armoured formations rolled into Amiens and Doullens just 64km (40 miles) from the coast. Guderian's 2nd Panzer Division reached the English

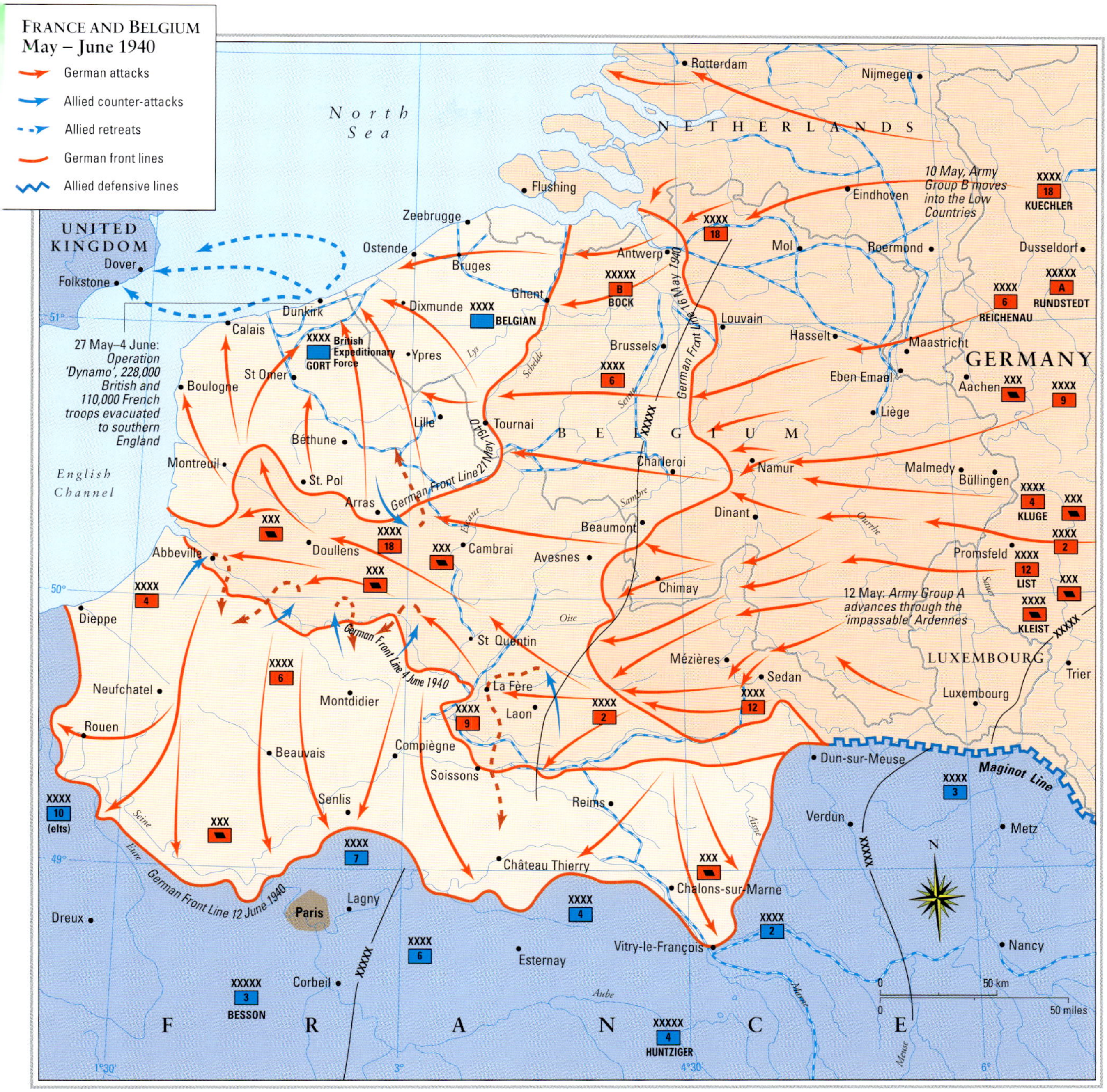

Channel on 20 May and captured Boulogne after a two-day fight with the 2nd Battalion, Irish Guards, a battalion of the Welsh Guards, three battalions of The Rifle Brigade, the Third Royal Tank Regiment and roughly 800 French infantry. Calais fell on 27 May. In a week of fighting Guderian had advanced 257km (160 miles).

Action at Arras

The most serious threat to the spectacular advance of the 7th Panzer Division occurred at Arras on 21 May as 74 Matilda I and II tanks of the 4th and 7th Royal Tank Regiments accompanied two battalions of the Durham Light Infantry in a spirited counterattack. Initially, the British made good progress and the Germans feared they had been assaulted by up to five divisions. Never far from the front, Rommel took personal command and ordered that 88mm (3½in) anti-aircraft and 105mm (4in) field cannon be deployed as far as possible and used in an anti-tank role. The British advance was halted, and a German counterattack was thrown back.

CASE RED
The resounding success of Case Yellow led to a follow-up operation, Case Red, to subjugate France. German troops surrounded Allied forces at Dunkirk and marched triumphantly into Paris. Within days, the French sued for peace.

WAR PRIZE
The Germans captured large stores of equipment during their conquest of France, pressing much of it into service. Here, a German soldier rides in the turret of a French Hotchkiss H39 tank emblazoned with a German cross.

However, the Allied position was untenable and Rommel pressed on to the English Channel.

Eventually, the remnants of the French armies in the north and the beleaguered British Expeditionary Force concentrated at the port of Dunkirk, where nearly 350,000 troops were rescued from the beach between 2 and 4 June by a motley collection of civilian and naval vessels. Hitler's decision to halt his Panzers short of Dunkirk to allow the Luftwaffe to deliver the coup de grace to the Allies has since been roundly criticized. With the escape of Allied forces across the Channel, Hitler allowed the Allies a breathing space, and missed his best chance of invading Britain.

FORWARD IN FRANCE
Panzerkampfwagen II tanks of the German 1st Panzer Division advance rapidly during their thrust into the interior of France. Within weeks, German armour and infantry had defeated French forces and compelled a new government to capitulate.

Case Red

As disaster loomed, the organized French Army had been reduced to about 50 divisions, less than half its original strength. Following a brief period of rest and resupply, the Germans turned their attention toward Paris, intent on completing their conquest of France. On 5 June, the offensive resumed with assaults along the River Aisne, while bridgeheads across the River Somme were secured by 8 June, and a major breakthrough in the area unhinged the French flank along the Aisne.

The French fought stubbornly for three days, and a counterattack by the 14th Infantry Division actually destroyed a German XXIII Corps bridgehead and resulted in the capture of 1,000 prisoners. Nevertheless, the exposed French flank compelled the defenders to retreat to the River Marne. On 12 June, four German Panzer divisions broke through the thin French lines and headed directly for Paris, entering the City of Light in triumph two days later. In short order, Fascist Italy declared war on Britain and France and invaded in the south.

Humiliating defeat

By mid-June, German tanks had rolled up to the Swiss border and the French government had fled, first to Tours and then to Bordeaux. The government of Prime Minister Paul Reynaud fell. Elderly Marshal Philippe Pétain was installed as the new French head of state and opened negotiations

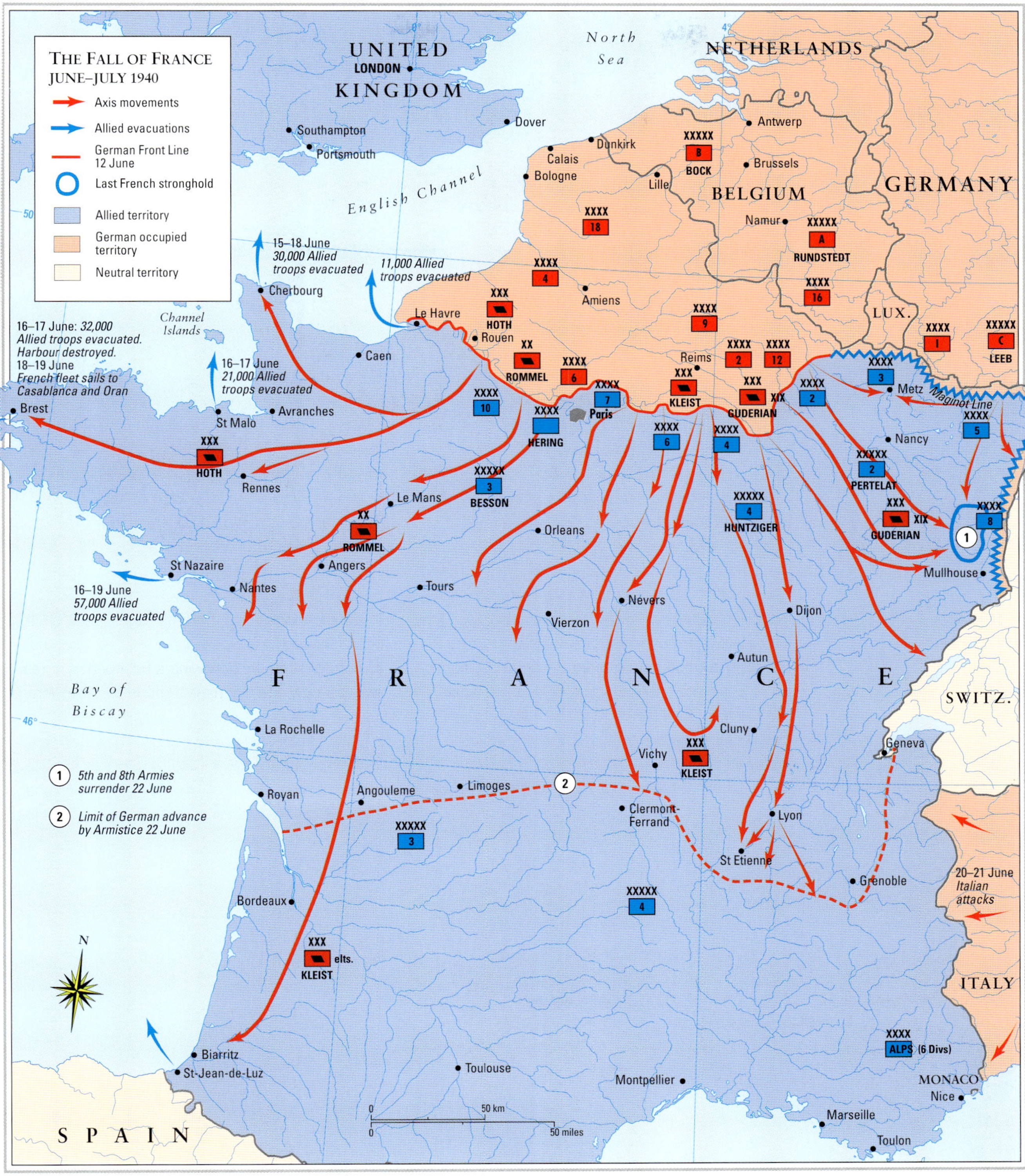

for an armistice. His Vichy government would later be vilified for its collaboration with the Nazis. On 25 June 1940, the Germans dictated surrender terms to the stunned French in the Forest of Compiègne. The surrender took place in the same railroad car used by the French to dictate terms to Germany in 1918, at the end of World War I.

The German victory was complete. In 42 days, France had been conquered. Hitler was master of Western Europe, and Great Britain stood alone.

CASE RED
German spearheads struck toward the Channel ports and Bay of Biscay, outflanked the Maginot Line, captured Paris and pushed south beyond Vichy.

WORLD TANK FORCES 1941–42

Military historians have acknowledged that 1942 was the year that doomed the Axis. In the Pacific, the United States had seized the land initiative with the assault on Guadalcanal and inflicted a decisive defeat on the Imperial Japanese Navy at Midway. On the Eastern Front, the Soviets had blunted the Nazi drive on Moscow and trapped the German Sixth Army at Stalingrad. In North Africa, the British Eighth Army had decisively defeated Panzer Armee Afrika at El Alamein in October. The following month Anglo-American forces had executed Operation Torch, landing at Oran, Algiers and Casablanca.

Behind this tremendous turning of the tide during World War II was the weight of Allied industrial capacity, particularly that of the United States and the Soviet Union. The production of vast quantities of war materiel was only part of the equation. Logistics and supply efforts were coordinated, and tanks and other weapons were shipped halfway around the world to combat areas.

Arsenal of democracy

Even before U.S. entry into the war, President Franklin D. Roosevelt had described his country as the great 'Arsenal of Democracy' and pushed the Lend-Lease Act through Congress. The provision allowed nations then fighting the Nazis, particularly Great Britain, to obtain the weapons of war on credit, rather than the 'cash and carry' basis that had been enforced previously. Later, the Soviet Union benefited from Lend-Lease and also accepted armoured vehicles of British and Canadian manufacture in its war against the invading Germans.

During the course of the war, Great Britain received no fewer than 17,000 M4 Sherman medium tanks, roughly 35 percent of the total number produced from 1941 to 1945. For Lend-Lease purposes and to equip its own armed forces the United States produced more than 29,000 tanks during 1941 and 1942, the bulk of these consisting of the M3 Stuart light tank, the M3 Grant and Lee medium tanks, and the ubiquitous M4 Sherman. US tank production peaked at 29,497 in 1943, and the total number of tanks that rolled off American assembly lines in World War II topped 100,000.

Great Britain continually improved its tank designs during the course of the war and steadily increased production of armoured vehicles. In 1939, fewer

DESERT COMBATANTS
On the eve of the decisive Battle of El Alamein, General Bernard Montgomery's Eighth Army had built its armoured strength to more than twice that of General Erwin Rommel's Afrika Korps, strained by equipment and supply shortages.

EIGHTH ARMY AND AXIS TANK STRENGTHS
OCTOBER 1942 – JANUARY 1943

		23 OCT '42	5 NOV '42	15 NOV '42	25 NOV '42	11 DEC '42	30 DEC '42	15 JAN '43
AXIS	**German**	238	35	35	54	c. 60	60	34
	Italian	279	0	45	42	c. 30	?	57
	TOTALS	**517**	**35**	**80**	**96**	**c. 90**	**60?**	**91**
BRITISH	**Frontline**	1.029	537	395	418	454	367	532
	Forward Res.	200	67	105	144	105	180	91
	TOTALS	**1,229**	**604**	**500**	**562**	**559**	**547**	**623**

than 1,000 tanks were manufactured in Britain, while production in 1940 increased to 1,399. More than 4,800 tanks were manufactured in Great Britain in 1941, and 8,600 the following year.

British production exceeded that of Nazi Germany, which depended upon the PzKpfw III and PzKpfw IV as its frontline tanks, producing a combined total of 7,400 of these vehicles in 1941 and 1942. While Allied production stressed quantity over quality, the German emphasis on highly engineered and technologically advanced models proved their undoing in wartime.

Soviet tank production accelerated rapidly following the German invasion of the country in June 1941. Although only 115 of the superb T-34 medium tank were built in 1941, production increased dramatically the following year, and by 1942 more than 15,000 had been built. In 1943 alone, production of the T-34 exceeded the previous three years combined at nearly 16,000. Production of the KV-1 heavy tank and heavy self-propelled assault guns totalled 4,300 from 1940 to 1942 and approached 14,000 by the end of the war.

Soviet armoured strength

Despite their experience in the Spanish Civil War in 1936-37, which demonstrated that the superior firepower of their tanks could be decisive, the Soviet military establishment nevertheless opted to disband the armoured corps and distribute tanks throughout infantry divisions. By 1939, however, General Georgi Zhukov, later to achieve fame fighting the Nazis on the Eastern Front, had defeated the Japanese in Manchuria decisively, and the effective use of armoured formations was acknowledged as primary factor in the victory. The Soviets had also witnessed the dismal performance of their tanks as deployed in the Winter War with Finland and observed the successes of the German Panzer divisions in Poland.

Therefore, by 1940, the Red Army had reconstituted nine mechanized corps, which included two armoured divisions and a motorized infantry division. Each armoured division consisted of two tank regiments, a motorized infantry regiment, and a motorized artillery regiment. Its complement of BT series, T-34 and KV-1 tanks totalled approximately 400. The motorized infantry division included two motorized infantry regiments, an armoured regiment, and a motorized artillery regiment. Plans to increase the number of armoured corps to 20 by the autumn of 1941 were derailed by the German invasion in June of that year.

By the spring of 1942, organizational changes had been almost continuous. Early in the year, the organization of the armoured corps included two tank brigades and a motorized brigade. The tank brigade included three armoured battalions of 23 tanks each, a motorized infantry machine gun battalion, anti-tank company, mortar company, reconnaissance battalion, and anti-aircraft battalion. In April, the corps strength was increased to three tank brigades and a motorized brigade. By the end of July, four tank armies, the 1st, 3rd, 4th and 5th, had been formed. They consisted of two armoured corps, an independent tank brigade, a rifle division, a light artillery division, a guards rocket launcher regiment, anti-aircraft and attached support units. In September, another mechanized corps was added, which included a tank brigade and three motorized brigades.

SOVIET BT-5
The Soviet BT-5 tank was fast, maneuverable and well armed and armoured for its day. During the Spanish Civil War, tanks of the BT series deployed with Republican forces outclassed German and Italian armour in action with the Nationalists.

VISCOUNT MONTGOMERY
General Bernard Law Montgomery achieved lasting fame as commander of the British Eighth Army in the North African desert. The victor of El Alamein, he went on to command Allied ground forces in Western Europe.

PANZERKAMPFWAGEN III
Intended to engage in tank-versus-tank combat, the PzKpfw III was developed in the 1930s and deployed widely with the German Army in World War II. It was progressively upgunned during the war.

British brute force

On the eve of his El Alamein offensive, General Bernard Montgomery had increased the strength of his Eighth Army to more than 1,000 tanks, easily double that of his adversary, the Panzer Armee Afrika under General Erwin Rommel. The 7th Armoured Division, nicknamed the 'Desert Rats', included four armoured brigades, the 4th Light, 8th, 22nd, and 131st Queens. Although both sides sustained heavy losses at El Alamein and during the ensuing desert pursuit, Montgomery was able to replace those tanks put out of service, while Rommel's successors watched their Panzer numbers dwindle and were powerless to remedy the situation. By the time of the Axis surrender in Tunisia in early 1943, combined German and Italian armoured strength totalled less than 100 tanks fit for service.

The organization of British armoured divisions in the desert remained fluid through 1942 but, in the spring of that year, the standard British armoured division was comprised of one armoured brigade, one infantry brigade, and a motorized battalion. These were supplemented by various self-propelled and towed artillery and reconnaissance formations.

American armour

In the spring of 1942, the United States had been at war only a few weeks; however, Army Chief of Staff General George C. Marshall and General Leslie J. McNair, commander of Army Ground Forces, had worked diligently for months to modernize the U.S. Army. On paper, the army included 16 armoured divisions. Tanks were largely considered infantry support weapons, and consequently were lightly armed and armoured – built for speed.

In 1942, the standard U.S. armoured division included a headquarters element, two tank regiments, a regiment of armoured infantry, three battalions of 105mm (4in) self-propelled howitzers, a reconnaissance battalion, and support units. The tank regiment included two medium tank battalions and one light tank battalion, while the armoured infantry regiment included three armoured infantry battalions. Within the division were two combat command headquarters. These units were brigade level and coordinated tank, armoured infantry, and artillery battalions formed as mission-oriented task forces according to orders from the divisional commander.

Panzers prominent

In 1938, the standard German armoured, or Panzer division included a combat element of a Panzer brigade that consisted of two Panzer regiments, each containing two or three Panzer battalions as available equipment and command priorities dictated. The Panzer battalion included three tank companies, each with three platoons of up to five individual tanks. With an armoured strength of nearly 400 tanks, the Panzer division was a potent force. Its strike capability was increased with a motorized rifle formation, often a brigade, an anti-tank battalion, an artillery regiment with two battalions of light field howitzers in three batteries, an armoured reconnaissance battalion, and support elements such as a light combat engineer company.

During the first year of World War II, the Panzer division's support capabilities were improved with an anti-aircraft battalion, the increase of the combat engineer, or pioneer, detachment to a full battalion, and enhanced reconnaissance capabilities. By the autumn of 1940, plans for Operation Barbarossa,

M3 STUART
The M3 Stuart light tank epitomized the U.S. doctrine involving fast moving reconnaissance and infantry support tanks. Armed with a 37mm (1½in) cannon, it was most effective against Japanese forces in the Pacific during World War II.

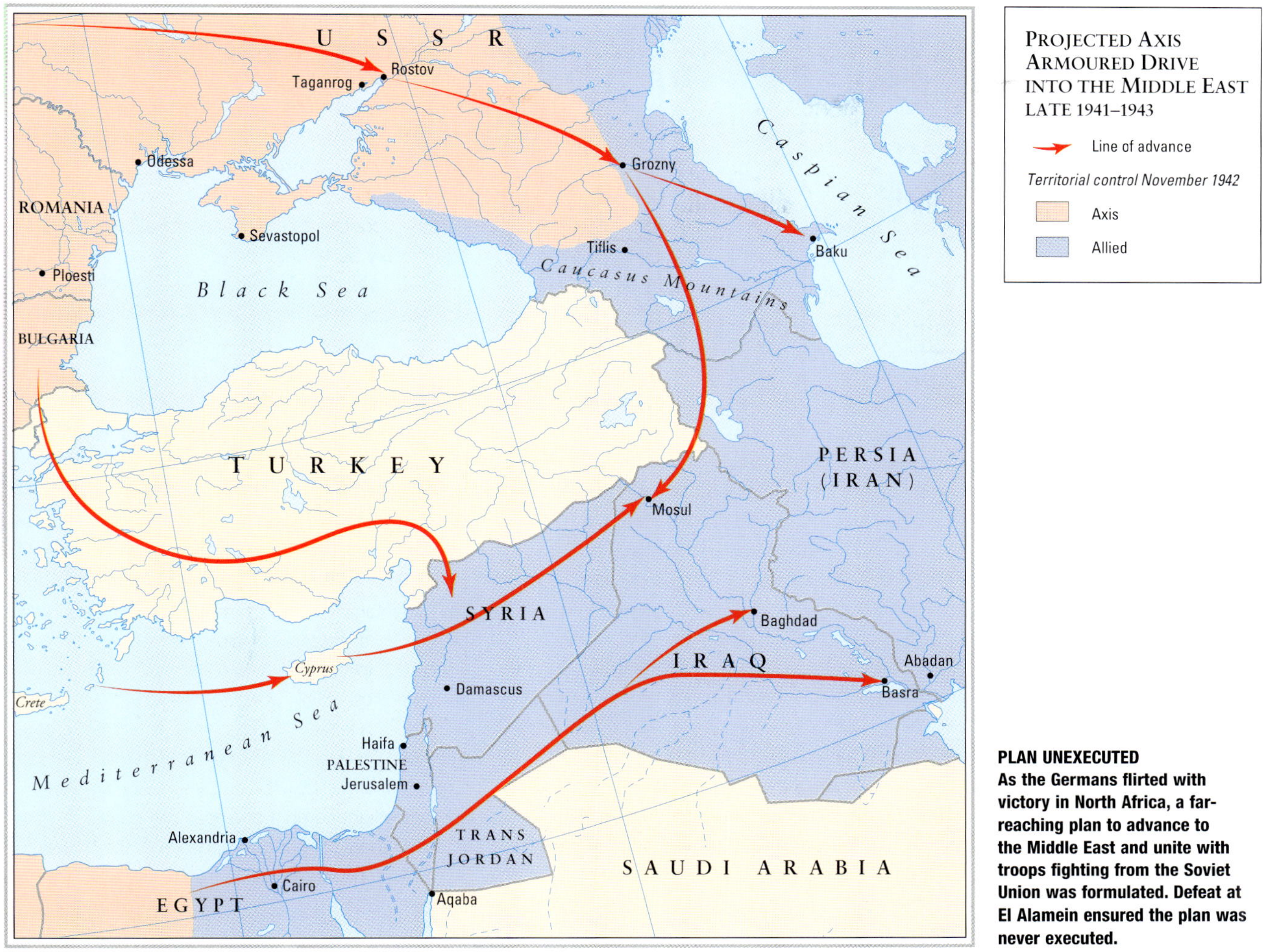

PLAN UNEXECUTED
As the Germans flirted with victory in North Africa, a far-reaching plan to advance to the Middle East and unite with troops fighting from the Soviet Union was formulated. Defeat at El Alamein ensured the plan was never executed.

the June 1941 invasion of the Soviet Union, were well under way. A significant reorganization occurred with the number of Panzer divisions doubling since 1938. At mid-year, the German Panzer forces had totalled 10 divisions and the 40th Panzer Abteilung, an independent battalion-sized formation. Tank strength, including the most up-to-date PzKpfw III and IV types, totalled 3,465.

Although the number of Panzer divisions increased in the autumn of 1940, the number of tanks did not. The table of organization and equipment in the new 1940 Panzer division included between 150 and 200 tanks, approximately half the previous number. The change was initiated ostensibly to provide Panzer formations more freedom of movement across the vastness of Russia, particularly as more than 4,000 tanks were allocated to Barbarossa.

The organic troop strength of the Panzer division remained fairly constant throughout World War II at roughly 14,000. In the spring of 1942, the motorized rifle regiments of the Panzer divisions were redesignated as Panzergrenadier regiments, and the number of companies in each rifle battalion was decreased from five to four. Infantry capability was improved with the addition of machine-gun and light infantry gun companies, while more anti-aircraft units were added.

The German Army had fielded Panzergrenadier divisions as early as 1938, although these were originally known as motorized infantry divisions and included some rifle formations and a few units designated as motorized cavalry. In the summer of 1942, these divisions became officially known as Panzergrenadier divisions, which included two infantry regiments, support units, and a battalion of tanks or self-propelled assault guns. Their troop strength was comparable to that of a Panzer division, and these soldiers were primarily transported by truck or marched on foot as regular infantry. Routinely, only a single Panzergrenadier brigade within the division was supplied with half-tracks or other armoured vehicles for transportation.

NORTH AFRICA 1941–42

Senior British commanders had realized as early as the 1930s that Italian adventurism and troop concentrations in North Africa posed a threat to Egypt, including Alexandria, the Royal Navy's principal port in the Eastern Mediterranean, and the strategically vital Suez Canal. Therefore, Imperial strength in the region was bolstered somewhat given the continuing threat of a Nazi invasion of Great Britain and the débâcle that befell the British Expeditionary Force on the Continent in the spring of 1940. Despite these efforts, Italian forces in North Africa greatly outnumbered British troops and other assets.

Predictably, an Italian offensive in the spring and summer of 1940 resulted in the occupation of British Somaliland and several towns across the frontier in Kenya and the Sudan. In September, the Italians launched an offensive into Egypt. Against long odds, British forces, which had reached a peak strength of about 100,000 that year, achieved significant success during a counteroffensive initiated in East Africa in January 1941. Meanwhile, to the northwest, British forces launched a response to the Italian incursion into Egypt.

Operation Compass

Toward the end of 1940, British and Imperial reinforcements began to arrive in North Africa, and General Sir Archibald Wavell, Commander-in-Chief, Middle East, authorized Operation Compass. Troops of the 7th Armoured and 4th Indian Divisions, dubbed the Western Desert Force, were to undertake a five-day raid against the Italians, intended to slow their progress in Egypt. The 7th Armoured Division would go on to gain lasting fame as the 'Desert Rats'.

Commanded by Major General Richard O'Connor, the British force surged forward at Sidi Barrani on 9 December, gathering momentum as Italian formations wilted before them. Soon, the raid had developed into a full-scale offensive, and within three weeks the British had ejected the Italians from Egypt, captured Fort Capuzzo and won a pitched battle at Bardia. On 22 January, the British captured the port of Tobruk, and during the first week of February the retreating Italians were cornered at Beda Fomm and surrendered en masse. During Operation Compass, the Western Desert Force, which was renamed XIII Corps in

ITALIAN ARMOUR
Italian tank soldiers stand in ranks and listen to a presentation by their commanding officer. Behind the tankers are Fiat-Ansaldo M13/40 medium tanks. Influenced by the British Vickers 6-ton tank design, these armoured vehicles mounted 47mm (1¾in) cannon.

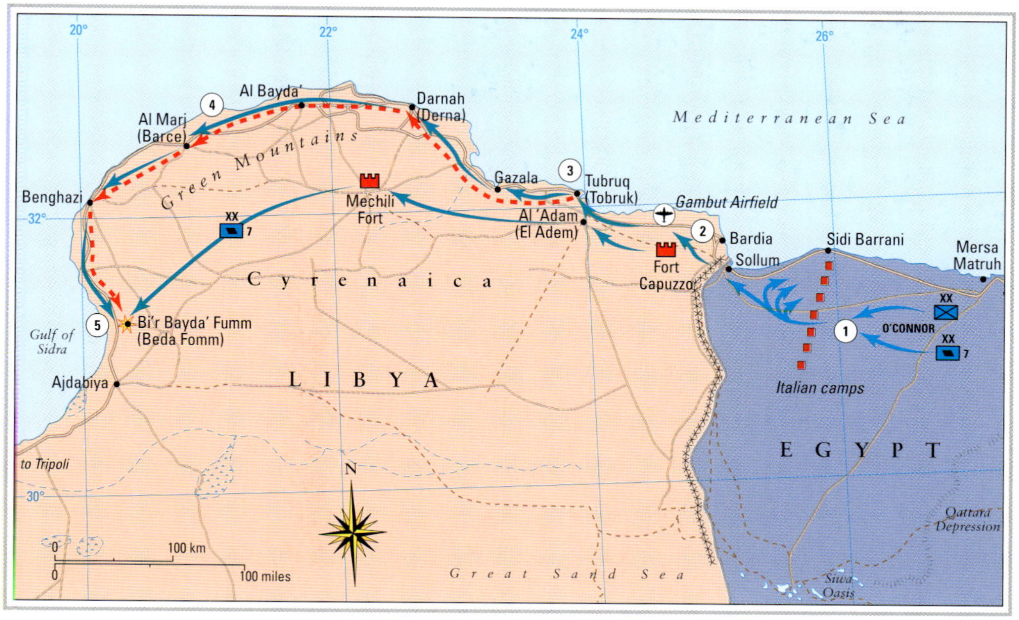

Operation Compass
9 December 1940 –
7 February 1941

- Allied army movements
- Italian army retreat
- Major battle site

1. 9 December 1940: *General O'Connor opens by hooking behind Graziani's camps. In three days he recaptures Sollum and Sidi Barrani taking 39,000 prisoners.*
2. 5 January 1941: *Captured by newly arrived Australians who replaces 4th Indian Division.*
3. 8 January: *Tobruk falls to Australians who take 25,000 prisoners.*
4. *Italians retreat along via the Balbia coastal highway pursued by the Australians.*
5. *7 February: Italians cut off and ambushed by 7th Armoured Division at Beda Fomm. In ten weeks O'Connor has captured 130,000 prisoners.*

January and became the nucleus of the British Eighth Army, captured 130,000 Italian prisoners, including 22 generals, while suffering fewer than 1,800 casualties.

Rommel arrives

As the rout of Italian forces in North Africa unfolded, Hitler committed German troops to the theatre in support. A few days after the triumphant conclusion of Operation Compass, Italian reinforcements came ashore at Tripoli. Along with them, the German 90th and 5th Light Divisions arrived with their commander, General Erwin Rommel, who would soon become known as the 'Desert Fox'. On 19 February, the German formations were formally named the Deutsches Afrika Korps, while the entire Axis military presence in North Africa became known as Panzer Armee Afrika.

Rommel stabilized the situation rapidly, turning back a series of British offensive actions during the spring and summer of 1941. In April, he launched an offensive of his own. During Operation Sonnenblume, Rommel's Panzer formations struck swiftly, pushing imperial forces out of Libya as the 5th Light Division took 2,000 prisoners at Fort Mechili and besieged Tobruk.

ITALIAN ROUT
Intended as a five-day raid into Italian territory, Operation Compass succeeded beyond the wildest expectations of British planners and continued as a full-scale offensive effort.

ROMMEL ARRIVES
The Germans dispatched the initial elements of the Afrika Korps to Tripoli, Libya. Arriving in February 1941, General Erwin Rommel wasted no time mounting an effective counteroffensive against the British.

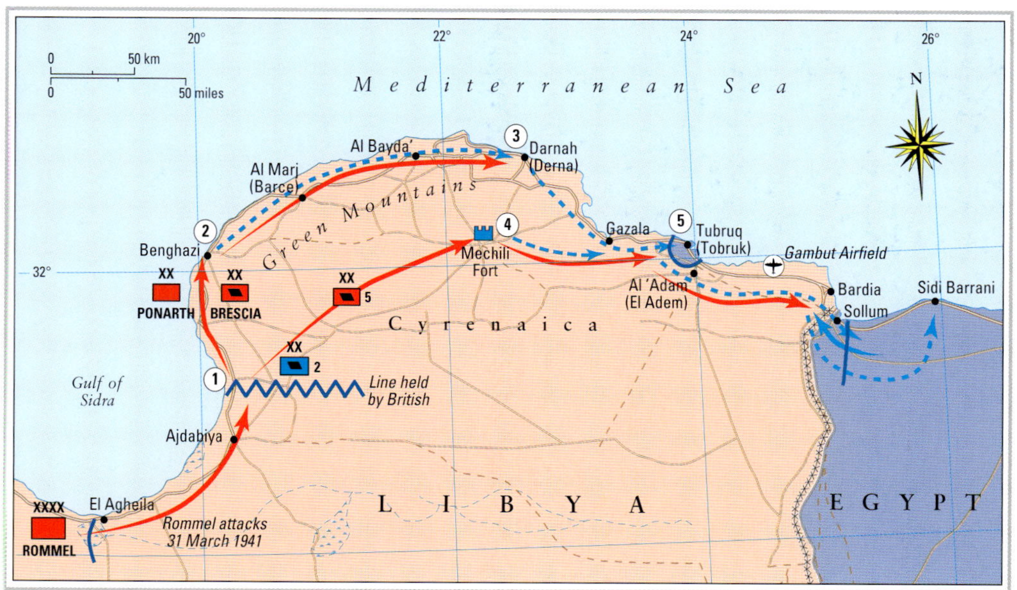

Operation Sonnenblume
April 1941

- German army movements
- Alllied army retreat

1. *Rommel divides his forces sending a Division under Brescis plus a small battle group under Ponarth along the coast road. He sends 5th Light Division along the desert route to attack Fort Mechili.*
2. 3 April1941: *British withdraw allowing Rommel's forces to walk straight into Benghazi.*
3. 7 April: *Battle group under Ponarth captures Generals O'Connor and Neame near Darnah.*
4. 8 April: *Major-General Gambier-Parry and 2000 British troops are captured at Fort Mechili after failing to escape. Few soldiers reach Tobruk.*
5. 11–13 April: *Siege of Tobruk commences. During this Easter weekend the Australian and British troops repel three Panzer attacks.*

OPERATION CRUSADER
NOVEMBER – DECEMBER 1941

Allied army movements
Italian army retreat
Major battle site

1. 18 November 1941: *Auchinleck launches Operation Crusader.*
2. 19 November: *British attempt to breakout from Tobruk thwarted by German 90th Light Division. This German division is attacked from the rear by New Zealand Infantry and 7th Armoured Division who are, in turn, attacked by Panzers moving towards Sidi Rezegh.*
3. 20 November: *Ariete Armoured Division repulses 22nd Armoured Brigade attack.*
4. 22 November – 7 December: *Confused tank battles rage at Sidi Rezegh, Rommel withdraws from Cyrenaica.*
5. 7 December: *End of 242-day siege of Tobruk.*
6. 30 December: *Rommel's withdrawal stops at Mersa el Brega.*

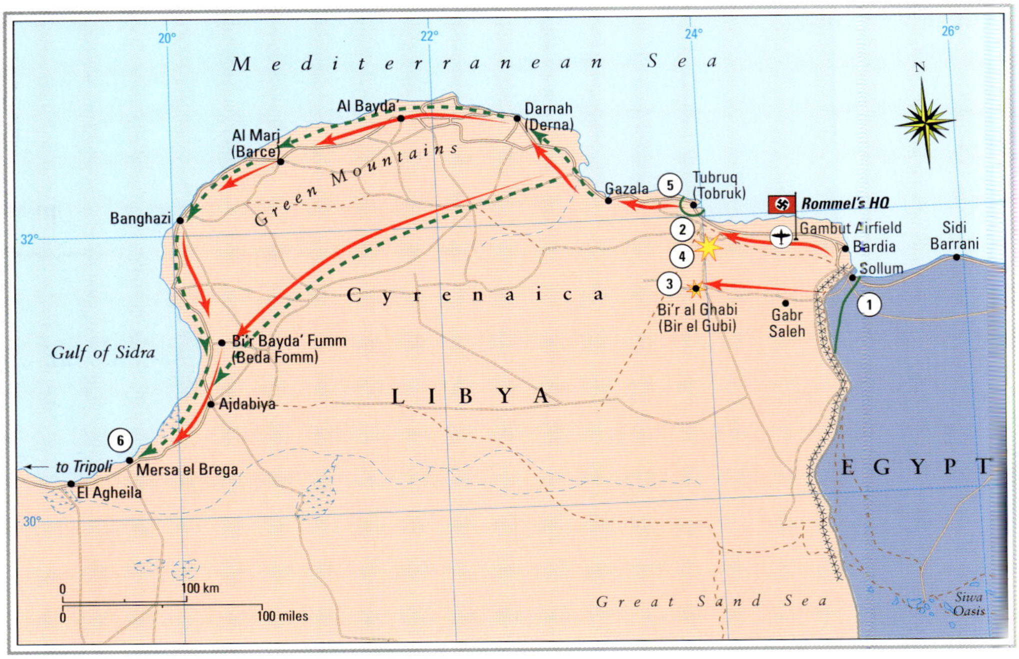

TOBRUK SAVED (ABOVE)
Operation Crusader raised the siege of Tobruk in the spring of 1941 as British and German armour clashed at Halfaya Pass.

DESERT DENIZENS (BELOW)
The composition of the British armoured divisions in the desert was somewhat separate from those organising to fight in Europe. Its combat elements consisted of two powerful armoured brigades, artillery and motorised infantry.

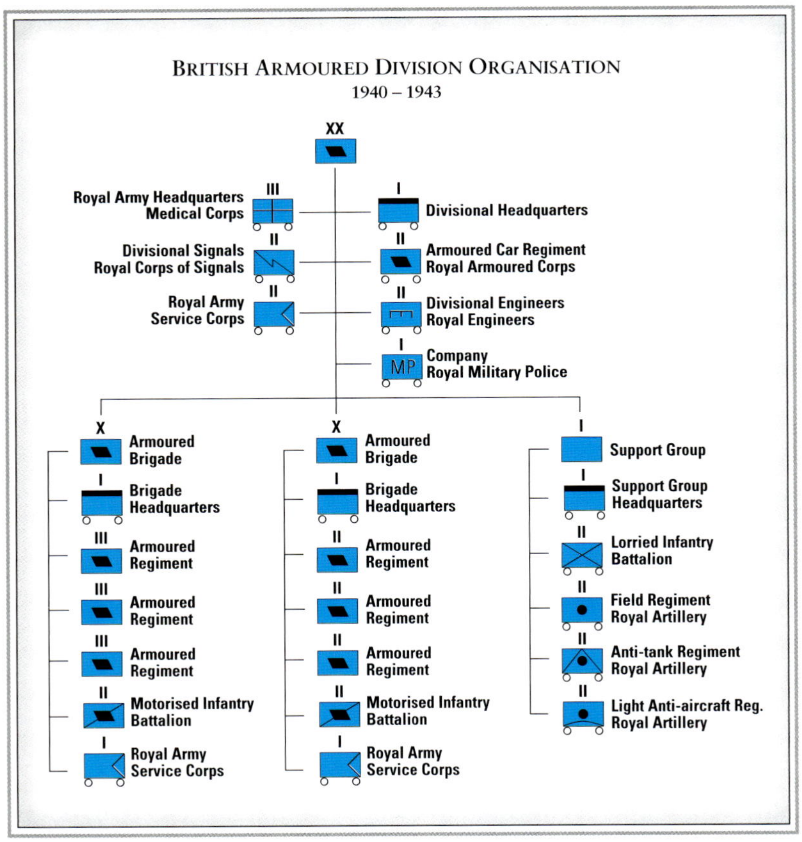

Operation Crusader

For seven months, beleaguered Australian and British troops held the port of Tobruk against the Germans. On 18 November, the British launched Operation Crusader with the intent of raising the siege and bringing German armour to battle. A furious two-day engagement was fought at Sidi Rezegh between the veterans of the 7th Armoured Division and the 15th and 21st Panzer Divisions. Under the command of General Sir Arthur Cunningham, the XIII and XXX Corps of Eighth Army moved along the Mediterranean coast through the Halfaya Pass and southwestward against Rommel's flank intending to cut his lines of communication and beat back the German armour.

The armoured strength of XXX Corps totalled more than 450 vehicles, while the 1st Army Tank Brigade attached to XIII Corps included 135 new Matilda and Valentine tanks. At Bir el Gubi, the 22nd Armoured Brigade destroyed 34 tanks of the Italian Ariete Division, but lost 25 of its own to enemy fire, while 30 others were disabled by mechanical failure. Heavy losses occurred on both sides, and Rommel cancelled an offensive to deal with the British advances, finally withdrawing as his supply lines came under enemy fire. Rommel was surprised by the strength of the British opposition and his attempt to relieve hard-pressed troops near the Halfaya Pass was abandoned. During the first week of December troops of the British 70th Division broke out of the Tobruk encirclement and linked up with New Zealanders, ending the 242-day siege

Tobruk revisited

Despite the successful relief of Tobruk, the British Eighth Army had suffered serious casualties during Operation Crusader, and now under the command of General Sir Claude Auchinleck, was seriously overextended. Rommel regrouped and received badly needed reinforcements as 1941 ebbed, and in January Panzer Armee Afrika renewed offensive operations, driving the British eastward 480km (380 miles) to the Gazala Line.

The Desert War settled into a four-month lull as both British and Axis forces were augmented. Eighth Army armoured strength grew to more than 850 tanks, a few of which were the U.S.-built Grant with a hull-mounted 75mm (3in) cannon and a 37mm (1½in) gun mounted in a traversing turret. British defences at Gazala consisted of six well-defended strongpoints, located within mutually supporting distance of one another. At the end of May, Rommel attacked. Two of the boxes, defended by Indian troops, fell rapidly. At Sidi Muftah, the 150th Brigade Group held out against six German armoured assaults before being virtually wiped out.

To the south at Bir Hacheim, 3,600 soldiers of the Free French Brigade fought off the Italian Trieste Division and held out for 16 days. A full 2,700 of the defenders evaded capture; however, the British situation worsened significantly in an area of heavy fighting known as the Cauldron. By the middle of June, the 201st Guards Motor Brigade had lost the Knightsbridge Box and in a single day's fighting the Eighth Army had suffered 6,000 killed, wounded and captured with 150 tanks destroyed.

As the Eighth Army withdrew eastward, Rommel brushed aside a series of delaying actions and small-scale counterattacks. On 21 June, elements of Panzer Armee Afrika rolled into Tobruk and captured huge stockpiles of fuel and materiel.

ABANDONED MATILDAS
Destroyed in combat with German armour and anti-tank guns, British Matilda II tanks lie along a desolate stretch of desert. Both sides lost huge numbers of tanks in North Africa.

FALL OF TOBRUK
In the spring of 1942, General Erwin Rommel captured his prize, the port city of Tobruk on the Mediterranean.

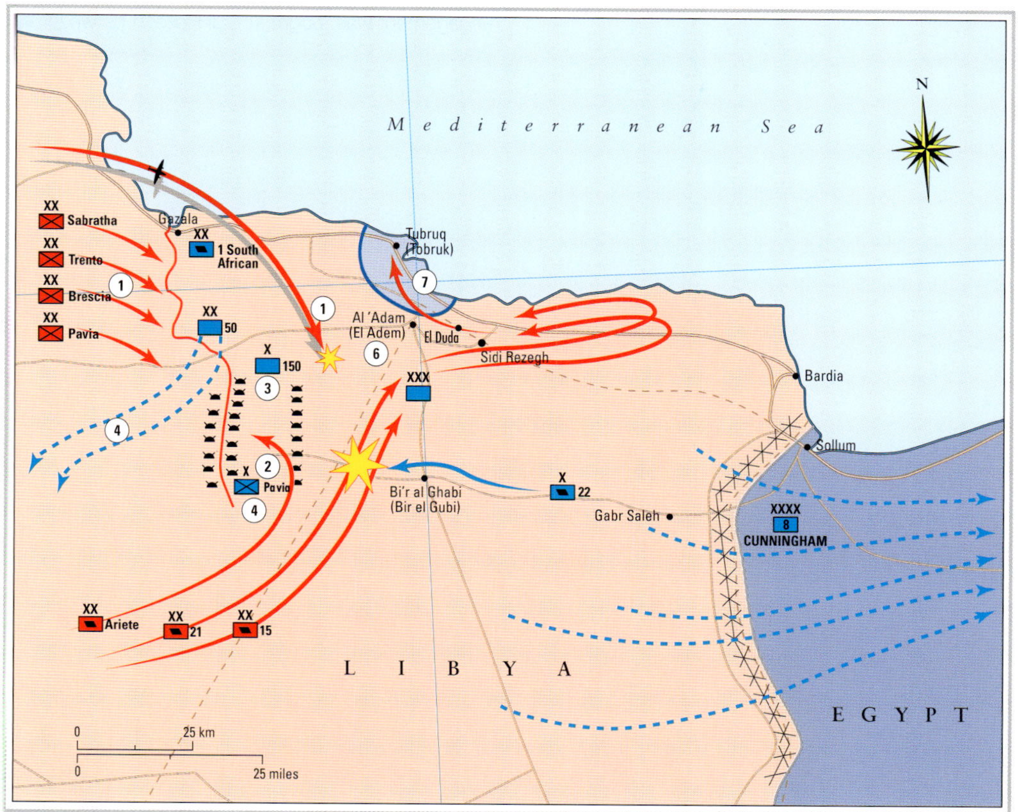

GAZALA AND THE FALL OF TOBRUK
26 MAY – 21 JUNE 1942

- Allied front lines
- Allied attack
- Allied retreat
- Axis movements
- Major battle site
- Minefields

1. 26 May 4.00 pm.: *Offensive starts, General Crüwell feints in the north with mainly Italian divisions.*
2. 26–27 May: *Rommel's real attack. His armour hooks around the French box at Bir Hacheim.*
3. 2 June: *150th Brigade falls, 3000 prisoners taken.*
4. 10 June: *After two weeks of siege, Koenig's Free French Brigade withdraws from Bir Hacheim box.*
5. 14 June: *British 50th Division escapes by first heading west and then south-west breaking through Axis lines.*
6. 14 June: *Scots Guards and South African anti-tank gunners suffer heavy casualties delaying German advance.*
7. 21 June: *Rommel smashes through Tobruk perimeter and the port is captured with 35,000 prisoners taken.*

BARBAROSSA (22 JUNE – 1 OCTOBER 1941)

PANZERS EAST
German tanks roll across the flat, dusty landscape of Russia during the opening phase of Operation Barbarossa. The 22 June 1941 invasion of the Soviet Union was initially successful, destroying large Soviet troop formations and capturing vast territory.

OPERATION BARBAROSSA (OPPOSITE)
The Germans launched their invasion of the Soviet Union with a million troops on a 1,500km (1,000-mile) front. Initially, Soviet forces were routed, and the Germans threatened Moscow before the onset of winter.

In the early hours of 22 June 1941, some 161 Axis divisions commenced Operation Barbarossa, the invasion of the Soviet Union. The campaign's intent, finalized in the 17 December 1940 plan, was the destruction of the bulk of the Red Army located in the 400km-deep (250-mile) border region up to the Dnepr and Dvina Rivers. The German-led Axis invasion would initiate an ideologically-driven war of racial annihilation that would utterly obliterate the scourge of Communism and enslave the Soviet peoples for the benefit of the Reich. Once the German Army had destroyed the Soviet forces in the border regions, it would to push on against meagre resistance until it reached a line that stretched from Arkhangelsk on the White Sea in the north, down through the Ural Mountains to the Caucasus region. To achieve this victory, the German Army would undertake bold and rapid operations that involved deep strategic penetrations by armoured (Panzer) spearheads. Four Panzer groups would conduct a series of envelopments and double envelopments that would encircle and destroy large parts of the Red Army until the latter collapsed.

This final Barbarossa plan envisaged a three-pronged invasion undertaken by three Army Groups. In the north, Army Group North would attack from East Prussia through the Baltic States to capture Leningrad and link up with the Finnish forces pushing south toward the Karelian isthmus In the centre, Army Group Centre would advance through White Russia to Smolensk and then Moscow, the Soviet capital, before pushing on through Kazan to the Urals. In the southern sector Army Group South's mixed Axis forces would advance through the Ukraine, via Kiev and Kharkov onto Rostov and beyond into the oil-rich Caucasus. The scale of the invasion was immense: The initial start line, excluding the Finnish front, was 1,464km (910 miles) long, while Moscow was 990km (616 miles) distant and the Urals a staggering 2,188km (1,360 miles) away.

Four armoured groups (Panzergruppen), with a combined total of 17 armoured and 12 motorized divisions that fielded 3,517 tanks, spearheaded this invasion. In total, some 4,431 Axis AFVs participated in the campaign. The Fourth Panzer Group, commanded by Colonel-General Hoepner, spearheaded Army Group North's attack. Colonel-General Heinz Guderian's Second and

COMMAND STRUCTURES (RIGHT)
The German High Command was divided into three potent army groups totaling 121 divisions for Operation Barbarossa. Stavka, the Soviet high command, was caught unprepared for the German onslaught.

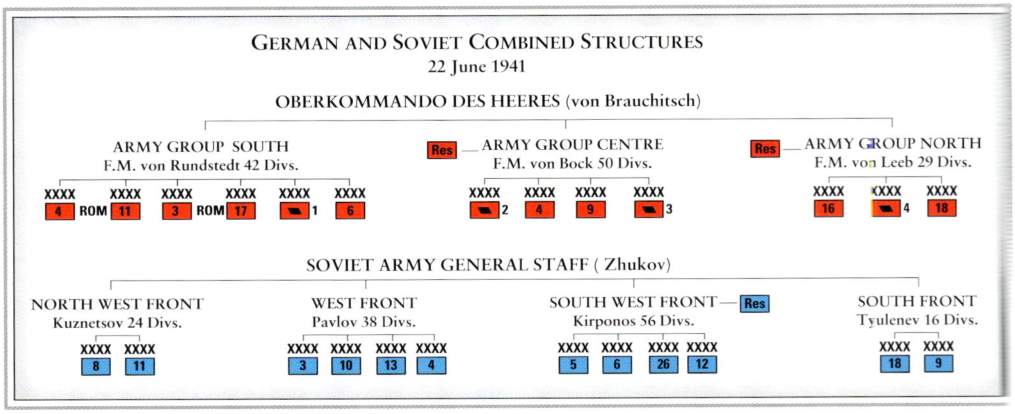

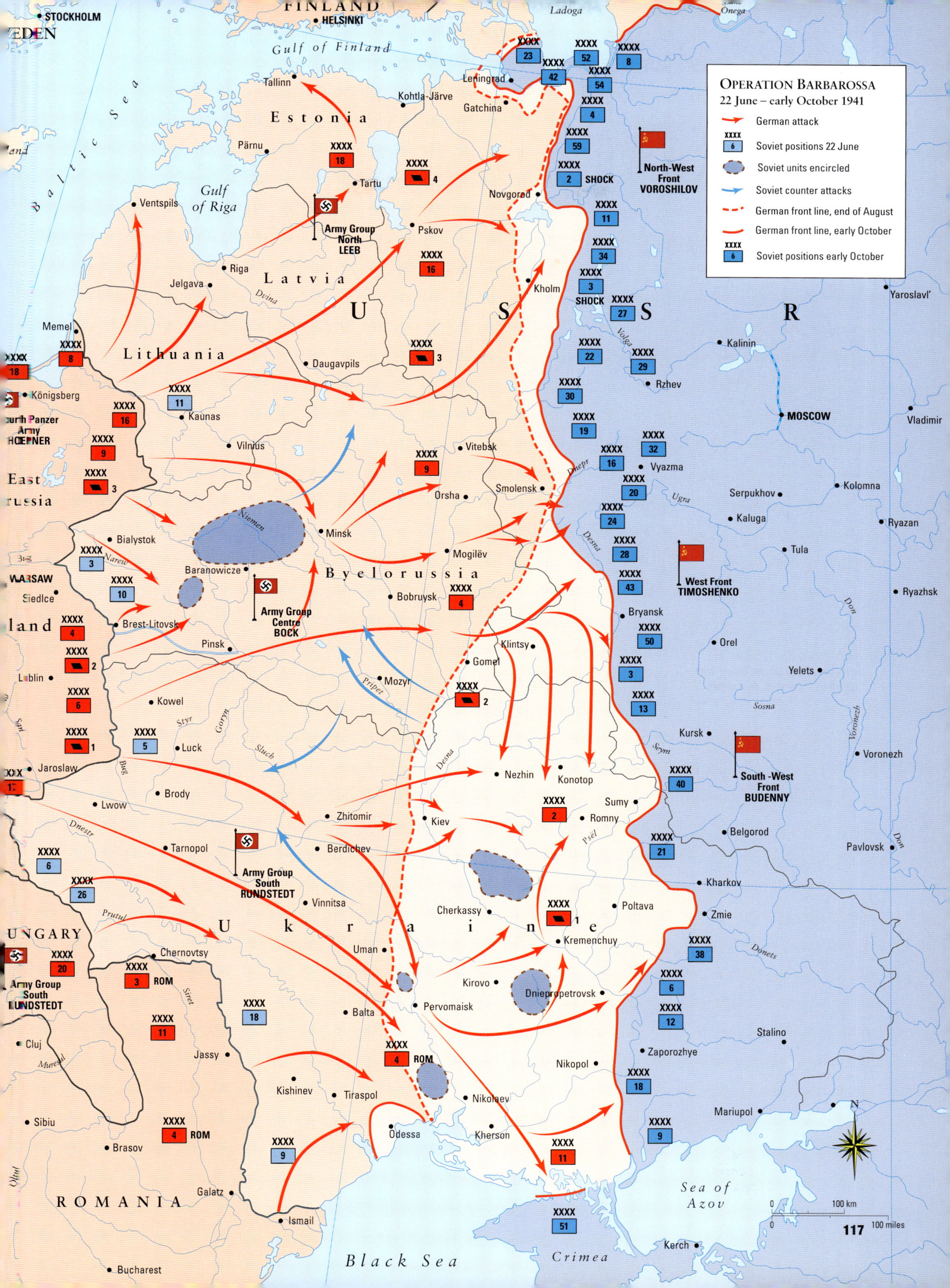

OPERATION BARBAROSSA
22 June – early October 1941
German attack
Soviet positions 22 June
Soviet units encircled
Soviet counter attacks
German front line, end of August
German front line, early October
Soviet positions early October
FINLAND
HELSINKI
STOCKHOLM
Gulf of Finland
Ladoga
Onega
Baltic Sea
Leningrad
Tallinn
Kohtla-Järve
Gatchina
Estonia
Pärnu
Tartu
Gulf of Riga
Ventspils
Army Group North LEEB
Pskov
Novgorod
North-West Front VOROSHILOV
Riga
Jelgava
Latvia
Dvina
Kholm
Memel
Lithuania
Daugavpils
U S S R
Yaroslavl'
Kalinin
Rzhev
MOSCOW
Vladimir
Königsberg
Kaunas
Vilnius
East Prussia
Vitebsk
Vyazma
Smolensk
Orsha
Dniepr
Niemen
Minsk
Bialystok
Narew
Baranowicze
Byelorussia
Mogilëv
Bobruysk
Army Group Centre BOCK
WARSAW
Siedlce
Brest-Litovsk
Pinsk
Lublin
Kowel
Pripet
Mozyr
Gomel
Klintsy
Bryansk
Orel
West Front TIMOSHENKO
Serpukhov
Kaluga
Kolomna
Ryazan
Tula
Ryazhsk
Ugra
Desna
Don
Yelets
Sosna
Kursk
Voronezh
Seym
South -West Front BUDENNY
Styr
Goryn
Luck
Sluch
Bug
Jaroslaw
Brody
Lwow
Nezhin
Konotop
Sumy
Romny
Zhitomir
Kiev
Berdichev
Psël
Belgorod
Pavlovsk
Kharkov
Tarnopol
Dnestr
Army Group South RUNDSTEDT
Vinnitsa
Cherkassy
Poltava
Zmie
Prutul
Ukraine
Kremenchuy
Donets
HUNGARY
Chernovtsy
Uman
Kirovo
Dniepropetrovsk
Siret
Balta
Pervomaisk
Cluj
Jassy
Stalino
Zaporozhye
Mures
Nikopol
Kishinev
Tiraspol
Nikolaev
Mariupol
Sibiu
Odessa
Kherson
Brasov
Sea of Azov
Olt
Galatz
ROMANIA
Ismail
Black Sea
Crimea
Kerch
Bucharest
SHOCK
ROM
Fourth Panzer Army HOEPNER
100 km
100 miles

PANZER PAUSE
With smoke from destroyed Red Army tanks billowing in the background, the command element of a German armoured unit pauses to scan the horizon during offensive operations in the Soviet Union.

Colonel-General Hoth's Third Panzer Groups led the operations undertaken by Army Group Centre. Finally, Colonel-General von Kleist's First Panzer Group formed the vanguard of Army Group South's attacks. In addition to these four armoured groups, eight German, two Romanian and one Finnish (infantry) armies advanced behind the armour, occupying ground and mopping-up. The German forces involved in Barbarossa amounted to 3,060,000 troops, back by 7,189 artillery pieces. Against this Axis invasion, the Red Army deployed 203 divisions in European Russia with 11,000 tanks, although just 1,440 of these were modern. Throughout the entire Soviet Union, the Red Army fielded 304 divisions, with 4.5 million troops and 20,000 tanks, while the Soviet air force fielded 14,600 aircraft. The Soviets also possessed immense mobilizational capabilities, calling-up some 5.5 million reservists during the 1941 campaign alone.

Invading Russia

In the early hours of 22 June 1941, Hitler's forces crossed the Soviet border. The initial armoured assaults achieved overwhelming surprise, mainly because Stalin had refused to accept the intelligence that suggested an invasion was imminent. During 22-23 June, massed Luftwaffe strikes destroyed 2,100 Soviet aircraft. The Panzer divisions quickly smashed through the border defences and raced deep into enemy territory. By 2 July, the 11th day of the invasion, Army Group North's panzer spearheads had advanced 328km (204 miles) to cross the River Dvina in Estonia. This command's northern armoured wedge, XLI Motorized Corps, now aimed to advance directly toward Leningrad via Ostrov and Sabsk, while LVI Panzer corps would advance toward Leningrad along a more southerly and easterly route via Soltsy and Novgorod; in so doing it provided protection for Army Group Centre's exposed northern flank. Along the central axis, by 2 July Army Group Centre's two Panzer groups had crossed the River Beresina, having advanced

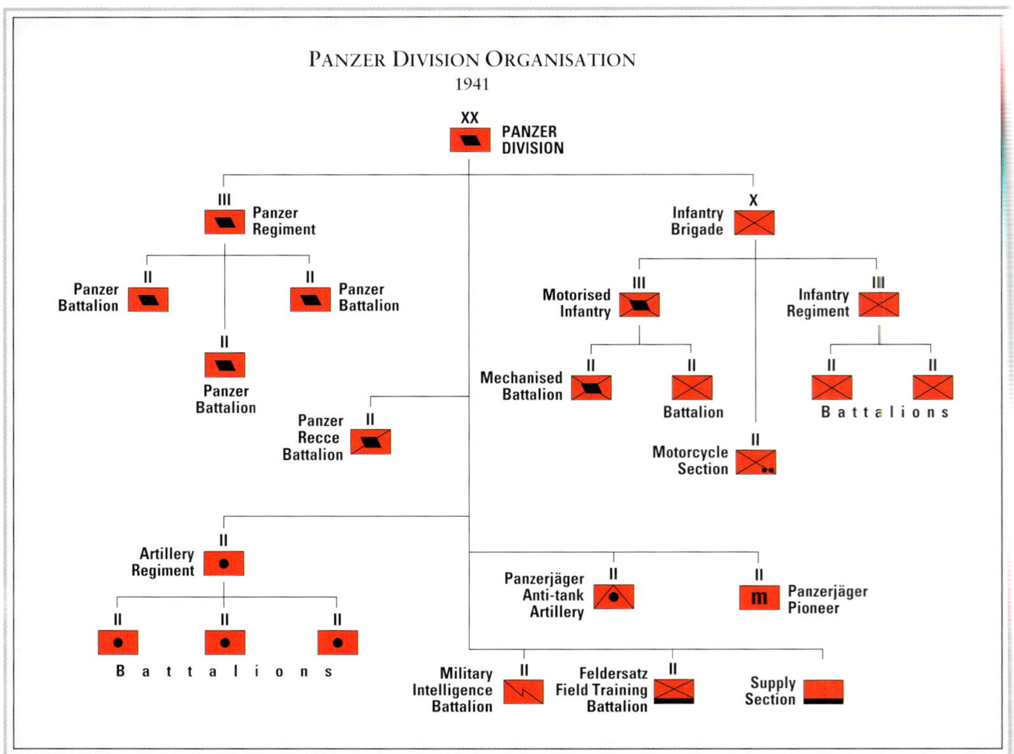

ARMOUR ORGANIZATION
The German Panzer division of 1941 included powerful tanks, armoured infantry also referred to as Panzergrenadiers, artillery, anti-tank, engineer and other support units. During World War II, the composition of German armoured formations changed numerous times.

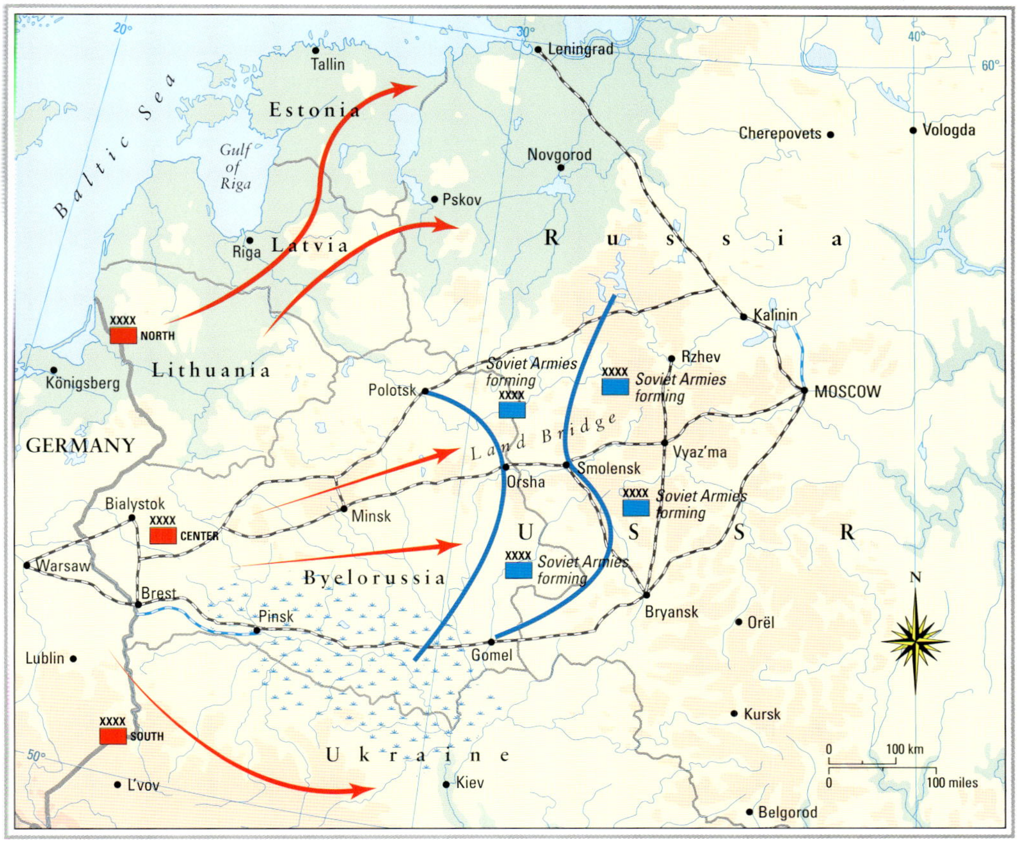

THE PANZER DRIVE TO MOSCOW
JUNE – SEPTEMBER 1941

Main German thrusts
Soviet defence lines
Strategic railways on the central front

TOWARD MOSCOW
During their drive on the Soviet capital of Moscow, German forces attempted a giant pincer movement. However, favorable weather waned and Soviet resistance stiffened as one Panzer formation reached within 12 miles of the city.

395km (246 miles) deep into Soviet territory. In the process, the Panzers had created two successful encirclements at Bialystok and Minsk, which bagged 29 Soviet divisions. Army Group South, however, had encountered much stronger Soviet resistance and thus made less significant advances than the other two army groups.

Directive 33

These Panzer spearheads, moreover, continued to make good progress during 2-15 July. In the north, Fourth Panzer Group advanced a further 253km (157 miles) beyond the River Dvina to Sabsk and Soltsy, just 160km (100 miles) short of Leningrad; this rapid advance, however, overstretched the German logistic system, and the advance now stalled. In the central sector during 4-20 July, the Second and Third Panzer Groups advanced another 285km (177 miles) toward Moscow, virtually encircling another 285,000 Soviet troops at Smolensk before logistic problems also stalled their advance. In the south, First Panzer Group made steady advance east toward the River Dnepr in western Ukraine. On 19 July, Hitler issued Directive 33. The Führer was concerned about the threat to Army Group Centre's southern flank posed by the still cohesive Soviet forces located in the Ukraine. He thus ordered the German Army to defeat the Red Army in the Ukraine before Army Group Centre resumed its advance on Moscow. Hitler instructed Army Group Centre's Second Panzer Group to divert its advance south behind the Soviet defences in the Ukraine, where it would link up with the eastward advance of First Panzer Group from Army Group South. He also ordered Army Group Centre's remaining armour, Third Panzer Group, to thrust

DEFENSIVE PLANNING
In this staged photograph, Red Army troops discuss defensive plans before Moscow as some of the weary soldiers rest atop an armoured vehicle. The civilian population of the Soviet capital was mobilised during the defensive effort.

T-34 CONCENTRATION
Columns of the highly successful Soviet T-34 medium tank take up positions for a Red Army counteroffensive against the invading Germans.

north to help the German advance on Leningrad. The crucial drive on Moscow was postponed until later in the campaign.

As Army Group Centre attempted to ready its forces to implement Directive 33, the Soviets launched repeated powerful counterattacks at Yelnya and Yartsevo during late July and August, which actually drove back the German forces some miles. Consequently, the two Panzer groups of Army Group Centre only managed to commence their redeployments on 21 August. Meanwhile, Fourth Panzer Group from Army Group South had successfully thrust east to reach Uman, in the process encircling 200,000 Soviet troops. Even as the German pincers closed neared Uman, other elements of First Panzer Group continued to advance east, until on 12 August they captured Kremenchug on the River Dnepr. These Panzer spearheads were now 182km (113 miles) behind the main Soviet defensive line that ran from north to south from Gomel, through the eastern Pripyet Marshes and down to Kiev. If Army Group South's Panzers could push north to meet the intended thrust south by Guderian's Second Armoured Group, the ensuing encirclement would probably trap no fewer than five Soviet armies. Consequently, after a pause to replenish their supplies, on 26 August Army Group South's armour thrust north toward Guderian's Panzers as the latter advanced south from Roslavl and Krichev. As the two armies approached one another it seemed that the anticipated German encirclement of the Soviet forces strung north-south along the Chernigov-Kiev-Cherkassy line would work.

On 22 August, Guderian had ordered his armour to advance south from Roslavl deep into the rear of the five Soviet armies lined-up north-to-south in front of Kiev. After a 26-day advance southwards, Second Panzer Group reached Lokhvitsa, 400km (248 miles) south of Roslavl. Here at Lokhvitsa on 15 September Guderian's Panzers linked up with the northward advance of First Panzer Group. By mid-

FACTORY FORTITUDE
Even as German armoured spearheads advanced perilously close to their locations, Soviet factory workers continued to turn out tanks for the defending Red Army. Soviet industrial capacity was a key factor in winning World War II.

September, therefore, these two Panzer groups had surrounded a staggering 500,000 Soviet troops in the Chernigov-Kiev region. The Axis infantry armies that had been following in the wake of the Panzers now moved to form a tight perimeter around the pocket to prevent the encircled Soviet forces from escaping. The last enemy forces in the region had been mopped-up by late September. The stunning operational-level success achieved by the Axis forces in the Ukraine removed any risk the enemy posed to the southern flank of Army Group Centre.

Meanwhile, Army Group North's advance north-east toward Leningrad had continued in the face of stiffening enemy resistance. From late August Fourth Panzer Group received additional mobile reinforcement as much of Third Panzer Group from Army Group Centre arrived, after being redeployed from the Moscow axis along the lines of Directive 33. By late August these forces had neared the southern outskirts of Leningrad. In the ensuing bitter assaults, the German forces fought their way to the shores of Lake Lagoda, located northeast of Leningrad, by 7 September. These advances cut the last land route from Leningrad to the rest of the Soviet Union, and the German siege of the city began. On 8 September, moreover, Hitler issued new orders for the campaign. He instructed Army Group South to exploit the operations then unfolding in the Ukraine by advancing on Kharkov and the Crimea. At the same time, he ordered Army Group North to continue its siege of Leningrad. Crucially, on 8 September Hitler also ordered that Army Group Centre simultaneously resume its attack on Moscow, where the front had stalled since July. This new drive on Moscow was codenamed Operation Typhoon and the designation of a new codename effectively announced the unspoken reality that the original German Barbarossa plan had already failed.

The renewal of the advance on Moscow by Army Group Centre was supposed to represent the main German point of effort in the East, but the need to support the ancillary attacks in the north against Leningrad, and in the south against Rostov, would starve it of badly-needed resources. During 9-30 September, Army Group North continued its siege operations around encircled Leningrad, supported by Finnish forces, which were holding the northern sector of the Axis perimeter in the Karelian isthmus. In the south, meanwhile, Axis forces advanced southeast from Pervomaisk down to the northwestern coast of the Sea of Azov, thus cutting off the Soviet 51st Army in the Crimea; these forces then began preparing for a renewed thrust east along the Azov coast toward Rostov. On the central axis of the Army Group South sector, Axis forces continued to drive east toward Kharkov in the face of stiffening enemy resistance. These continuing operations, however, were increasingly overshadowed during September by German preparations for Operation Typhoon, the renewed thrust toward Moscow. By 30 September, therefore, Operation Barbarossa had secured vast territorial gains in the East, with an advance of up to 1,000km (630 miles) deep into Soviet territory. Despite capturing over two million prisoners and inflicting immense personnel casualties and equipment losses on the Soviets, however, Barbarossa had failed to bring the enemy to strategic defeat. Substantial numbers of newly-mobilized Red Army recruits and newly-produced tanks continued to reach the front, replacing the immense Soviet losses. At this point in time, the Germans gambled everything on one final, last-ditch attempt to capture Moscow and thus win the war.

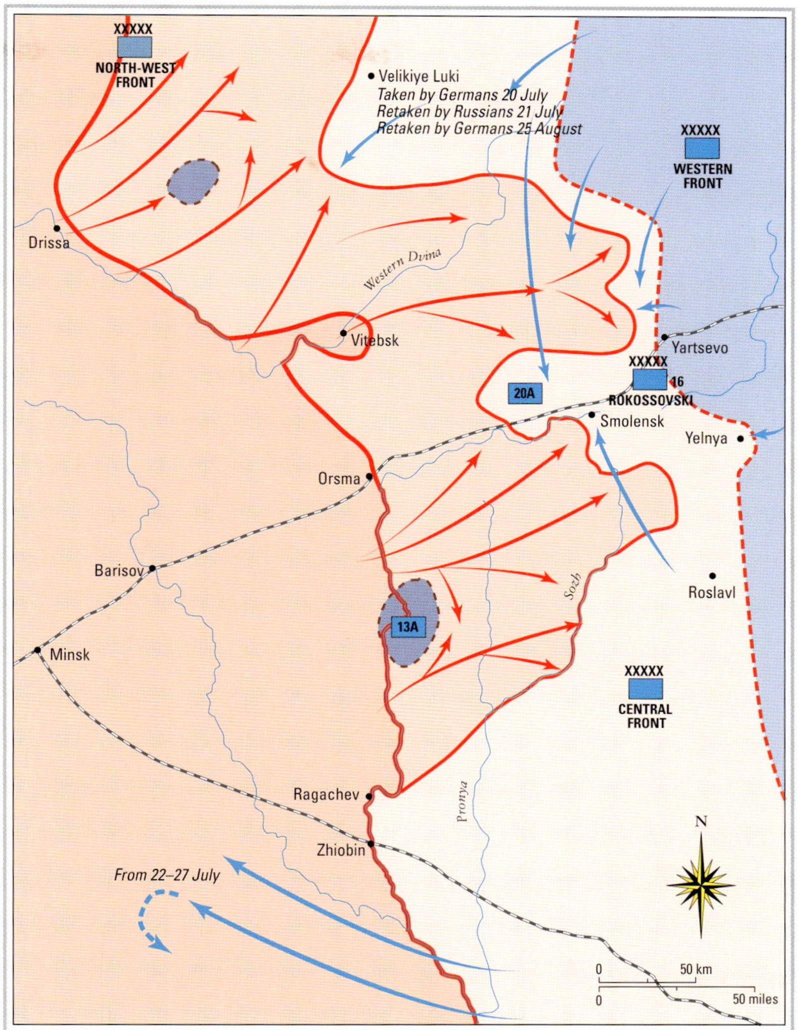

BATTLE FOR SMOLENSK
10 July – 10 September 1941

- German advances
- Soviet counter-attacks
- Soviet retreat
- German front line, 3 July
- German front line, 22 July
- German front line, 10 September
- Soviet troops surrounded

SMOLENSK STROKE
Within two weeks of the German invasion of the Soviet Union, Second Panzer Group, under General Heinz Guderian, and Third Panzer Group, under General Hermann Hoth, threatened the city of Smolensk, encircling elements of four Soviet Army fronts.

OPERATION TYPHOON - WOTAN 1941

OUTCLASSED PANZER
Despite the fact that it was already functionally obsolescent, the PzKpfw I remained in action during Operation Barbarossa. Lightly armed and armoured, it was vulnerable to Soviet tanks and anti-tank guns and subsequently withdrawn.

To implement the Führer's order on 8 September 1941 to decisively renew the attack on Moscow, which had been stalled since mid-July, the Second and Third Panzer Groups redeployed during the second half of September from the eastern Ukraine and the Leningrad area, respectively. Initiating an offensive against Moscow so late in the year was an immense gamble, based mainly on meteorological predictions that the usual intense autumn rains might come later than normal at the end of October. This gave the Germans a scant four weeks, at best, to complete the conquest of Moscow, and thus, they believed, achieve strategic victory over the Soviet Union. On 30 September 1941, Army Group Centre's 78 divisions – now spearheaded by 1,350 tanks – commenced their preliminary attacks, with the offensive proper commencing only on 2 October. During 2-14 October, the German armoured divisions out-manoeuvred the static Soviet defences in front of Bryansk and Vyazma to create two pockets that contained no fewer than 660,000 Soviet troops; it would take another 11 days to digest these two vast encirclements, however. By mid-October, therefore, these tremendous German successes had destroyed the heart of the Soviet

SOVIET TANK DIVISION
The Soviet tank division of 1941 included the heavy KV tanks, light BT series tanks, artillery and anti-tank units, and armoured cars. Its armoured element was significantly upgraded with the introduction of the T-34.

SOVIET TANK DIVISION ORGANISATION
JUNE 1941

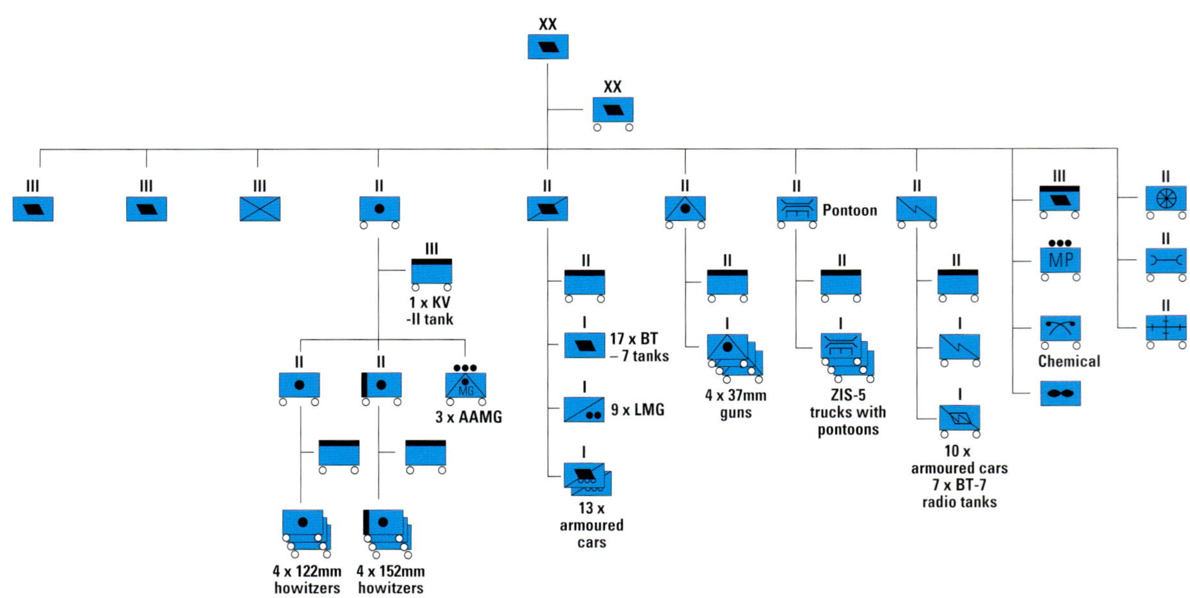

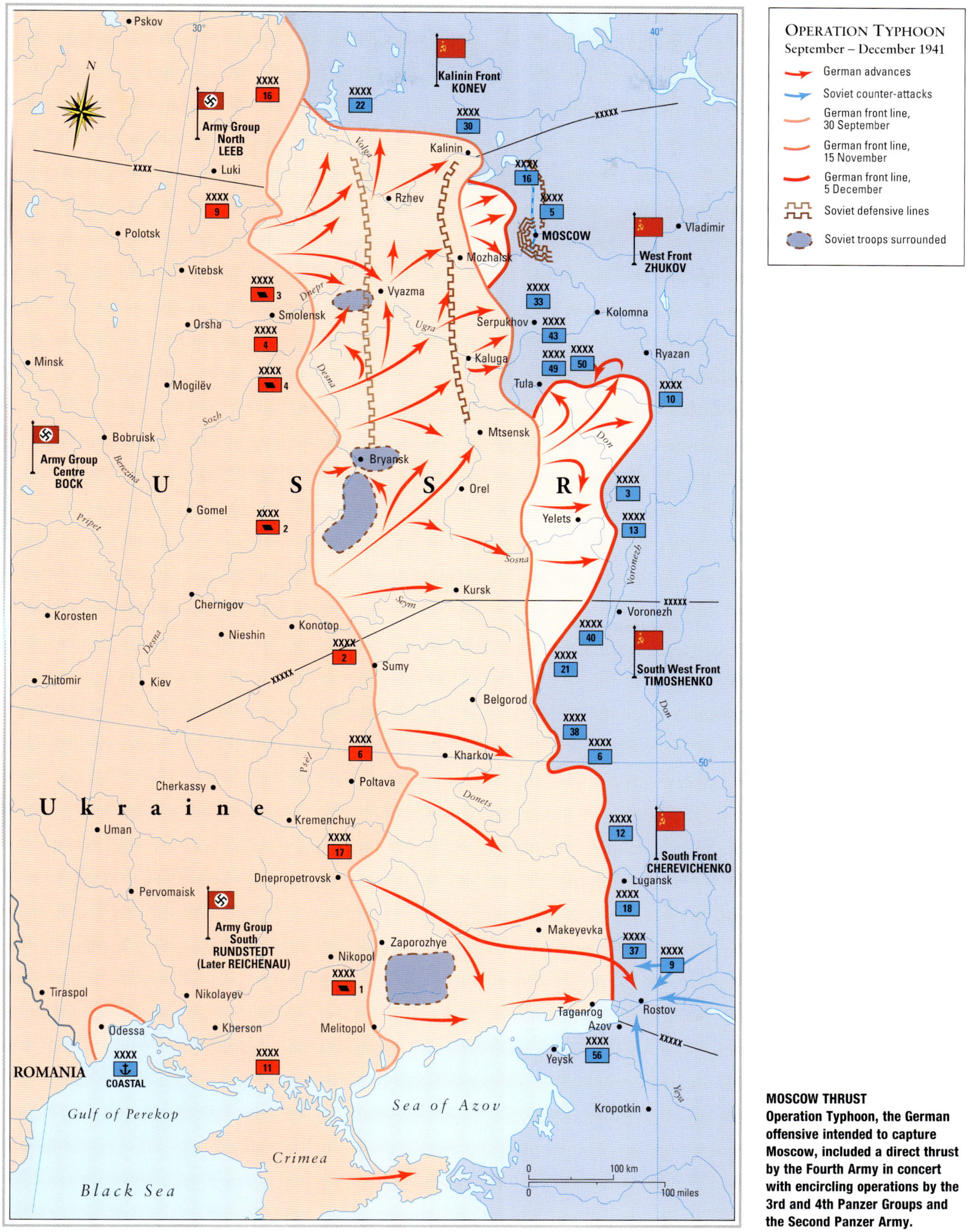

MOSCOW THRUST
Operation Typhoon, the German offensive intended to capture Moscow, included a direct thrust by the Fourth Army in concert with encircling operations by the 3rd and 4th Panzer Groups and the Second Panzer Army.

ABANDONED PANZER
Its hatch open, a German PzKpfw III lies abandoned on a street during Operation Typhoon and Operation Wotan, a failed armoured thrust launched against Moscow in September 1941. Worsening weather contributed to the failure to capture the Soviet capital.

defensive system in front of Moscow. Just as the German armour was poised to race into Moscow, however, the torrential autumn rains arrived on time, not late as predicted, and the appalling mud that ensued slowed the Panzers' advance to a crawl. In the meantime, the Soviets flung in newly mobilized half-trained recruits and newly produced equipment to hastily improvise a makeshift defensive position – the Mozhaisk Line – in front of Moscow. In the face of the fanatical resistance offered by these scratch forces the Germans had to temporarily halt Typhoon on 25 October to reorganize their logistical system.

During early November, the first severe frosts of the approaching winter hardened the boggy ground, and thus the Germans recommenced Typhoon on the 15th. With just 900 operational tanks, the 1.2 million troops of Army Group Centre now faced much-reinforced enemy forces, who were also better-equipped to deal with the extreme cold. The Panzers developed two thrusts – one north of Moscow via Kalinin, the other to the south via Tula – to outflank and encircle Moscow. Beset by serious logistic problems, the German armour made only slow progress against determined Soviet resistance. By 1 December the northern thrust had crawled forward to within 18km (11 miles) of Moscow, but now only had 40 tanks still operational; by the 4th its advance had come to a permanent halt. The next day, Army Group centre went over to the defensive. The 1941 Axis invasion of the Soviet Union had failed.

The Soviet counteroffensive

On 6 December 1941, in the wake of the failure of Typhoon, 30 Soviet divisions counterattacked Army Group Centre's exhausted troops. These counterattacks forced the Germans back from Kaluga and Kalinin, thus opening up gaps in the German front at Livny, Mtsensk, and Juchnow, through which Soviet armoured reserves poured during the latter half of December. Hitler insisted that his forces stand and fight where they were because he feared that any strategic withdrawal would turn into a rout. This insistence, and the inability of the Soviets to exploit their earlier successes, led to the German front assuming a bizarrely-contorted shape, with a long German-held salient extending east from Smolensk through to Vyazma and Rzhev. In addition, German counterattacks, notably at Vyazma, had surrounded several of the Soviet spearheads, creating enemy-held pockets behind the German lines that were reinforced by partisans; the hard-pressed German forces remained too weak, however, to eliminate these pockets. Subsequently, between 5 January and late February 1942, an over-confident Stalin widened these counterattacks into a general counteroffensive across several much-wider sections of the front. This over-ambition meant that most of these counterattacks achieved only modest success. During spring 1942 in the Moscow sector the small beleaguered German salients or pockets at Cholm, Demyansk and Velikye-Luki continued to hold out against Soviet attacks, despite appallingly dire tactical situations. By June, however, the Soviet counterattacks against Army Group Centre's front had petered out after their troops had become exhausted. Even though the German army group had seemed on the point of disintegration on several occasions, the German front somehow held until the enemy's offensive power had been exhausted. The German High Command's thoughts now turned back to offensive operations for the summer 1942 campaign season.

BLAZING TANK
Camouflaged with treebranches to no avail, a Soviet tank belches smoke and flame on the Russian steppe. Following the failure of Operation Typhoon, 30 Red Army divisions launched counterattacks against the Germans.

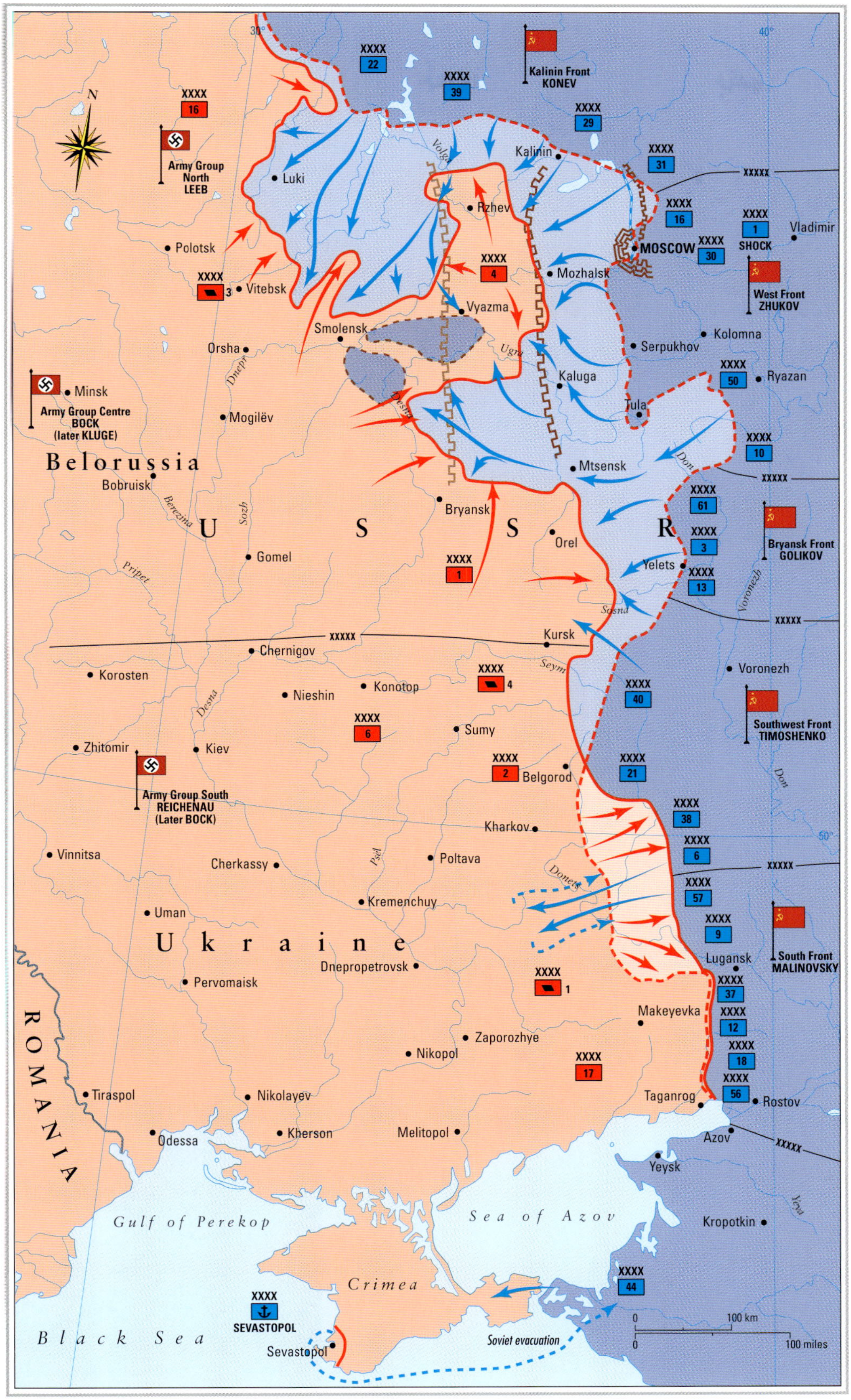

Battle for Moscow
January – June 1942

- German advances
- Soviet counterattacks
- German front line end May
- German front line June
- Soviet defensive lines
- Soviet Partisans operating behind enemy lines

MOSCOW BATTLE
The Soviet effort to capture Moscow failed due to stiffening Red Army resistance and the onset of autumn rains and winter cold. German spearheads reached within a dozen miles of the Soviet capital but could advance no further.

T-34 IN ACTION

T-34 TERROR
The Soviet introduction of the T-34 medium tank came as a shock to the Germans and prompted the development of the PzKpfw V Panther in response. Here, a column of T-34s and supply trucks pauses somewhere in embattled Russia.

Ironically, the tank which many historians and military theorists consider the most effective of World War II, traces much of its original design to the West. Yet the iconic T-34 came to symbolize the Soviet victory over the Nazis in the Great Patriotic War.

In the early 1930s, the Soviet Union imported prototypes of a tank designed by American engineer Walter Christie, who had hoped to sell his idea to the U.S. military establishment. When this did not happen, it was the Soviets who purchased Christie's prototypes and used many of their components as the basis for their BT series of light tanks that debuted in the mid-1930s and performed well against the Japanese at Nomonhan and elsewhere in Manchuria in 1939.

Engineers of the Kharkiv Morozov Machine Building Design Bureau began working on a successor to the BT series in 1937 and immediately began to incorporate the best features of the Christie-inspired series into a heavier tank that combined speed with increased firepower. The early T-34 weighed 29.5 tons and was powered by a 12-cylinder V-2 engine capable of reaching a top speed of 55km/h (34mph). It retained sloped armour, a Christie innovation that increased the effectiveness of its average armour thickness of 52mm (2in), and sometimes caused enemy shells to deflect harmlessly off the tank. The main weapon was a 76mm (3in) cannon, adequate to take on German tanks such as the PzKpfw III and IV.

The T-34 entered production in 1940, and by the end of the following year nearly 3,000 had been manufactured. For all the innovations in the early T-34, it was plagued by inadequate communications as only command tanks were initially equipped with radios, and the tank commander was also responsible for operating the main gun in combat.

The Great Patriotic War

When Hitler launched Operation Barbarossa, the speed and power of the onslaught sent the Red Army reeling eastward with tremendous losses in men and machines. As the Germans advanced, Stalin ordered Soviet manufacturing facilities dismantled, moved to safety east of the Urals, and reassembled to produce the armaments necessary to turn the tide. One factory west of the Urals at Stalingrad on the River Volga continued to produce tanks throughout the autumn of 1942 as the Germans approached. Some of these were reportedly manned by crews of factory workers and were driven directly off the assembly line into combat without paint or other finishing work.

As Soviet industrial capacity shifted to a complete war footing, productivity soared, and in 1942-43 nearly 30,000 T-34s were produced. In 1944, the Soviets introduced the T-34/85, mounting a high-velocity 85mm (3⅓in) gun capable of penetrating

PRODUCTION PROWESS
Soviet industrial capacity easily outstripped that of Germany as tanks rolled off assembly lines in factories east of the Ural Mountains, a safe distance from attack by the invading Germans.

ARMOURED FIGHTING VEHICLE STRENGTHS ON EASTERN FRONT
1941 –1945

	JUNE '41	MAR '42	MAY '42	NOV '42	MAR '43	AUG '43	JUNE '44	SEPT '44	OCT '44	NOV '44	DEC '44	JAN '45
SOVIET	28,800	4,690	6,190	4,940	7,200	6,200	11,600	11,200	11,900	14,000	15,000	14,200
GERMAN	3,671	1,503	3,981	3,133	2,374	2,555	4,470	4,186	4,917	5,202	4,785	4,881

the armour of a new generation of German tanks, including the PzKpfw V Panther and PzKpfw VI Tiger, at moderate distances. The T-34/85 also included a larger turret to accommodate three men, including a gunner and loader, relieving the commander of the responsibility of servicing the main gun in combat.

By the end of World War II, more than 57,000 T-34 and T-34/85 tanks had been produced by the Soviets. In sharp contrast, fewer than 6,500 of the PzKpfw V Panther and 2,000 of the PzKpfw VI Tiger and Tiger II had been manufactured in Germany.

T-34 AND PANZER V PRODUCTION STATISTICS
1940 –1945

TYPE	1940	1941	1942	1943	1944	1945	TOTAL
T-34	115	2,800	12,553	15,812	3,500		34,780
T34-85					10,449	12,110	22,559
PANTHER V				1,122	3,958	1,398	c. 6,472

Panzer parity

The Panther had actually been developed in response to the appearance of the T-34 on the battlefield in the autumn of 1941. When confronted by the T-34, it was readily apparent to General Heinz Guderian, the foremost German authority on tank strategy and tactics, that the Soviets possessed a tank capable of taking on the German PzKpfw III and IV and winning. Guderian prepared a detailed report describing the best attributes of the new Soviet tank and spurred German designers to redouble their efforts.

When the Panther was introduced, it proved comparable to the T-34 in several respects. Every bit as fast and even more heavily armoured, its 75mm (3in) cannon was capable of defeating the T-34 in single combat. The performance of the T-34 was solid enough to allow a seasoned crew to even the odds, while overwhelming numbers of Soviet tanks took their toll of German armoured vehicles and strained German industrial capacity.

Soviet armoured formations suffered staggering losses during World War II, primarily because of the superiority of German armour early in the conflict, and the lack of training for the hastily replaced Red Army tank crews. Soviet armoured tactics evolved slowly, and massive tank charges were often thrown against strong enemy positions with disastrous results. Nevertheless, Stalin's factories made good the heavy losses, enabling the mass-produced T-34 to prevail against the costly and over-engineered German tanks.

The T-34 continues in service with some armed forces today, more than 75 years since its inception, a testament to the enduring excellence of its design.

QUALITY AND QUANTITY
German forces could not cope with the combination of quality and quantity evident in the Soviet T-34 medium tank. The T-34 was perhaps the best tank of World War II and was produced in greater numbers than any other.

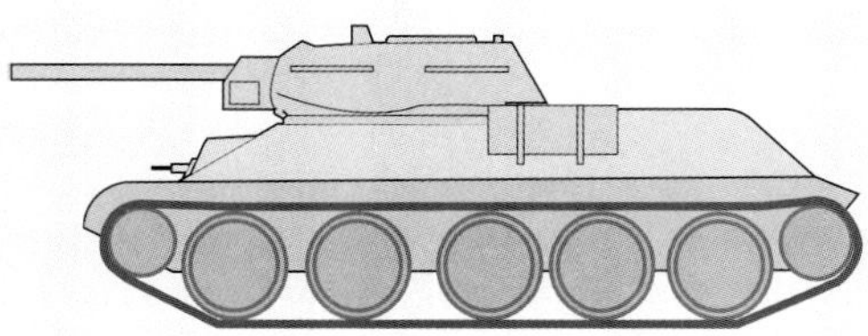

T-34

Manufactured 1940

No. built 57,339 (inc. T34/85)
Weight 26.5 tonnes
Length 6.6 m
Crew 4
Armour 15–60 mm
Main Armament 76.2 mm F-34
Secondary Armament 2 x 7.62 MG
Speed 33 mph (53 kph)

Panther V

Manufactured 1943

No. built 6,472
Weight 44.9 tonnes
Length 6.87 m
Crew 5
Armour 15–120 mm
Main Armament 1 x 75 mm KwK 142L/70
Secondary Armament 2 x 7.92 MG
Speed 34 mph (55 kph)

T-34 VERSUS PANTHER
The German response to the T-34 was the potent PzKpfw V Panther with a 75mm (3in) main gun. The Panther was more heavily armoured and slightly faster than the T-34 but suffered from mechanical breakdowns early in its deployment.

AIRCRAFT VS TANK

FLYING TANK
The Soviet Ilyushin Il-2 ground attack aircraft was produced in larger numbers than any other aircraft of World War II. It was heavily armoured and highly adept at destroying German tanks exposed in open country.

While the tank proved lethal in its own right, the required combination of speed, firepower and armoured protection presented challenges for designers, particularly when it came to keeping these elements in balance. Each slight variance in one key element resulted in a weakening of performance in another. Although tanks were formidable battlefield opponents, it was understood that they were not impervious to attack from a number of quarters, including other tanks, anti-tank guns, infantry weapons, mines, artillery, and certain aircraft that could target the tank's relatively thin top armour, drop bombs or fire rockets.

Flying artillery

Prior to the outbreak of World War II, the German military establishment recognized that its Luftwaffe would serve as flying artillery, devastating enemy troop concentrations, disrupting communications, and destroying armoured columns exposed on roads or in open fields in daylight. In Poland, the Junkers Ju 87 dive bombers of the Luftwaffe, sirens blaring, destroyed Polish armour and helped pave the way for ground forces to overrun the country in three weeks.

During the war on the Eastern Front, Stukas became even more specialized as tank-busters. Some were fitted with a pair of under-wing 37mm (1½in) cannon capable of blasting Soviet tanks to pieces. The most prolific Stuka pilot was Colonel Hans-Ulrich Rudel, who flew more than 2,500 combat missions and destroyed over 500 Soviet tanks. Rudel was the most highly decorated member of the German armed forces during the war and was wounded five times. He flew several variants of the Ju 87, including the cannon-equipped G-2.

SHTURMOVIK ATTACK 1

Aircraft attacked from the side of the enemy tank formation taking advantage of the thinner armour on the sides of the tanks. The weaving line of flight meant that it was more difficult for flak guns to track and engage the aircraft.

SHTURMOVIK ATTACK
The Soviet Ilyushin Il-2 Shturmovik often flew low and attacked targets of opportunity as German armoured columns advanced through Russia. The Shturmovik took a heavy toll in German tanks and contributed significantly to the Soviet victory in the East.

Shturmovik menace

The most effective ground attack aircraft of World War II was the Ilyushin Il-2 Shturmovik, which was heavily armed with a pair of 23mm (1in) VYA-23 cannon, two 7.62mm (⅓in) machine guns, a 12.7mm (½in) machine gun, a payload of bombs up to 600kg (1,320lb), and deadly RS-82 or RS-132 rockets. The Shturmovik was heavily armoured, difficult to shoot down, and its pilot and rear gunner were well protected by a steel tub. With its armour accounting for almost 15 percent of its weight, not for nothing was the aircraft nicknamed the 'Flying Tank'.

More than 36,000 Shturmoviks were produced from 1941 to 1945, and including its successor the Il-10, more than 42,000 were built. It remains the most highly produced aircraft in history. The Il-2 made such a significant contribution to the Soviet military during the Great Patriotic War that Premier Josef Stalin once remarked that it was as essential to the Red Army as air or bread.

The Il-2 was so successful that German tank crews came to dread its appearance above the battlefields of the Eastern Front. The Soviet pilots flew low and initially made horizontal approaches directly at ground targets from an altitude as low as 50m (164ft). Later, refined tactics dictated that a target kept to the pilot's left could be attacked in a shallow 30-degree dive, with cannon, rockets or bombs capable of blowing a Tiger tank apart with a direct hit. Shturmoviks were most effective attacking in echelon, with groups of four to 12 aircraft diving together and then climbing, circling and executing the manoeuvre over and over again.

One combat report of the Il-2 claims that a flight destroyed as many as 70 tanks of the German 9th Panzer Division in just 20 minutes on the Kursk battlefield in July 1943.

Wings in the West

Following the landings in Normandy on June 6, 1944, Allied efforts to break out of their beachheads were painstakingly slow. By late summer, however, German forces were in headlong retreat. Resisting Allied spearheads became problematic for German Panzer divisions, since moving during daylight hours invited attack by American and British fighter bombers such as the Republic P-47 Thunderbolt and the Hawker Typhoon. The Typhoon could carry up to 454kg (1,000lb) of bombs and four 27kg (60lb) RP-3 rockets mounted under each wing. Although the rockets were highly inaccurate, a direct hit was lethal, and the bombs were powerful enough to obliterate enemy tanks.

The ground-attack version of the sturdy Thunderbolt, which was nicknamed the 'Jug', carried two 227kg (500lb) bombs, mounted eight 12.7mm (½in) machine guns firing armour-piercing rounds that could penetrate the deck and turret armour of most German tanks, and later in the war was equipped with M8 110mm or 130mm (4 or 5in) HVAR rockets. During the German retreat from Normandy, approximately 50,000 troops and a large number of armoured vehicles were surrounded in the Falaise Pocket. These formations were repeatedly subjected to air attack and artillery bombardment, and estimates of tanks and assault guns lost in this theatre run as high as 500 units.

SHTURMOVIK ATTACK 2

When attacking dispersed armour vehicles, the aircraft would adopt a roughly circular air path attacking with anti-tank rockets and canon.

ECHELON ATTACK
Soviet tank-killing Ilyushin Il-2 Shturmovik ground attack planes sometimes attacked German armour in staggered formations, banking to the left and approaching targets from different altitudes to maintain pressure on the enemy on the ground.

ARMOUR VS FIREPOWER

ARCHER ASSAULT
The open-turreted Valentine Mk I Archer tank destroyer entered service with the British Army in September 1944. Mounting the 17-pdr anti-tank gun, it was deployed in limited numbers during World War II. Fewer than 700 were manufactured.

The multiple roles that the tank was required to play in World War II placed different demands on tank designers. For reconnaissance or exploitation of a breakthrough, speed was usually paramount. For tank-versus-tank combat or infantry support, firepower and armour protection most often dictated the outcome of the contest. As the capabilities of modern tanks increased, the Allies and the Axis pursued opposing viewpoints on tank development. The Germans emphasized highly engineered tanks with heavier firepower and armour protection, while the Allies, particularly American designers, favoured lighter armour and main armament. In the end, it was the prevalence of Allied tanks on the battlefield that contributed to the defeat of German forces as the greater numbers of M4 Shermans fielded by the Western Allies and the Soviet T-34 overwhelmed the PzKpfw V Panther and the PzKpfw VI Tiger.

Tank destroyers

During World War II both sides developed a new generation of armoured fighting vehicles designed specifically to defeat enemy tanks. Early versions, such as the German Marder, open-turreted and armed with a 75mm (3in) cannon, were complemented by a variety of later tank-hunting armoured vehicles, including improved variants of the Marder, the Jagdpanzer IV, the heavy Nashorn with its 88mm (3½in) cannon, and the massive Panzerjäger Tiger, mounting an 88mm (3½in) cannon and weighing nearly 72 tons.

The US M10, also with an open turret and mounting a 75mm (3in) gun, was produced for a limited period in the autumn of 1942. The M18 Hellcat was one of the fastest armoured vehicles of the war, and was capable of reaching a top speed of 88km/h (55mph). However, like many other tank destroyers, its open turret was vulnerable to attack and it sacrificed armour for speed.

In late 1944, the British introduced the innovative Archer tank destroyer, a variant of the Valentine infantry tank, with units of the Royal Artillery. Its 17-pdr main gun (that fired a 76mm/3in shell) faced to the vehicle's rear, and the Archer proved a success, particularly lying in ambush, firing at a target and then withdrawing from the action with its weapon still pointed toward the enemy.

Anti-tank guns

The most famous anti-tank gun of World War II was the German 88mm (3½in) gun, originally an anti-aircraft weapon that was pressed into anti-tank service during the campaign in France in May 1940 and in the desert of North Africa. The dreaded '88' was capable of dominating the battlefield, and the towed version of the gun was adapted for mounting in the turret of the PzKpfw VI Tiger tank.

Early British anti-tank guns were adequate against the thin skin of the German PzKpfw I and II, but heavier German tanks were often impervious to the QF 2-pdr that fired a 40mm (1½in) shell. The

SWIFT HELLCAT
The U.S. M18 Hellcat tank destroyer was the fastest tracked vehicle of World War II, capable of a top speed of 97 km/h (60mph). The Hellcat mounted a 76mm (39in) gun in an open turret.

QF 6-pdr (with a 57mm/2¼in shell) was introduced in late 1941 and served with distinction in North Africa. However, the arrival of the Panther and Tiger tanks rendered it ineffective. The QF 17-pdr reached combat units in late 1942 and became one of the most best towed anti-tank guns of the war.

The Soviet Red Army fielded a 45mm (1¾in) anti-tank gun that was later supplanted by the 76mm (3in) field gun, which was pressed into an anti-tank role on the Eastern Front. Although the 45mm was considered extremely light by contemporary standards, near the end of the war it was still present with Soviet forces in large numbers. The U.S. Army relied on the antiquated 37mm (1½n) M3 weapon at the outbreak of war, but this was rapidly replaced by the M5, a 76mm (3in) weapon that proved capable although cumbersome and difficult to move.

Anti-tank guns were sometimes placed in fixed defensive positions on the battlefield, often dug-in and camouflaged against detection by air. These positions were then ringed by infantry dugouts, bunkers or foxholes with interlocking fields of fire. At times, the infantry positions were connected by trenches to allow concealed movement to a threatened area of the defensive perimeter.

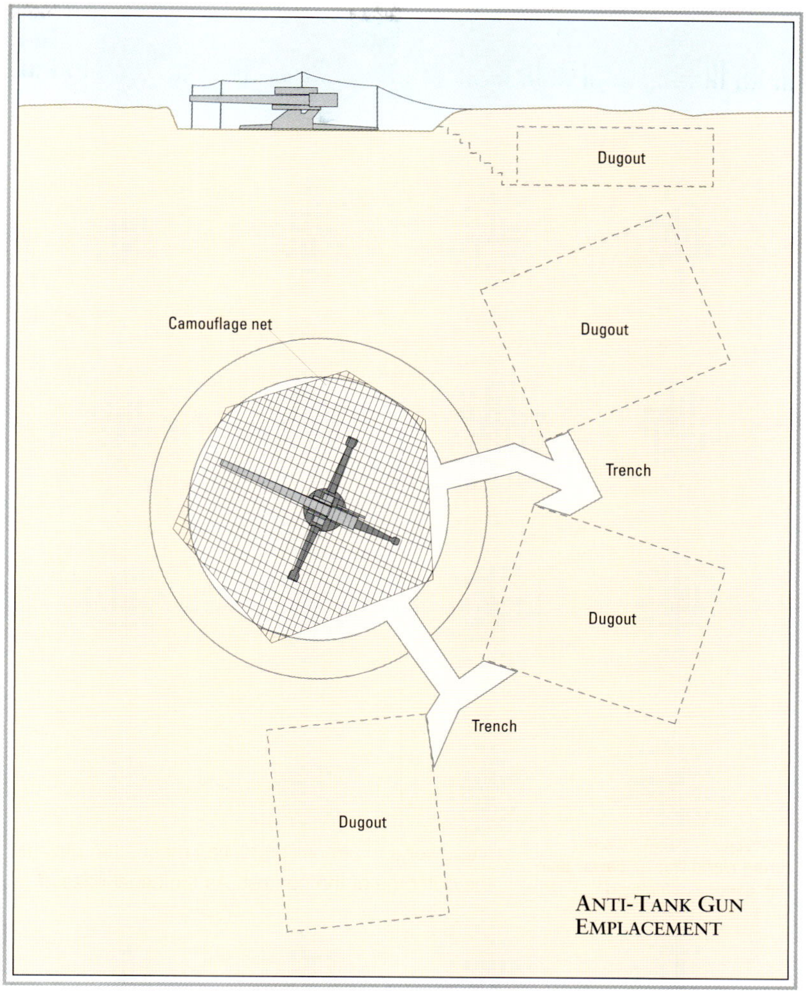

TANK KILLER
The anti-tank gun emplacement typically included a camouflaged position surrounded by dugouts connected by trenches with interlocking fields of fire. Often, an advancing tank's best defence was its screening infantry, capable of neutralising such a position.

Infantry anti-tank weapons

For the combat infantryman in World War II, the threat of approaching enemy armour was terrifying. It was readily apparent that infantrymen, particularly lightly armed airborne troops, needed some type of portable, shoulder-fired anti-tank weapon. Heavy-calibre anti-tank rifles were deployed with infantry units early in the war, but their effectiveness diminished markedly with the improved armour that began to appear on the battlefield.

In 1933, the U.S. Army began testing a shoulder-fired anti-tank weapon that became popularly known as the 'bazooka'. The weapon was essentially a steel tube, open at both ends, into which a 60mm (2⅜in) rocket was inserted. The operator gripped the bazooka with two wooden handles, and the weapon was fired with a trigger on the rear handle. The bazooka entered service in North Africa in the autumn of 1942, and its effective range was limited to about 91m (100 yards).

Rather than a rocket projectile, the British fielded the PIAT (Projector, Infantry, Anti-Tank), which was actually a spring-loaded spigot mortar that fired a hollow-charge bomb to an effective range of 110m (115 yards). The PIAT was serviced by a two-man crew. Although it was produced in great numbers (more than 115,000), and served throughout the war, it was generally unpopular with the troops because of its weight and bulkiness.

The best known German infantry anti-tank weapon of World War II was the shoulder-fired Panzerfaust. Cheap to produce and firing an effective hollow-charge warhead weighing 3kg (6.6lb), it was operated by a single soldier and its launch tube was discarded after firing. Capable of penetrating up to 140mm (5½in) of armour, the weapon was highly effective against enemy tanks. However, one disconcerting aspect of the weapon's operation was its short range, particularly with its early variants. At a distance of only 30m (33 yards), the soldier armed with the Panzerfaust was often exposed to immediate return fire.

German infantrymen became particularly reliant on the Panzerfaust as they made the transition from offensive to defensive combat. Late in the war, some units of the Volkssturm, or home guard, faced the advancing Allies armed only with the Panzerfaust.

CAUCASUS – SOUTH RUSSIA 1942

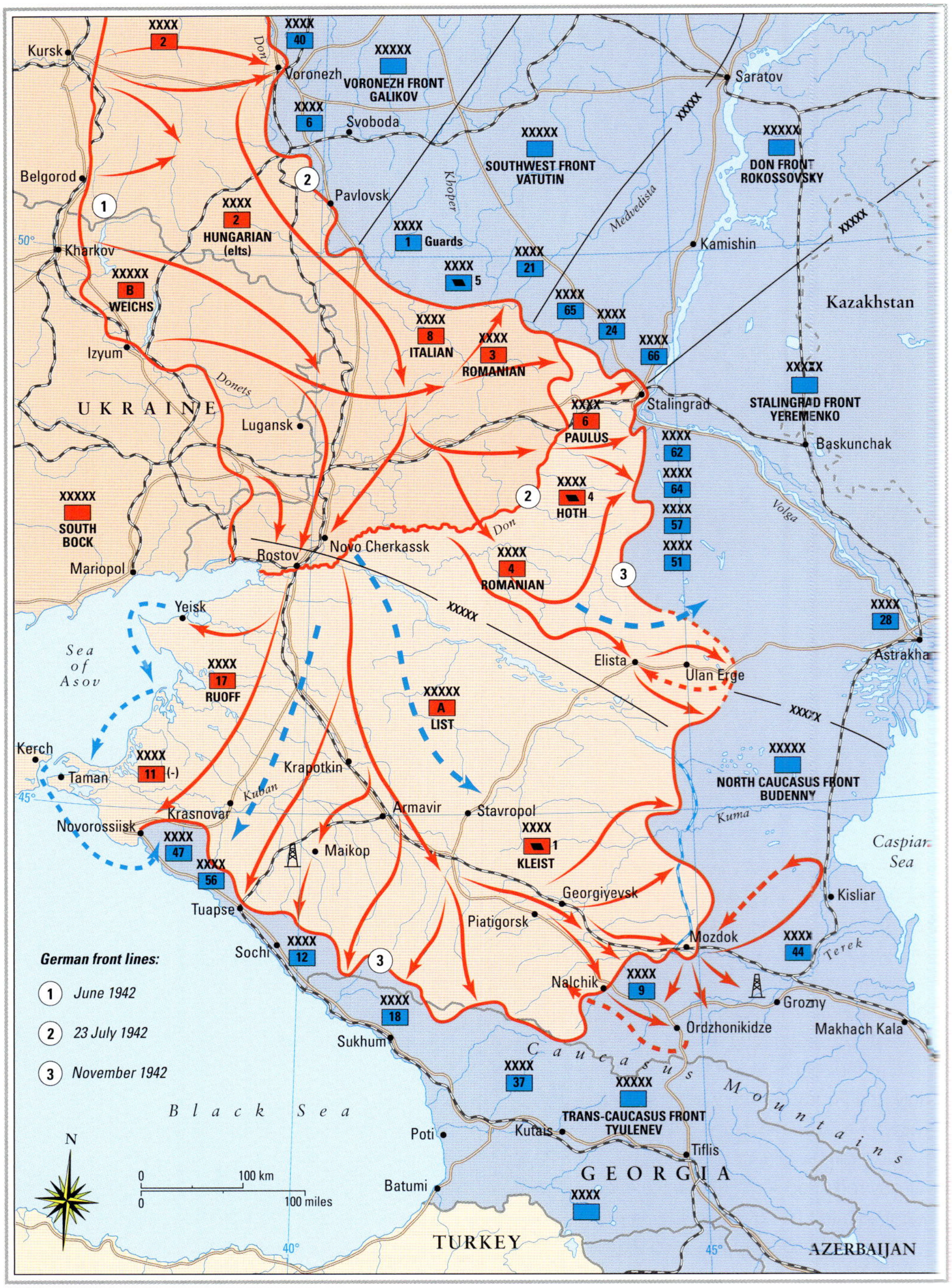

INTO THE CAUCASUS
In the spring of 1942, Hitler sent his powerful army groups south and east toward the oil fields of the Caucasus and the industrial center of Stalingrad on the Volga River.

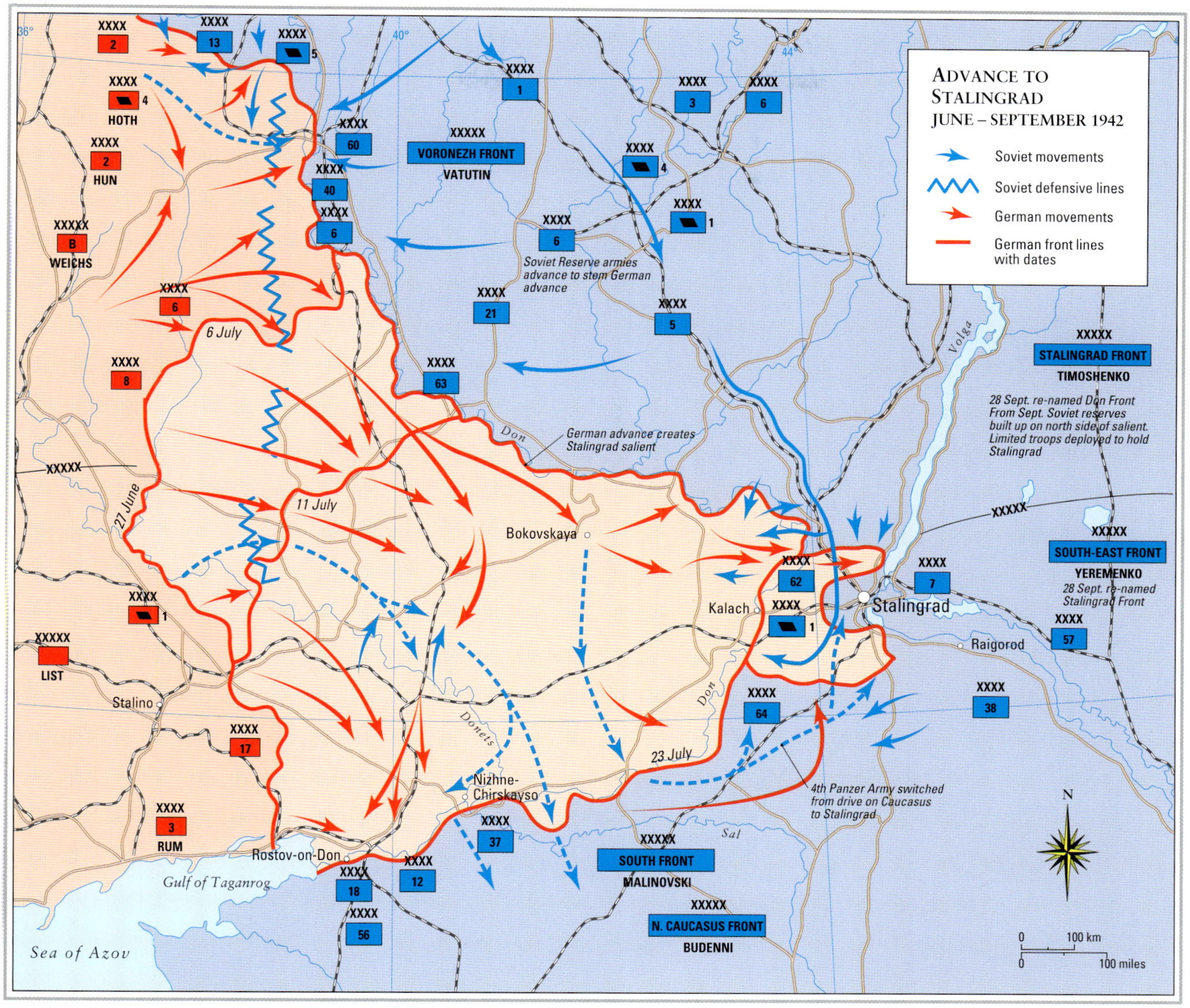

On 28 June 1942, the northern elements of German Army Group South commenced Operation Blue (or Case Blue) with an assault launched from the Kursk-Belgorod sector that aimed to capture the key city of Voronezh located on the River Don. Army Group North fielded 68 divisions with 1.4 million troops and 1,495 armoured fighting vehicles (AFVs). The operation was a continuation of Operation Barbarossa and was intended to force the surrender of the U.S.S.R. once and for all. By 6 July, the Fourth Panzer Army had successfully advanced east to reach the suburbs of Voronezh. On 9 July the remainder of Army Group South (that day renamed Army Group A on Hitler's orders), located along the Isyum-Taganrog sector, struck east and southeast toward Rostov and the Don estuary.

Next, on 25 July Army Group A's 20 divisions thrust south and southeast across the Don estuary into the northern Caucasus region. By then Hitler had set new objectives for Operation Blue: Army Group B was to capture Stalingrad, while Army Group A was to capture the Baku oilfields on the coast of the Caspian Sea. During August, Army Group A advanced rapidly southeast deep into the Caucasus before its advance stalled. Meanwhile, the Sixth Army fought its way east to reach Stalingrad, and surrounded it up to the River Volga from the north, west and east. During September and October 1942 some 20 divisions of the German Sixth and Fourth Panzer Armies fought through Stalingrad's ruined streets against intense resistance; if the city fell, Hitler now believed, the Soviets would never recover from such a psychological defeat.

STALINGRAD ASSAULTED
As autumn turned to inhospitable winter, the initially successful German drive into the Caucasus had stalled and Wehrmacht troops were engaged in bitter house-to-house fighting with stubborn Red Army defenders for control of Stalingrad.

PZKPFW IV
Staring across the vast expanse of the Russian landscape, a German commander pauses in a wheatfield to scan the horizon. His tank, a PzKpfw IV, was the workhorse of the Panzerwaffe during World War II.

Operation Uranus

But even as the German forces slowly fought their way through Stalingrad's ruined streets toward the River Volga during October and early November, the Soviets were husbanding powerful reserves to the northwest and south of the city, which were being readied for a decisive Soviet strategic counteroffensive. By 18 November the Axis forces had continued to drive back the Soviet troops that defended Stalingrad, until all they held was a 1.5km (1-mile) deep and 13km (8-mile) long strip of the city that ran north-south adjacent to the River Volga. Their continued, albeit painfully slow, progress convinced the Germans that one last push would deliver to them the great prize of the capture of the city. However, to the northwest, on the eastern side of the River Don around Melokletski, Marshal Vatutin's southwestern front had assembled powerful armoured forces, including two elite Guards armies. In the south, Marshal Yeremenko's Stalingrad front had similarly concentrated powerful reserves, including the 51st, 57th and 64th Armies.

On 19 November the Red Army initiated this long-planned counteroffensive, codenamed Operation Uranus. This riposte caught by surprise the thinly-held Axis fronts located to the northwest and south of the densely-packed battle-space around Stalingrad. In the north, Marshal Vatutin's armour attacked from the remaining Soviet bridgeheads on the western bank of the Don between Ust-Khoperski and Melokletski, bridgeheads that the Germans ought to have reduced in previous weeks. In the south, Marshal Yeremenko's armour struck along the Andreyevka-Plodovitoye sector. Both of these attacks quickly broke through the weak Axis positions in these locations – most of the

SOVIET RESISTANCE
As Army Group A advanced on the Terek river barrier north-west of Grozny, Soviet reistance increased. Their objectives were to capture Baku in Azerbaijan, which produced 80 per cent of Soviet oil.

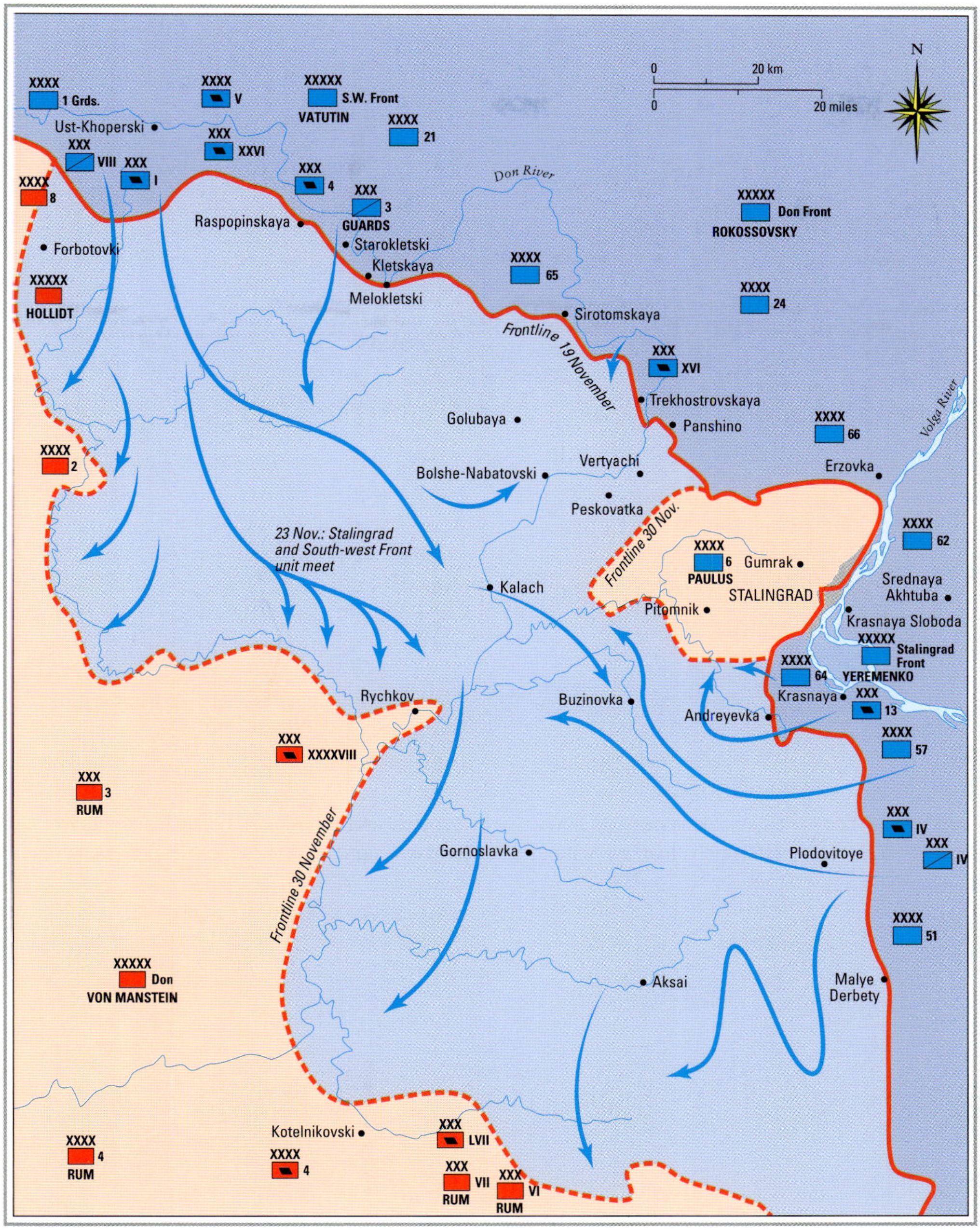

Soviet Counter Offensives
19 November – 1 December 1942

German front line 19 November

German front line 30 November

Major Soviet attacks

SURROUNDING STALINGRAD
In November 1942, the Soviets unleashed Operation Uranus, throwing fresh divisions transferred from Siberia against the Germans at Stalingrad and surrounding the Sixth Army in the city. By February 1943, the Germans had capitulated.

Axis forces had been sucked into the battle for control of Stalingrad's ruined streets. During 20-22 November, the Soviet armour quickly charged south and west deep into the German rear areas to converge on one another. On 23 November 1942, the converging Soviet pincers duly met up at Kalach to form a shallow encirclement that successfully trapped some 255,000 troops of the German Sixth Army at Stalingrad, together with a few elements of the Fourth Panzer Army. The Soviets' Uranus counteroffensive had inflicted a classic double-pincer German-style encirclement operation on the Germans themselves. By sheer force of numbers it had also imposed the most significant setback that the German forces had so far faced in the entire war. By 30 November, continuing Soviet attacks had driven Sixth Army's western flank back toward Pitomnik airfield, while its eastern flank remained locked on Stalingrad itself. In the meantime, the Soviet forces located west of Kalach drove back the forces of the newly-renamed Army Group Don to the River Chir, although XXXXVIII Panzer Corps clung onto a small salient that jutted east toward the surrounded Sixth Army.

SOVIET T-26 AND T-34 TANKS
Soviet infantrymen accompanied by a T-26 tank and its improved replacement within the Red Army armoured divisions, the legendary T-34, pause along a roadway. Transitioning to updated armoured vehicles was a logistical challenge in wartime.

Operation Winter Storm

In the aftermath of the Soviets successfully surrounding the German Sixth Army at Stalingrad, Hitler ordered Army Group Don to mount a relief attempt east from the River Chir toward Stalingrad, while the Sixth Army was to form a staunch defensive hedgehog around the city of Stalin's name, and it was not to be abandoned, however dire the tactical circumstances, the Führer admonished. Hitler envisaged that this relief effort would not only restore a land corridor to the Sixth Army and allow it to maintain its grip on large parts of the city itself, but would also permit a successful subsequent capture of the remaining parts of the city. This was woefully over-optimistic. By contrast Field Marshal Erich von Manstein, commanding Army Group Don, believed that the Sixth Army's only hope of salvation was to break-out westward from Stalingrad when the relief attack west was launched, but Hitler expressly forbade this. Manstein, however, was unable to initiate this relief effort (codenamed Operation Winter Storm), until 12 December 1942. Manstein had originally intended that the German LVII and XXXXVIII Panzer Corps would simultaneously develop twin thrusts toward Stalingrad. In the north, XXXXVIII Panzer Corps, part of the Army Detachment Hollidt was to attack from the Rychkov salient. The Corps was to force a crossing of the River Don and then drive 61km (38 miles) to Sixth Army's southwestern perimeter at Marinovka, where it would link up with the advance of LVII Corps. The latter, from Colonel-General Hermann Hoth's Fourth Panzer Army, was to strike north-northeast from Kotelnikovo east of the Don to reach the southwestern corner of Sixth Army's perimeter, 145km (90 miles) away. Unfortunately for General Paulus' Sixth Army trapped at Stalingrad, fierce Soviet offensive pressure prevented Operation Winter Storm from even commencing as planned. Fierce Soviet attacks along the River Chir for example, sucked XXXXVIII Corps into defensive actions that prevented it from participating in

POWERFUL T-34
Clad in winter camouflage, Soviet infantrymen advance under covering fire from a T-34 medium tank. Early versions of the T-34 mounted a 76mm (3in) gun, upgraded later to 85mm (3⅓in) with greater armour-penetrating capability.

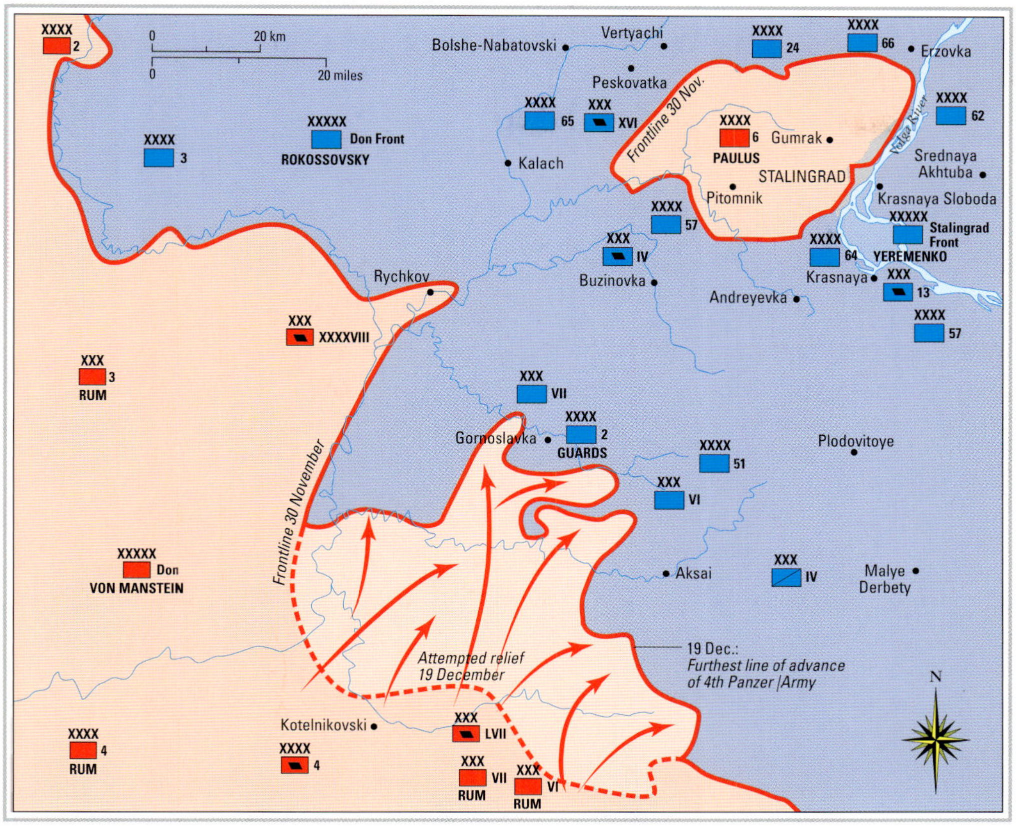

FOURTH PANZER ARMY ATTEMPT TO RELIEVE STALINGRAD TO 19 DECEMBER 1942

DESPERATE RELIEF MEASURE
Following Hitler's refusal to allow the Sixth Army to attempt a breakout, elements of the Fourth Panzer Army under General Hermann Hoth launched Operation Winter Storm but failed to relieve besieged German forces in Stalingrad.

the offensive at all. Consequently, Manstein was compelled to begin the relief effort on 12 December by mounting only a single-axis advance east toward Stalingrad, supported by flanking operations to the immediate north and south.

The German offensive made steady, but relatively modest, progress during the first four days, as fierce Soviet resistance slowed the advance. Finally, the operation stalled on 19 December 1942 in the face of the fierce opposition offered by the newly committed elite Soviet Second Guards Army. By then, Winter Storm had managed to advance northeast some 90km (56 miles) to reach the Myshkova River. From this location, the spearheads of LVII Panzer Corps remained just 58km (36 miles) short of the southwestern point of the Sixth Army's perimeter around Stalingrad. Unfortunately for Paulus' troops, LVII Panzer Corps would never get any closer to this perimeter. For back on the 16th, the Soviets had initiated Operation Little Saturn, an offensive into the northern flank of Manstein's Army Group Don. By the 19th the Soviets' spearhead had penetrated deep into the Axis rear areas, and this advance threatened to turn the northern flank of the Fourth Panzer Army, which thanks to its involvement in Winter Storm, remained in exposed positions on the eastern side of the Don.

The launch of the Soviet Little Saturn counteroffensive condemned the German Winter Storm relief operation to failure. Before accepting this reality, however, Manstein undertook one final desperate attempt to rescue the encircled Sixth Army. During 19-24 December, he tried to persuade Hitler to grant Paulus permission to break out toward the positions held by LVII Panzer Corps on the Myshkova River. Hitler, however, mistakenly believed that the Sixth Army could easily hold on for a number of weeks, whereas in reality it was already so short of munitions and fuel that its combat power was significantly denuded. The Führer forbade any breakout that would mean relinquishing the German hold on Stalingrad, even though the Sixth Army was already too weak to either break out or hold the city. By 24 December, it was probably already too late for the Sixth Army, as it lacked sufficient fuel to fight its way west to reach the German forces on the Myshkova River, even assuming that it could have overwhelmed the fierce Soviet attempts to resist its bid for escape. The failure of Winter Storm condemned the encircled Sixth Army to a long defensive battle for the city in the ensuing weeks that would inevitably end in destruction or surrender, thus creating what would be the greatest German setback of the war to date.

STALINGRAD 1942–43

On 23 November 1942, as we have seen, Soviet armoured counterattacks encircled the 240,000 German and 15,000 Axis Romanian troops of General Paulus' German Sixth Army around Stalingrad. Hitler instructed that these forces defend resolutely the pocket's perimeter, while during mid-December German armour outside the pocket counterattacked east to re-establish a land link to them. However, this relief effort stalled just 58km (36 miles) short of the perimeter. In the meantime, the Luftwaffe attempted to resupply the Sixth Army by air with the ammunition, fuel, and rations it needed to fend off savage Soviet assaults. In the face of numerous airborne enemy attacks, however, the Luftwaffe only succeeded in delivering a proportion of the supplies Paulus' forces required, which obviously weakened significantly their defensive combat power.

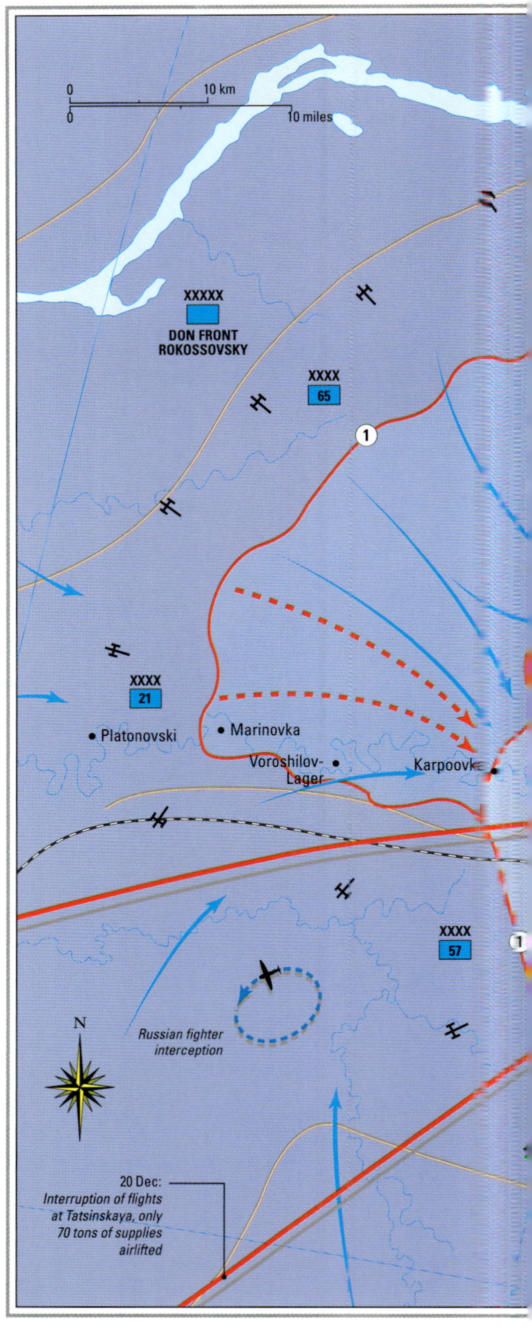

T-70 LIGHT TANK
Soviet troops rush past a disabled T-70 light tank as they advance toward German positions at Stalingrad. The lightly armed and armoured T-70 was introduced in 1942 primarily as an infantry support and reconnaissance vehicle.

During the first half of January 1943, continuing Soviet assaults on the pocket's western flank drove the defenders back 38km (24 miles). By 20 January the Axis pocket had shrunk further into a small enclave, just 40km (25 miles) wide, still anchored in the east amid Stalingrad's ruined streets. During 16-22 January the continuing Soviet advance from the west captured the vital Pitomnik and Gumrak airfields, depriving the already starving and under-resourced defenders of their last source of external re-supply. The next day, the Soviet advance east

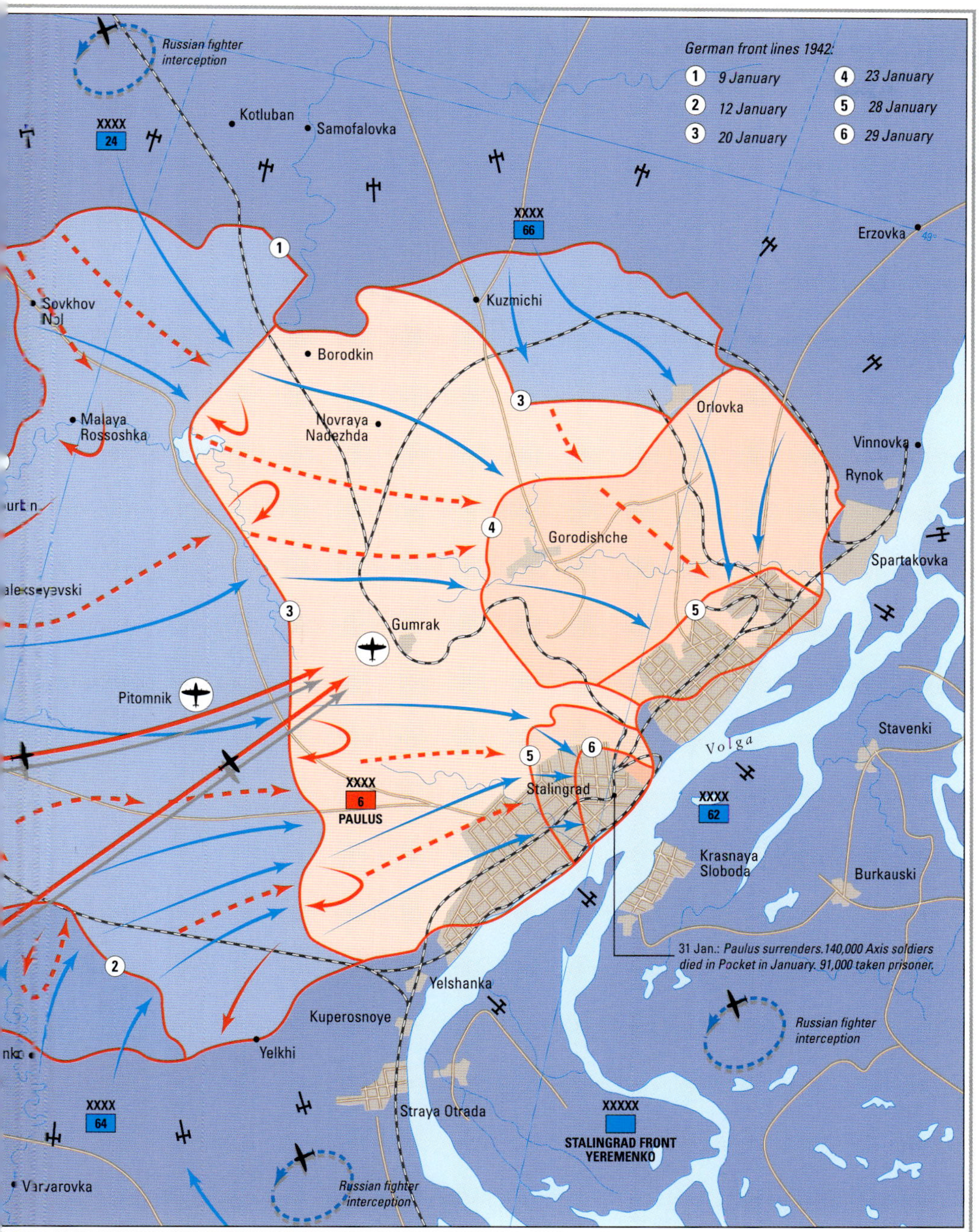

THE END AT STALINGRAD
JANUARY 1942 – FEBRUARY 1943

- Russian attacks
- German counter attacks
- German retreats
- German front lines
- Russian artillery concentrations
- Russian fighter interception
- Luftwaffe air lift

STRANGLING STALINGRAD
As re-supply efforts from the air proved woefully inadequate, the Red Army strengthened its grip on Stalingrad, reducing the Germans in the city to starvation rations and forcing a mass surrender of the Sixth Army.

successfully punched through the German positions in central Stalingrad to link up with those Soviet forces still holding on to the strip of the city adjacent to the River Volga's western bank. This advance thus cut the defending forces into two separate pockets, a larger enclave to the north and a smaller one to the south. In the desperate street fighting that ensued, the Soviet attacks drove the defenders, now virtually out of munitions and food, back into ever-shrinking pockets within the city itself. On 22 January, Paulus asked Hitler's permission to surrender. This was denied, with Hitler calling the battle 'an heroic drama of German history'.

From 29 January, the surviving defenders began to surrender anyway, with the last elements capitulating on 3 February 1943. Of the 91,000 Axis prisoners taken during this battle, just 5,000 survived Soviet captivity. One of the bloodiest battles in World War II, Stalingrad was marked by callous brutality on both sides. The might of the Wehrmacht was finally challenged and resisted, marking a turning point in the war.

TANK PRODUCTION 1941–45

ARMOUR PRODUCTION
As World War II progressed, the industrial might of the Allied nations, particularly the United States and the Soviet Union, resulted in the manufacture of thousands of tanks and self-propelled assault guns, far exceeding the output of Nazi Germany.

The production of tanks and armoured fighting vehicles remained a priority for the major belligerents throughout World War II, while the diversity and combat capabilities of these vehicles advanced rapidly with the real-time proving ground of the battlefield.

During World War II, the tank, self-propelled gun and armoured fighting vehicle were produced in great numbers as modern combat doctrine evolved. Conflicting perspectives as to the role of the tank on the battlefield were tested and proven either incredibly successful or fatally flawed. In the opening days of the war, the Nazi Blitzkrieg was ascendant, with its success depending on shock and speed. However, as the tide of war turned in favour of the Allies, the sheer weight of numbers doomed German armoured formations to defeat. The enormous industrial capacity of the Soviet Union and the United States proved overwhelming.

Arsenal of democracy

Although the United States was not at war in 1939, President Franklin D. Roosevelt recognized early that U.S. production capacity would likely make the difference between victory and defeat for Great Britain. U.S. factories produced only about 330 tanks in 1940. However, Roosevelt's Lend-Lease concept mobilized U.S. industry to arm the Allies.

Automotive assembly lines were adapted to tank production. In early 1941, the first production M3 tank rolled off the assembly line at a Chrysler production facility near Detroit, Michigan. Production of the M4 Sherman was started, and Chrysler's 500th tank rolled off the assembly line as the country went to war in December.

Other U.S. auto manufacturers, including General Motors and Ford, engaged in armoured vehicle production. By 1945, U.S. factories had delivered more than 88,000 tanks and assault guns.

TANK AND SELF-PROPELLED GUN UNIT PRODUCTION
1939–1945

	1939	1940	1941	1942	1943	1944	1945	TOTALS
USA	—	331	4,052	24,997	29,497	17,565	11,968	**88,410**
USSR	2,950	2,794	6,590	24,446	24,089	28,963	15,419	**105,251**
UK	969	1,399	4,841	8,611	7,476	4,600	?	**27,896**
CANADA	?	?	?	?	?	?	?	**5,678**
GERMANY	247	1,643	3,790	6,180	12,063	19,002	3,932	**46,857**
ITALY	40	250	595	1,252	336	—	—	**2,473**
HUNGARY	—	—	—		c. 500		—	**c. 500**
JAPAN	—	315	595	557	558	353	137	**2,515**
TOTALS	**4,206**	**6,732**	**20,463**	**66,209**	**74,187**	**70,649**	**31,456**	**227,235**

British influence

British tank development and production languished between the world wars. By the summer of 1941, however, the brilliant successes of German General Erwin Rommel in North Africa spurred the British military and industrial base to produce effective tanks. The majority of interwar tank production in Britain was undertaken by Vickers-Armstrong, and was later joined by others such as Nuffield. A team from Morris Motors was sent the U.S.A. to study the work of the American tank designer J. Walter Christie. By 1943, British tank production had grown from fewer than 1,000 annually to more than 8,600. By 1945, over 33,000 tanks and self-propelled guns had been built in Britain and Canada.

SOVIET PRODUCTION
On a crowded factory floor lined with tanks in various stages of completion, Soviet industrial workers await the placement of a hoisted turret atop the chassis of a T34 medium tank.

Soviet output

The Soviet Union produced the largest number of tanks and self-propelled guns during World War II. As German spearheads menaced Moscow, some Soviet tanks were driven directly off the factory floor by combat crews and into action. Entire factories were dismantled and reassembled east of the Ural Mountains.

As the combination of Red Army tenacity and the cruel Russian winter slowed the German tide, the Soviet design bureaus authorized production of the legendary T-34. The Soviet industrial colossus improved production from fewer than 3,000 tanks in 1939 to a peak of nearly 29,000 by 1944, and overall wartime production topped 100,000.

German precision

At the outbreak of hostilities, Germany was the most advanced military power in terms of armoured vehicle development, construction and tactics in the world. The Germans often showed design innovation with their stalwart PzKpfw IV, PzKpfw V Panther, and behemoth Tiger I and Tiger II tanks.

Additionally, the German penchant for quality over quantity meant that battlefield losses were not easily replaced, and that Allied tactical air supremacy and numerical superiority in vehicles would eventually win the war. Although the Tiger and Panther were fearsome foes, these tanks were plagued by breakdowns. German tank and self-propelled gun production peaked in 1944, when 19,000 were built despite intense Allied bombing.

STALWART SHERMAN
Nearly 50,000 of the M4 Sherman tank and its variants were produced during World War II, and its numerical superiority contributed to the attrition of German armoured vehicles that could not easily be replaced.

NORTH AFRICA 1942–43

EL ALAMEIN EVE
General Bernard Montgomery amassed an overwhelming superiority in tanks and artillery prior to launching the British offensive at El Alamein in October 1942. Commonwealth forces relentlessly battered German armoured strength to a shadow of its former self.

Employing brilliant tactical acumen and vital intelligence intercepts of top-secret Allied communications, Rommel was poised for final victory in the summer of 1942. His armoured spearheads had crossed the Egyptian frontier and mauled four British divisions at Mersa Matruh, 160km (100 miles) inside Egypt on 26 June. Astonishingly, the Germans had deployed only 60 tanks to achieve the one-sided victory.

Pressing on, the Germans approached the railway outpost at El Alamein. There, a scant 240km (150 miles) from the Egyptian capital at Cairo, the British Eighth Army would make its final stand in the desert. General Auchinleck had chosen well. In the north, the British position was anchored on the Mediterranean coast, while to the south the Qattara Depression, an impassable area of loose sand where vehicles could not operate, confronted the advancing Germans.

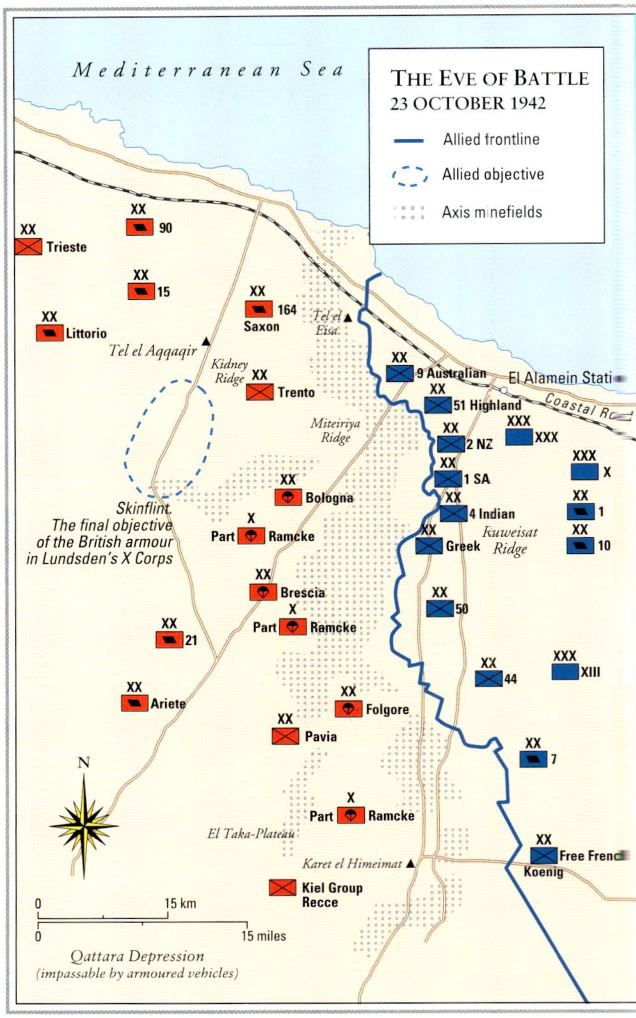

Thrust and parry

The defensive boxes of the Eighth Army near El Alamein were occupied by a polyglot force, including the 6th New Zealand Brigade, the 9th and 18th Indian Brigades, and the 3rd South African Brigade. Although the combined tank strength of the 1st and 7th Armoured Divisions amounted to just 155, the stout defenders managed to contain Rommel's probing attacks and then a general assault at Ruweisat Ridge on 21-22 June.

The action at Ruweisat Ridge slowed the momentum of Rommel's command, now known as Panzergruppe Afrika, but, British Prime Minister Winston Churchill had lost confidence in Auchinleck even though he had halted Rommel for the moment. Churchill appointed General Sir Harold Alexander to command imperial forces in the Mediterranean. Concurrently, General Bernard Law Montgomery was named commander of the Eighth Army.

Over-stretched supply lines and continuing

BREN GUN CARRIER
The ubiquitous tracked Bren gun carrier served to transport troops, ammunition, supplies and wounded for Commonwealth armies in every theater of operations during World War II. Through more than 25 years of production, over 113,000 were manufactured.

OPERATION LIGHTFOOT
In late October, General Montgomery launched Opertion Lightfoot, the opening phase of the Battle of El Alamein.

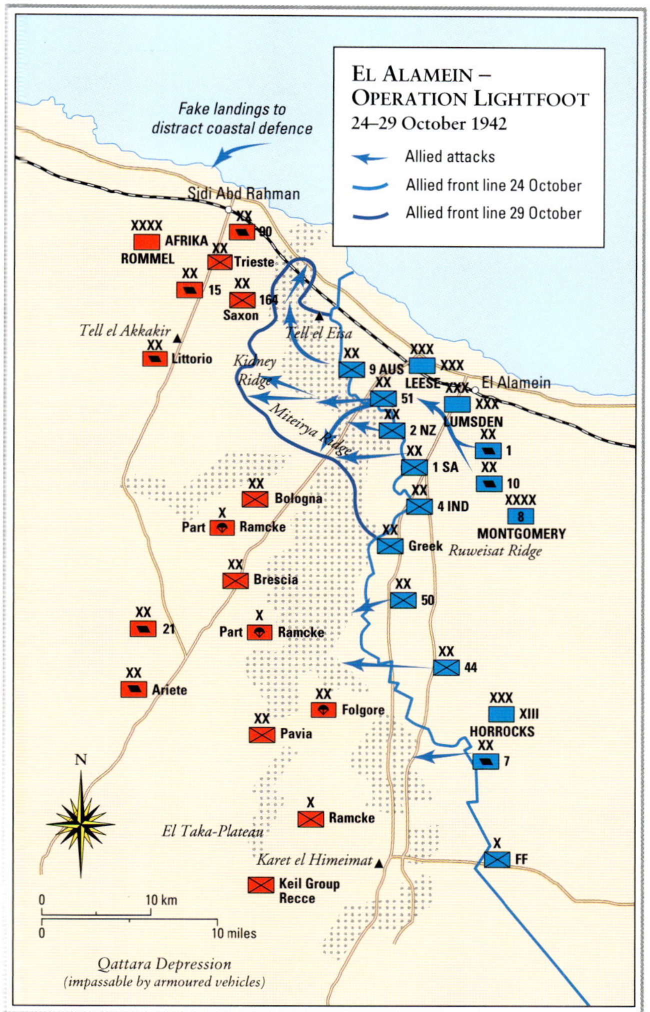

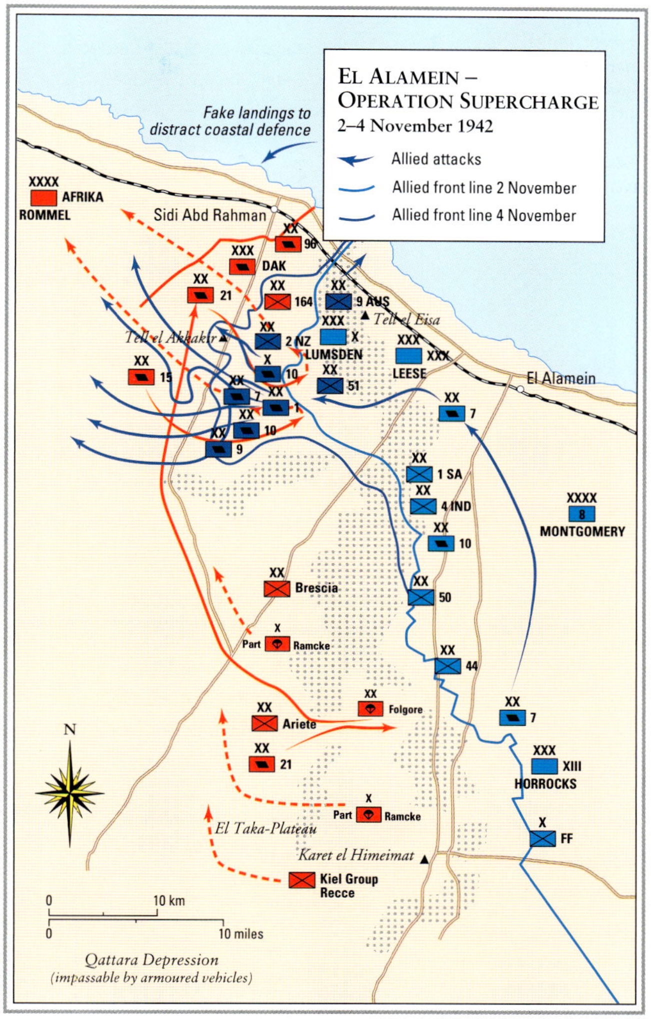

shortages of men and armoured vehicles strained Rommel's advance. Allied interdiction of supply convoys through the Mediterranean were having a telling effect, and the 'Desert Fox' was well aware that events beyond his control were conspiring against him. If he was to achieve final victory in North Africa, Rommel was compelled to act. Early in September, the offensive capability of his Panzer formations was further eroded at Alam el Halfa Ridge in action against the fresh 44th Infantry Division and 400 tanks of the 7th and 10th Armoured Divisions. Still, the British suffered mightily in action against the PzKpfw III tanks and anti-tank guns of the 15th and 21st Panzer Divisions, as a single armoured brigade lost 96 tanks.

Decision at El Alamein

As Rommel's combat efficiency waned, that of Montgomery and the Eighth Army increased significantly in the weeks following the fight at Alam Halfa. More than 300 new M4 Sherman tanks were due to arrive from the United States, and the Eighth Army grew to 200,000 troops, more than 1,000 tanks and 2,300 artillery pieces. In contrast, Panzergruppe Afrika could muster only 115,000 troops and 560 tanks, with little or no immediate prospect of re-supply from Germany. Montgomery completed plans initiated by Auchinleck for a decisive battle at El Alamein.

In the early hours of 23 October 1942, Montgomery unleashed the fire of 900 heavy guns

OPERATION SUPERCHARGE
On 2 November 1942, Operation Supercharge, the decisive blow at El Alamein, was initiated. Heavy attacks by more than 200 British tanks forced the Germans to abandon defensive positions at Kidney Ridge and elsewhere along the Mediterranean coast.

as British, South African and New Zealand troops crept forward and sappers removed thousands of German mines which had been sown in an area nicknamed the 'Devil's Gardens'. Tanks of the 1st and 10th Armoured Divisions followed to exploit any breaches in the German lines and consolidate the gains won by the infantrymen.

Operation Lightfoot

Operation Lightfoot was named for the fact that the infantry, not tanks, made up the initial advance. Lighter than machines, infantrymen did not trigger the mines, and they were followed by engineers who cleared a path for the armour. The opening phase of the British offensive made good progress, although the German minefields channelled Allied infantry and tank formations into predesignated fields of fire. German machine guns and anti-tank weapons slowed the British to a crawl after a few hours.

CRUSADER CRAWL
British tank soldiers replace a section of tread to restore their Crusader tank to battle readiness. The harsh climate of the North African desert took its toll on both men and equipment.

When the 10th Armoured Division finally cleared the German minefields, it became bogged down in heavy fighting at Miteirya Ridge. Eventually, troops of the 2nd New Zealand Division captured portions of Miteirya Ridge, suffering heavy casualties in the process. British troops occupied Kidney Ridge but were forced onto the defensive against ferocious German counterattacks. At Outpost Snipe, a battalion of British 6-pdr anti-tank guns ambushed two columns of advancing German and Italian armoured vehicles, destroying 33 tanks and five self-propelled guns, and damaging as many as 20 more. These were losses that Rommel could not replace.

Undeterred by the slow pace of Operation Lightfoot, Montgomery was determined to continue slugging it out with the Germans. Rather than pausing, he renewed his push during the first week of November.

Operation Supercharge

On 2 November, the British launched the decisive blow at El Alamein, Operation Supercharge. More than 200 tanks of the 1st Armoured Division spearheaded the effort, followed closely by the 7th and 10th Armoured Divisions, the 9th Armoured Brigade, and New Zealand infantry positioned to take advantage of any tactical gain. Heavy British attacks forced the Germans to abandon defensive positions between Kidney Ridge and the Mediterranean coast, while the 1st Armoured steadily advanced toward Tel el Aqqaqir.

Rommel knew that the end was near as his armoured strength dwindled to just 35 serviceable

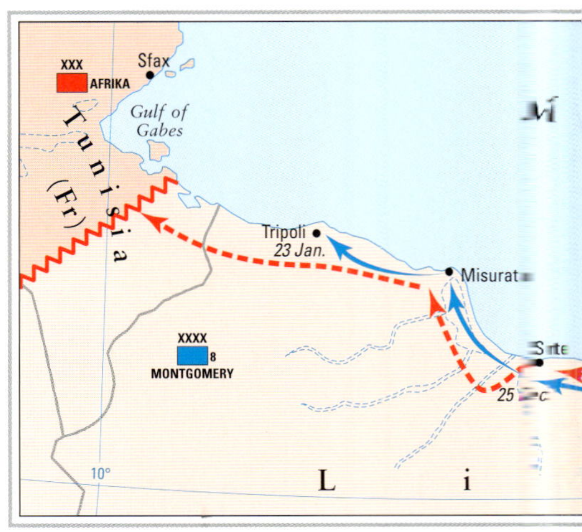

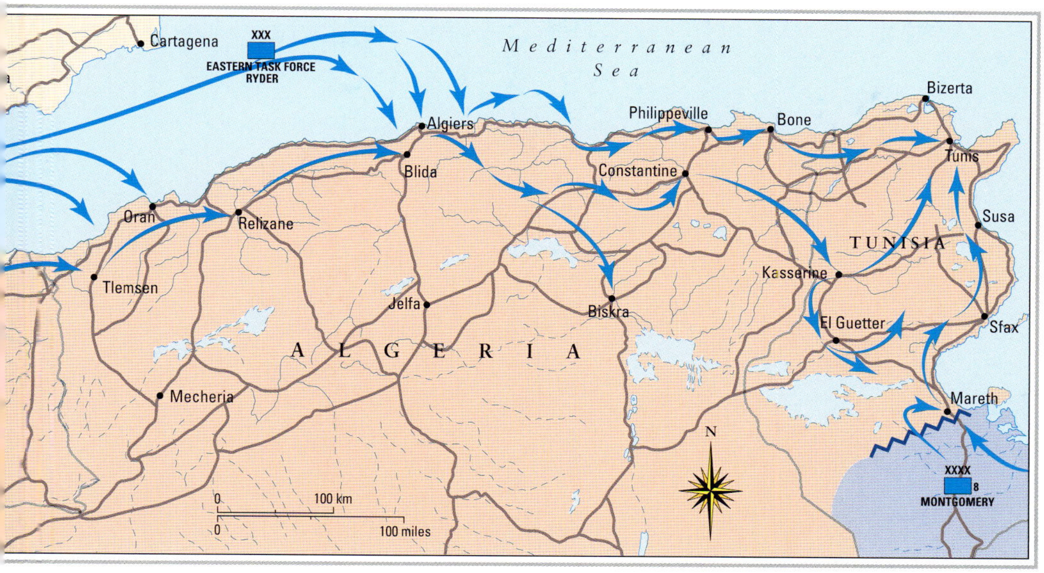

OPERATION TORCH
11 NOVEMBER 1942
Allied attacks

OPERATION TORCH
As the British Eighth Army pursued Panzerarmee Afrika from the west, Allied forces landed at Oran, Algiers and Casablanca in the east, forcing the Germans and Italians to fight on two fronts in North Africa.

tanks by 3 November. He requested permission for an orderly withdrawal in order to save what he could of the Axis forces as British tanks had broken through and threatened to cut off his line of retreat. After 11 days of fighting, the Eighth Army had destroyed 450 German tanks and 1,000 artillery pieces. More than 50,000 German and Italian troops were killed, wounded or captured. British losses amounted to 13,000 casualties and 500 tanks destroyed or disabled.

The Germans were compelled to begin a fighting retreat to Tunisia, a staggering 2,250km (1,400 miles) across the desert to the west.

Operation Torch

As Rommel began the long trek toward Tunisia, Allied forces mounted Operation Torch on 8 November 1942, landing nearly 100,000 Allied troops on the beaches at Oran, Algiers and Casablanca in the west. Rommel was thus forced to fight on two fronts, and U.S. ground troops entered combat against the Germans for the first time during the North African campaign.

Pressed by the British and U.S. First Armies in the west and the Eight Army advancing from the east, Rommel sensed that the Allied ring was closing. In mid-February, he turned on the Americans at Kasserine Pass and inflicted a stinging defeat on these untried troops. More than 6,300 US soldiers were killed, wounded or captured. However, the setback at Kasserine could not prevent the vanguard of the US II Corps and the 12th Lancers of the Eighth Army from linking up at Sfax on 7 April, 1943. In May, Rommel was recalled to Germany.

Surrounded and pressed against the sea, Axis forces under General Wilhelm Ritter von Thoma surrendered on 12 May 1943. The Allies had won a complete victory in North Africa.

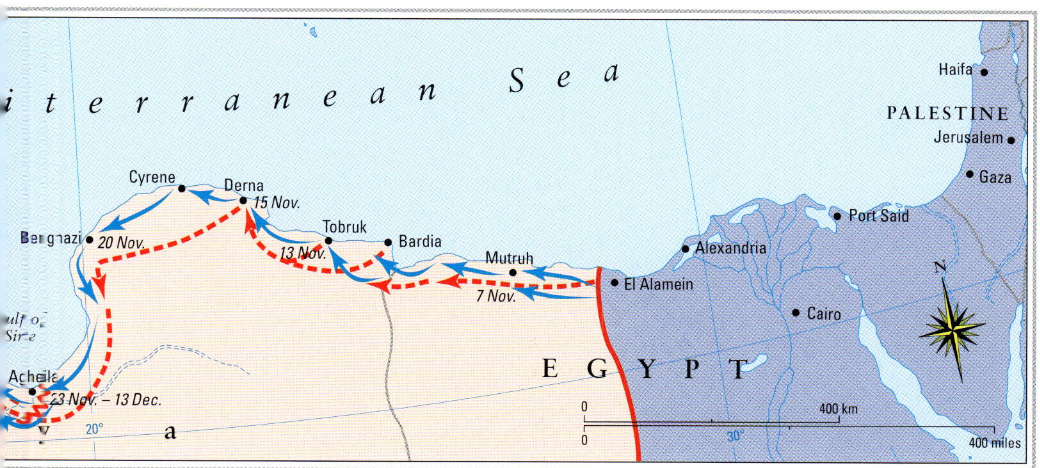

THE LONG ADVANCE
NOVEMBER 1942 – FEBRUARY 1943
Allied advance with dates
Axis retreat
Axis defence lines

BITTER RETREAT
Following their defeat at El Alamein, Axis forces retreated across hundreds of miles of desert toward Tunisia. Rommel was recalled to Germany, and by the spring of 1943, Allied victory in North Africa was complete.

VELIKIYE LUKI OPERATION 1942–43

VIGILANT WATCH
Red Army soldiers ride atop T-34 medium tanks in the vicinity of Velikiye Luki, where German troops were trapped in the city as they defended a vital rail link to Leningrad.

While the Red Army noose tightened around the German Sixth Army at Stalingrad to the south, elements of another German army, the Ninth, were fighting for their lives against encirclement at the strategically vital town of Velikiye Luki, on the Lovat River. Nearby, the rail line that ran through Novosokolniki supplied German Army Group North, besieging Leningrad. In November 1942, four elite Red Army guards divisions, supported by several armoured regiments with both the superb T-34 medium and KV-1 heavy tanks, undertook a pincer operation that would cut off the garrison at Velikiye Luki and threaten to drive a wedge between Army Group North and Army Group Centre, possibly unhinging the siege of Leningrad and liberating the city of Smolensk to the west.

Rapid encirclement

Within three days, the motorized infantry and tank units of the Red Army had successfully encircled Veliki Luki and nearly trapped a considerable number of German troops operating to the south. Characteristically, Hitler refused to allow an attempt to break out, or a tactical withdrawal that would continue to hold the rail line and place the flanks of advanced Soviet forces in jeopardy. As at Stalingrad, Hitler ordered a relief effort as troops of the 83rd Infantry Division and 3rd Mountain Division held the city that had been turned into a fortress during months of German occupation.

No relief

For several weeks, the defenders at Velikiye Luki successfully resisted all efforts to take the city, while few German troops were available to mount an effective relief effort, particularly as the situation at Stalingrad deteriorated. Nevertheless, the Soviets were denied their immediate objective – full control of the rail line at Novosokolniki – although they had cut it early in the offensive. By late December, a combined relief effort by elements of no fewer than five German divisions had ground to a halt.

The Germans launched a second breakthrough attempt on 4 January 1943, and came within 8km (5 miles) of the city. Among the German forces that attempted to relieve Velikiye Luki was the under-strength 8th Panzer Division, fielding only 32 operational tanks, many of them older Czech models. The following day, a Soviet counterattack succeeded in splitting the city in half. Two weeks later, the bulk of German troops in Velikiye Luki surrendered. A relative few had managed to escape the Soviet ring. The campaign resulted in a somewhat pyrrhic Soviet victory, with the Red Army sustaining more than 30,000 killed and achieving only limited success.

ARMOURED TRAFFIC
A column of PzKpfw IIIs led by a PzKpfw III Ausf. J with a short-barreled 50mm (2in) gun makes its way toward the fighting during the winter of 1943.

NORTHERN TRAP SPRUNG (OPPOSITE)
Hitler refused to allow a withdrawal from Velikiye Luki, and efforts to relieve the city's garrison were futile.

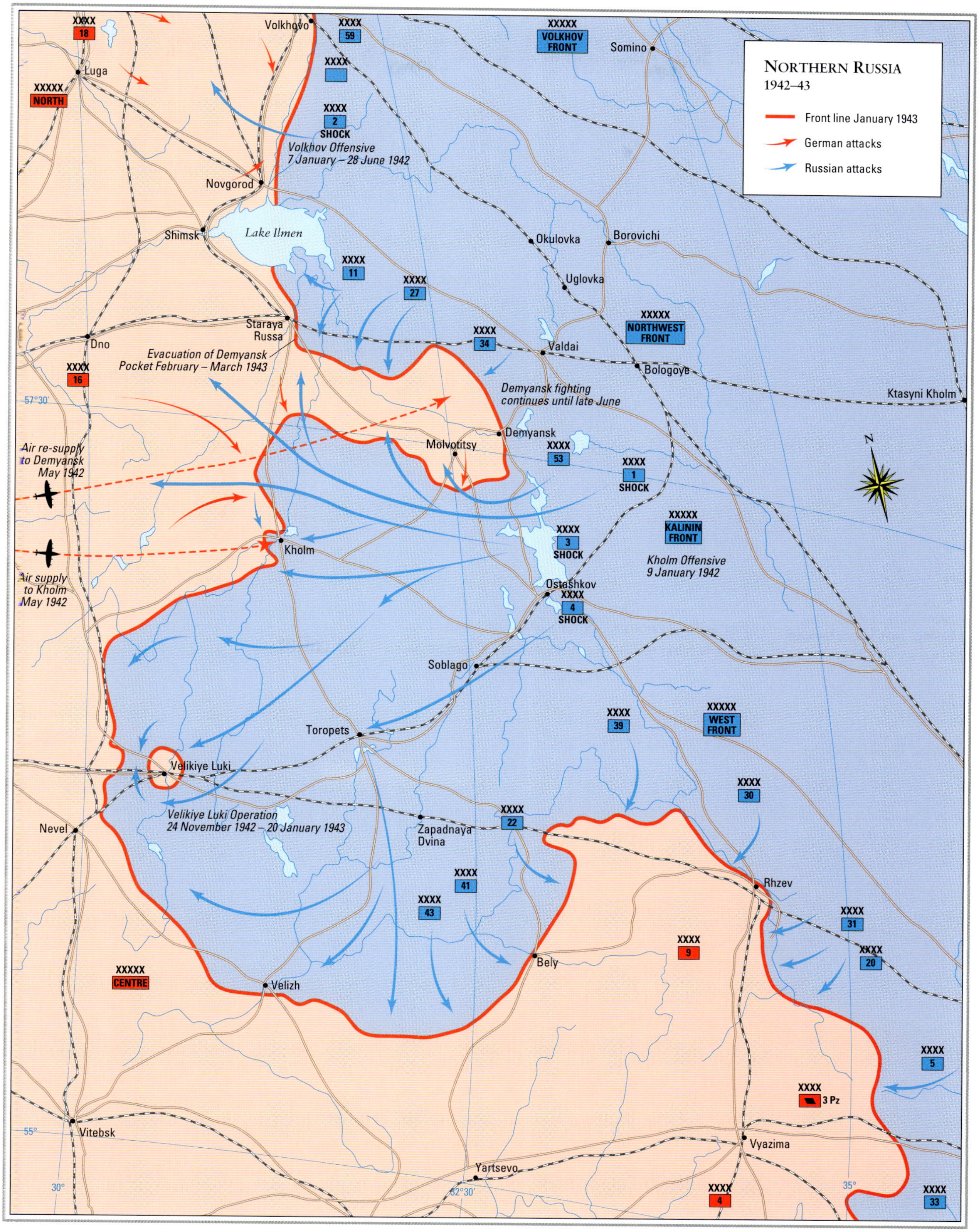
NORTHERN RUSSIA
1942–43
Front line January 1943
German attacks
Russian attacks
XXXXX NORTH
XXXX 18
XXXX 16
XXXXX CENTRE
XXXX 9
XXXX 3 Pz
XXXX 4
XXXXX VOLKHOV FRONT
XXXXX NORTHWEST FRONT
XXXXX KALININ FRONT
XXXXX WEST FRONT
XXXX 59
XXXX 2 SHOCK
XXXX 11
XXXX 27
XXXX 34
XXXX 53
XXXX 1 SHOCK
XXXX 3 SHOCK
XXXX 4 SHOCK
XXXX 39
XXXX 30
XXXX 22
XXXX 41
XXXX 43
XXXX 31
XXXX 20
XXXX 5
XXXX 33
Volkhov Offensive 7 January – 28 June 1942
Evacuation of Demyansk Pocket February – March 1943
Demyansk fighting continues until late June
Air re-supply to Demyansk May 1942
Air supply to Kholm May 1942
Kholm Offensive 9 January 1942
Velikiye Luki Operation 24 November 1942 – 20 January 1943
Volkhovo
Somino
Luga
Novgorod
Lake Ilmen
Shimsk
Okulovka
Borovichi
Uglovka
Staraya Russa
Dno
Valdai
Bologoye
Ktasyni Kholm
Demyansk
Molvotitsy
Kholm
Ostashkov
Soblago
Toropets
Velikiye Luki
Nevel
Zapadnaya Dvina
Rhzev
Bely
Velizh
Vitebsk
Vyazima
Yartsevo
57°30'
55°
30°
32°30'
35°
N

TIGERS ON THE BATTLEFIELD

TIGER TRIAL
Massive German Tiger I tanks advance along a snowy road. The Tiger mounted a menacing main gun but was prone to mechanical breakdowns. Many were lost to air attack or abandoned due to lack of fuel.

By the summer of 1944, the German PzKpfw VI Tiger tank had been in service little more than a year, and it had already become the stuff of legend to Allied soldiers in the West and earned the grudging respect of Red Army veterans in the East. The Tiger, with its formidable 88mm (3½in) cannon, is perhaps the best known tank of World War II. The main armament, capable of destroying Allied armour from more than a mile away, had already been proven effective in an anti-aircraft and anti-tank role. Coupled with the massive 57-ton turret and chassis of the Tiger, and later, the more than 60 tons of the Tiger II, the mere presence of such a heavy weapon on the battlefield often resulted in Allied tactical planning revisions. It was a formidable fighting vehicle that displayed tremendous potential and was capable of dominating the battlefield.

TIGER PROTECTION
The armour of the PzKpfw VI Tiger tank was up to 120mm (5in) thick in areas such as the turret and flanks. Allied tanks often positioned themselves for a shot at the rear, near the powerplant and exhaust areas.

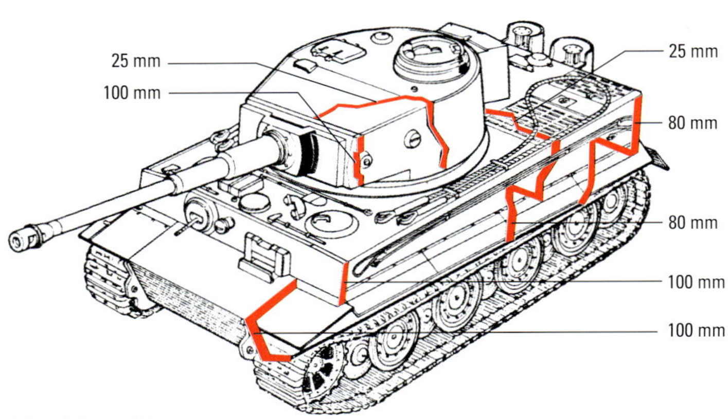

THE TIGER TANK
ARMOUR CONFIGURATION MODEL AUSF.E

Mammoth undertaking

Following a design competition between the Henschel and Porsche firms, the Henschel-designed Tiger went into production in August 1942. The exigencies of war resulted in the vehicle's deployment to combat in North Africa and on the Eastern Front after only limited trials, so the tank was plagued by mechanical difficulties, particularly early in its career. In addition, numerous design innovations made production slow and difficult. Due to its significant weight, concerns surfaced that the Tiger was initially underpowered, and the 12-cylinder Maybach engine that powered the first 250 Tigers was replaced with a more substantial V-12 HL 230 P45 engine. The Tiger was also costly to manufacture, with a price tag twice that of the proven PzKpfw IV.

While the Tiger was the epitome of German armoured might, it was also indicative of the German desire to produce quality over quantity. Its success was only limited by an arduous production protocol and the overwhelming numbers of Allied tanks, particularly the Soviet T-34 and the American M4 Sherman, that could be thrown against it. During the course of the war, 1,347 units of the Tiger I were produced, and from 1943 to 1945, a mere 492 of the follow-on Tiger II were manufactured.

Tiger in action

The Tiger was initially deployed with German forces in North Africa in late 1942 and quickly outclassed its opponents in terms of firepower and survivability, with armour plating as thick as 120mm (4⅘ in) on the frontal and turret areas. At

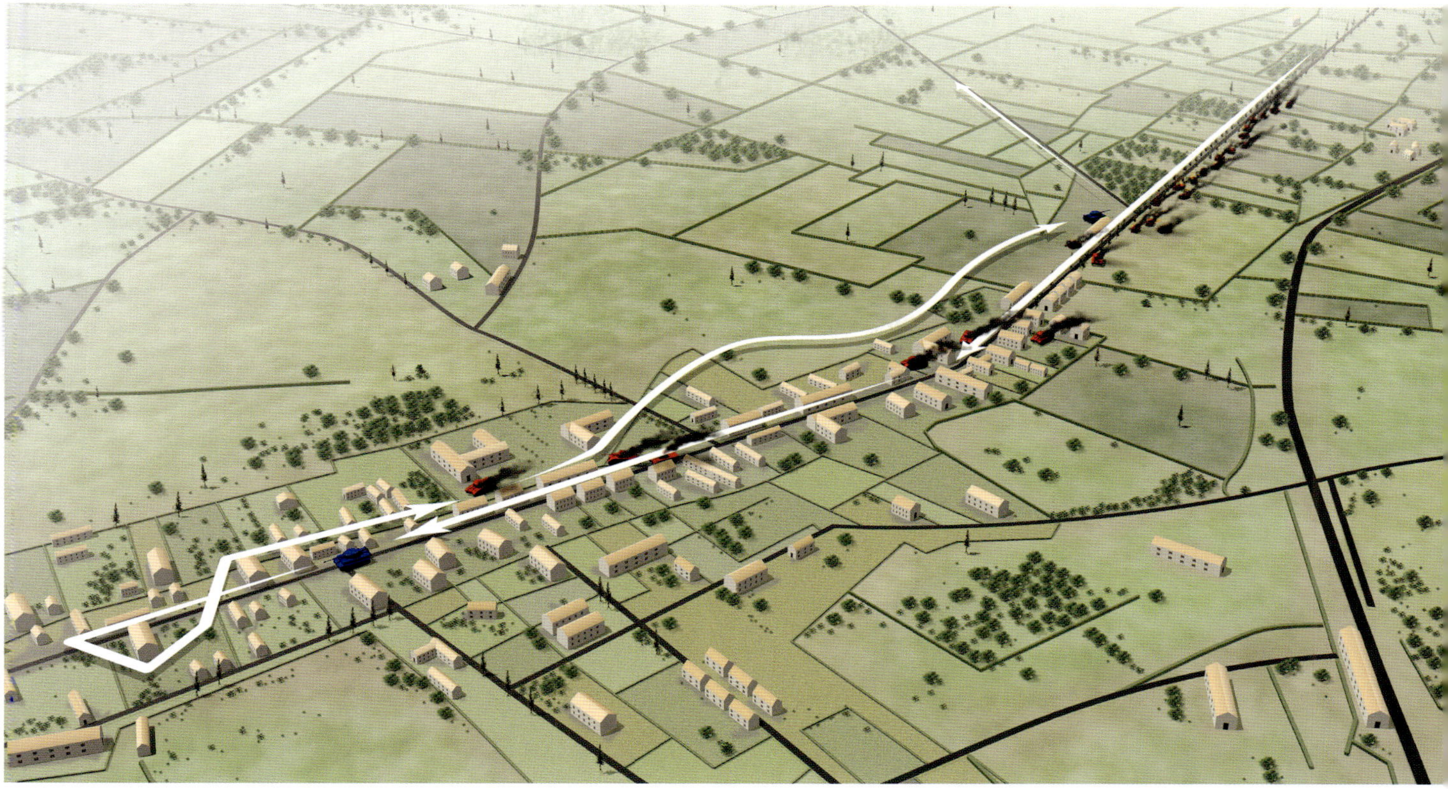

WITTMAN'S ONSLAUGHT
The single Tiger tank commanded by SS Hauptstürmfuhrer (Captain) Michael Wittman wreaked havoc among vehicles of the British 7th Armoured Division at Villers Bocage. During the action, he was credited with destroying numerous tanks and personnel carriers.

times, only concentrated artillery fire or direct air attack from tactical fighter bombers could stop the behemoth. On the ground, it was typical for a platoon of Allied tanks to engage a single Tiger, attacking simultaneously from multiple directions and acknowledging that some loss would be incurred. On the Eastern Front, Soviet tankers developed the tactic of closing rapidly with the Tiger to somewhat neutralize its long-range gunnery advantage and perhaps gain an advantageous firing position at the vulnerable rear of the German tank.

In combat, the Tiger was estimated to have achieved a kill ratio of six to one against Allied tanks, and its precision optics coupled with the 88mm (3½in) cannon were a deadly combination. Often, however, the Tiger reached the battlefield in insignificant numbers due to difficulty in transporting the heavy tank by rail, conducting routine maintenance in the field, and its complex overlapping wheels and suspension system that were prone to failure. Chronic fuel shortages and a ravenous powerplant limited its range.

Villers-Bocage

The battlefield prowess of the Tiger tank was readily apparent in the hands of numerous aces who achieved astounding combat successes against Allied armoured vehicles. The most famous Tiger ace, SS Hauptsturmführer (Captain) Michael Wittman, demonstrated just how deadly the Tiger could be at the French town of Villers-Bocage on 13 July 1944. A week after D-Day, Wittman encountered a column of the British 7th Armoured Division and within 15 minutes was credited with destroying as many as 14 tanks, 15 troop carriers and a pair of anti-tank guns. Some of these may, in fact, have been destroyed by other Tigers of Heavy Panzer Battalion 101; however, Wittman is known to have been the foremost tank commander in the engagement.

Two months after the fight at Villers-Bocage, Wittman was killed in action near the town of Cintheaux, France, presumably by the fire of a 17-pdr cannon mounted on an upgunned British Sherman tank known as a Firefly. During his brief career, Wittman was credited with destroying as many as 138 Allied armoured vehicles. At least a dozen Tiger commanders claimed to have destroyed more than 100 enemy tanks.

Considering the impressive combat record of the Tiger I and Tiger II tanks deployed during World War II, it is obvious that the weapon might well have altered the course of the war had significant numbers been manufactured, perfected and deployed. Quite simply, it was too costly in terms of resources and manpower for Germany's wartime economy.

SICILY AND ITALY
1943

Having liberated all of Axis-held North Africa by 12 May 1943, the Western Allies next planned to invade the Italian-held island of Sicily, located just southwest of the country's southwestern 'toe'. During 9-10 July 1943 some 140,000 troops Western Allied troops invaded the island. One Canadian and three British divisions from the British Eighth Army assaulted the island's southeastern coast in the vicinity of Pachino. Meanwhile, three divisions from the U.S. Seventh Army assaulted its southern coast between Licata and Scoglitti; both attacks were supported by complementary airborne landings. During 11-13 July the Allies successfully advanced to capture the port of Syracuse. Between 14 July and 5 August British forces cleared the island's central-eastern sector, while American forces secured all of western Sicily. By 6 August, therefore, these two Allied thrusts had forced the German and Italian defenders all the way back to the Etna Line, which protected Sicily's northeastern peninsula. During the next 10 days, the Allies mounted several successful amphibious landings to outflank the firm Axis defence line that covered this peninsula. These developments compelled Axis forces to begin evacuating the island, a process completed by the 17th.

Next, during 3-9 September troops from the British Eighth Army invaded Italy's 'toe' and 'heel'; on the 8th the Italians had surrendered and the Germans rushed to take control of much of the country. As the British pushed north, on 9 September U.S. Fifth Army forces landed at Salerno on Italy's western coast south of Naples, to outflank the enemy's defences. After fending off fierce German counterattacks during 14-17 September, the Fifth Army's forces

Invasion of Sicily
10 July – 17 August 1943

- Allied landings with dates
- Axis counterattacks
- Allied front line 11 July
- Allied front line 15 July
- Allied front line 23 July
- Axis retreat line
- Axis retreat line
- Axis retreat line
- Axis retreat route
- Airfields constructed by Allies
- Allied airborne landings

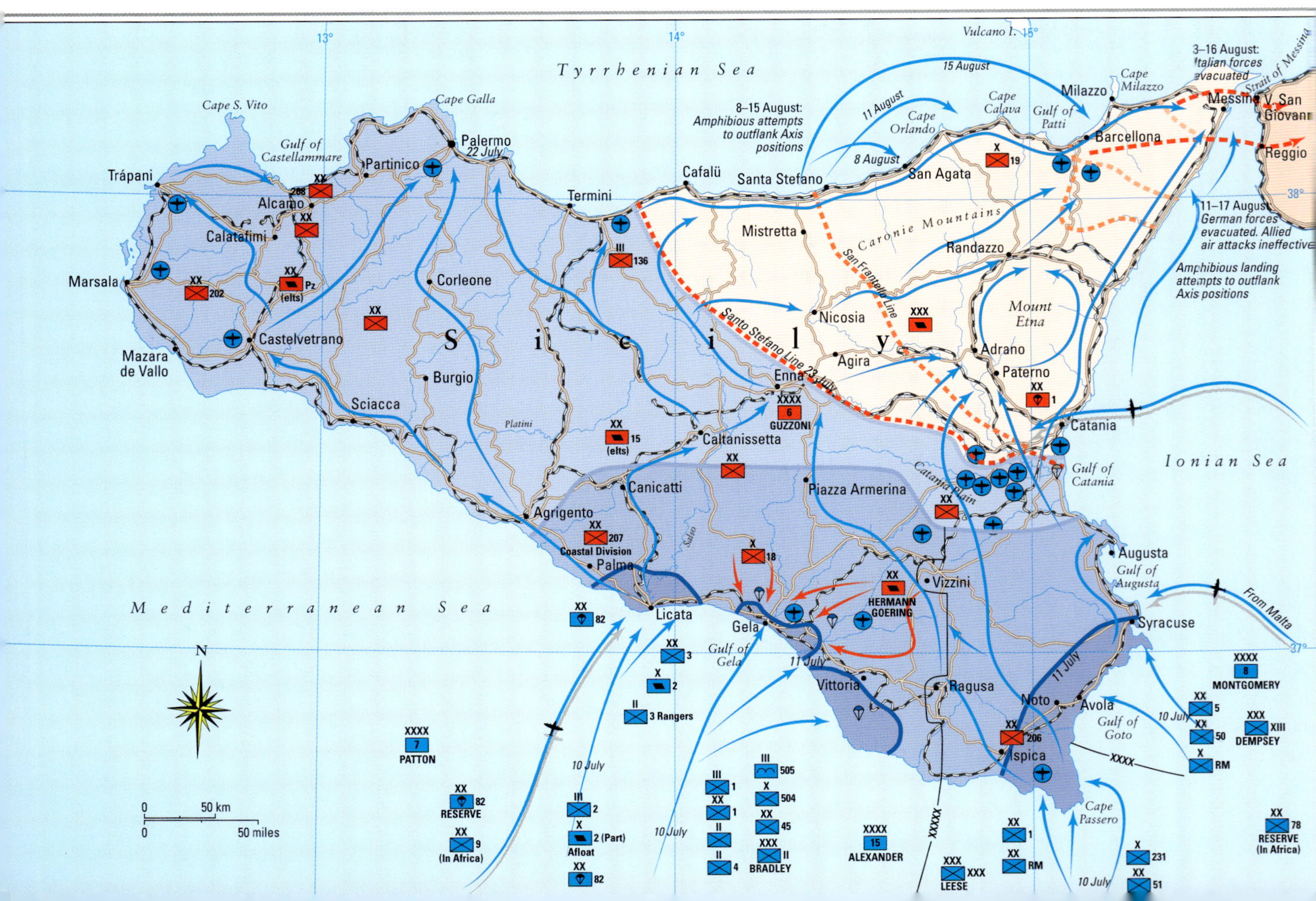

SOUTHERN ITALY
3 SEPTEMBER –
15 DECEMBER 1943

OPERATION HUSKY
The invasion of Sicily was conducted by the British Eighth and U.S. Seventh Armies, commanded by Generals Bernard Montgomery and George Patton respectively. While the British slugged northward against heavy resistance, Patton captured Palermo and raced to Messina.

subsequently advanced north to capture Naples and reach the River Volturno by 6 October. By then, the British Eighth Army had advanced up Italy's eastern (Adriatic) coast to link up with the Fifth Army. During October and November the Allies continued to advance north across the entirety of Italian front. By late December 1943, however, the Allied offensive had petered out along the front of a hastily constructed German defensive position, termed the Gustav (or Winter) Line. This position ran east across Italy from near Gaeta, on the western coast just south of Rome, through Monte Cassino, the Apennine Mountains and on to the estuary of the River Sangro on the eastern coast near Ortona.

INVADING ITALY (OPPOSITE)
In the late summer of 1943, the Allied Fifth and Eighth Armies landed at Salerno, Calabria and Taranto and began the arduous push northward up the 'boot' of Italy. Serious resistance temporarily threatened the beachhead at Salerno.

KURSK 1943

WATCHFUL GUNNER
A German soldier peers warily from behind the shield of an 88mm (3½in) gun. Originally designed as an anti-aircraft weapon, the '88' proved to be deadly against Allied armour across the European Theatre of Operations.

During the brief spring pause in operations on the Eastern Front in 1943, the German High Command decided to mount their next offensive, codenamed Citadel, against the obvious target of the Kursk salient; this was a large, broadly semi-circular bulge of Soviet-held territory that jutted west deep into the German line. By thrusting with two pincers into this salient from north and south, the Germans hoped to encircle and destroy a sizable Soviet force; by early July no fewer than 850,000 Soviet troops defended the bulge. On 4 July 1943, Field Marshal Erich von Manstein's Army Group South commenced its preliminary attacks from the salient's southern corner. Manstein's command controlled the 20 divisions and 1,405 armoured fighting vehicles (AFVs) fielded by General Werner Kempf's Army Detachment and Colonel-General Hermann Hoth's Fourth Panzer Army. The next morning General Walter Model's Ninth Army, subordinated to Army Group Centre, began its attack from the salient's northern shoulder. Model's command fielded six mechanized and 14 infantry divisions, together with 1,840 AFVs. The German commanders hoped that these two assaults would link up at Kursk to encircle over 670,000 enemy troops; if successful, Operation Citadel would deliver the largest bag of enemy prisoners from any of the many pockets formed on the Eastern Front during World War II.

Pre-emptive strike

Early on 5 July, the infantry spearheads of Model's Ninth Army's assaulted the first Soviet defence line along the northern corner of the salient. Soviet intelligence, however, had correctly identified the time when Model's forces would attack and so unleashed an intense artillery bombardment on the German positions to disrupt the assault. By mounting a series of intense all-arms assaults that day, Model's forces managed to advance 10km (6 miles) south into the Soviet defences. Along the salient's southern shoulder, meanwhile, the Germans also commenced their main attack on

TRUDGING GRENADIERS
German Panzergrenadiers, laden with heavy gear, begin advancing with armoured vehicles across the vast expanse of Russia. In the summer of 1943, the Battle of Kursk pitted German and Soviet tanks in the largest armoured battle of the war.

RUSSIAN FRONT
4 July 1943

Soviet Frontline, 4 July

KURSK SALIENT
In early July 1943, the Germans launched Operation Citadel, its objective to isolate the Kursk salient and trap approximately 850,000 Red Army soldiers inside. The German offensive failed following a decisive clash of armour at Prokhorovka.

WEHRMACHT THRUST
German soldiers and an open-turreted, self-propelled assault gun await the order to advance during the opening hours of Operation Citadel, an attempt to isolate the Kursk salient, a deep bulge into the German lines on the Eastern Front.

ARMOURED ONSLAUGHT (OPPOSITE)
General Hermann Hoth's 4th Panzer Army assaulted the southern shoulder of the Kursk salient, while General Walther Model's 9th Army attacked in the north. On the first day, however, neither managed to advance more than 10km (6 miles).

DETERMINED DEFENDERS
Lying in wait for the expected German assault on the Kursk salient, Soviet tanks have assumed hull-down positions as crewmen check their combat readiness. Armoured thrust and counterthrust marked the decisive Battle of Kursk.

5 July. By evening Hoth's forces had also advanced no more than 10km (6 miles) despite launching repeated intense attacks. Worse still, further to the east, Kempf's spearhead armour failed even to break through the first Soviet defensive line in the face of fanatical enemy resistance. Subsequently, during 6-9 July, Ninth Army's forces repeatedly attacked the enemy's second defensive line to seize the vital ground of the Olkhovtka Ridge. Vigorous enemy armoured counterattacks, however, prevented Model's forces from capturing the key hamlet at Ponyri. Then, during 10-11 July, yet more powerful Soviet counterstrikes halted the advance of Model's now exhausted and denuded forces. Thus, despite launching fierce assaults over an entire week, Model's drive south had only managed to advance just 16km (10 miles) deep into the enemy's layered defensive system. Yet worse was to come for the Germans, for from 12 July potent enemy counterattacks even forced Model's spearheads to give up ground that they had captured in the previous few days. The German northern thrust had been a dismal failure.

Battle of Prokhorovka

During 6-9 July, meanwhile, along the southern front, German armoured attacks had gradually driven the Red Army back toward the villages of Prokhorovka and Oboyan, despite encountering repeated Soviet ripostes. Here, Colonel-General Hausser's II-SS Panzer Corps secured the deepest advance, but to the east Kempf's armour was less successful,

and this exposed Hausser's flank to determined Soviet counter-blows. Next, on 10 July, the German armour penetrated parts of the third Soviet defensive line and closed on Oboyan and Prokhorovka. This setback led the concerned Soviets to commit their key reserve, the elite Fifth Guards Tank Army, to the area to stem the German advance. On 12 July, at Prokhorovka, a titanic armoured battle ensued when 805 Soviet tanks clashed with 495 Panzers in a series of bitter tactical encounters. Early that day, the Soviet armoured units smashed into the German II SS and III Panzer Corps as their divisions

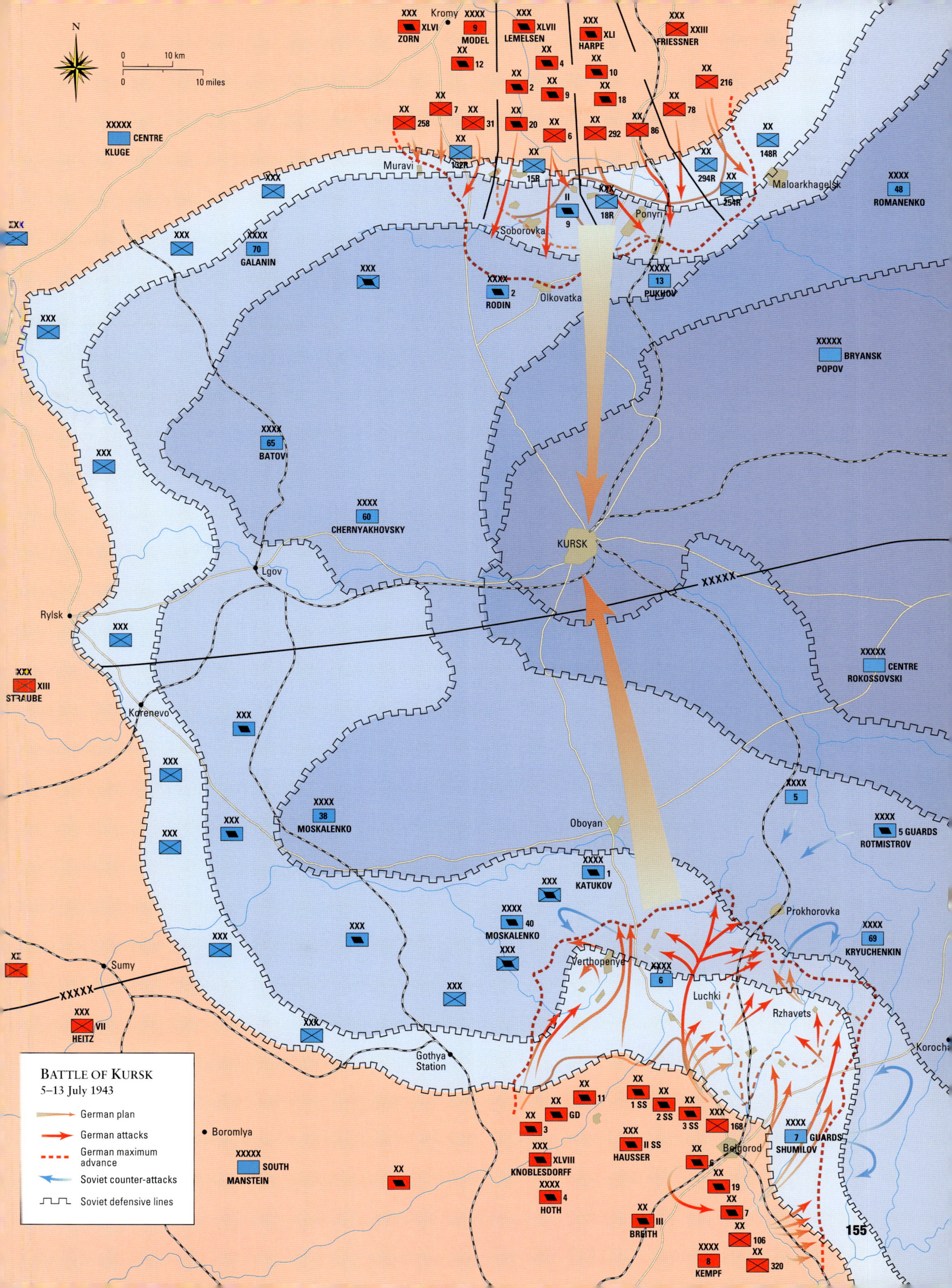
BATTLE OF KURSK
5–13 July 1943
German plan
German attacks
German maximum advance
Soviet counter-attacks
Soviet defensive lines
N
0 10 km
0 10 miles
XXX XLVI ZORN
Kromy
XXXX 9 MODEL
XXX XLVII LEMELSEN
XXX XLI HARPE
XXX XXIII FRIESSNER
XX 12
XX 4
XX 10
XX 216
XX 2
XX 9
XX 18
XX 7
XX 31
XX 20
XX 6
XX 292
XX 86
XX 78
XX 258
XXXXX CENTRE KLUGE
XX 132R
XX 15R
XX 148R
XX 294R
XX 254R
Muravi
Maloarkhagelsk
XXXX 48 ROMANENKO
II 9
XXX 18R
Ponyri
Soborovka
XXXX 70 GALANIN
XXXX 2 RODIN
Olkovatka
XXXX 13 PUKHOV
XXXXX BRYANSK POPOV
XXXX 65 BATOV
XXXX 60 CHERNYAKHOVSKY
KURSK
Lgov
Rylsk
XXXXX
XXX XIII STRAUBE
Korenevo
XXXXX CENTRE ROKOSSOVSKI
XXXX 38 MOSKALENKO
XXXX 5
XXXX 5 GUARDS ROTMISTROV
Oboyan
XXXX 1 KATUKOV
XXXX 40 MOSKALENKO
Prokhorovka
XXXX 69 KRYUCHENKIN
Verthopenye
XXXX 6
Luchki
Rzhavets
Sumy
XXXXX
XXX VII HEITZ
Gothya Station
Korocha
XX 11
XX 1 SS
XX GD
XX 2 SS
XX 3 SS
XX 3
XX 168
XXXX 7 GUARDS SHUMILOV
Boromlya
XXX II SS HAUSSER
XXX XLVIII KNOBLESDORFF
XXXXX SOUTH MANSTEIN
XX 6
XX 19
Belgorod
XXXX 4 HOTH
XX 7
XX III BREITH
XX 106
XXXX 8 KEMPF
XX 320

KURSK – SOVIET COUNTER OFFENSIVE
16 JULY – 27 AUGUST 1943

Russian counter attacks
German attacks
Soviet frontlines with dates

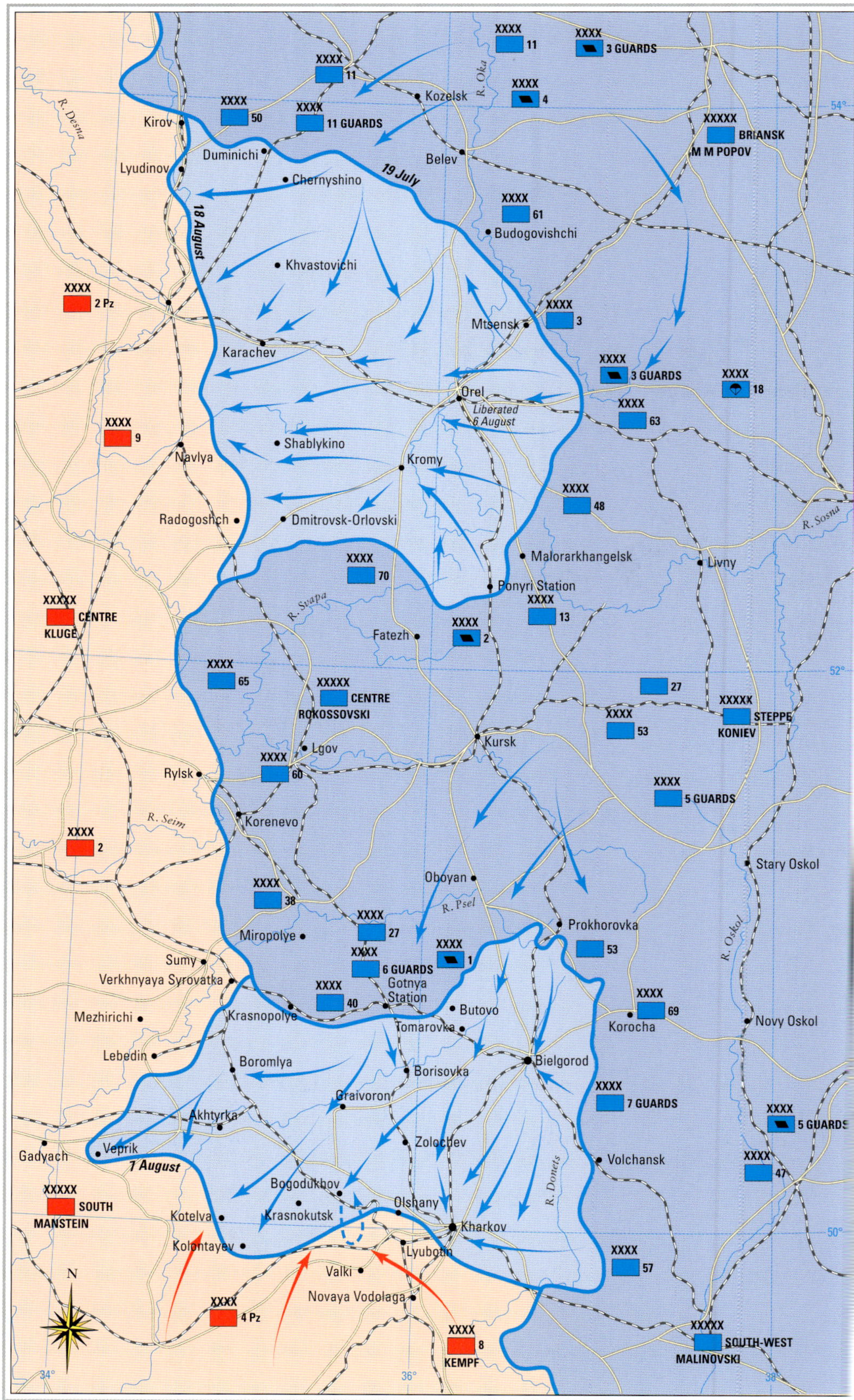

SOVIET COUNTERSTROKES
As the Red Army defenders grudgingly gave ground, German armoured spearheads made limited progress within the Kursk salient. Then, a series of Soviet counterblows, particularly at Prokhorovka, doomed the German offensive to failure.

FURIOUS FIGHT
Soviet tanks are directed toward the fighting at Kursk by a signalman. Realising that they were confronted by Panther and Tiger tanks, the Soviets closed rapidly with the German armour to negate the long-range advantage of the enemy guns.

formed up for the intended advance on the village of Prokhorovka. Knowing that the German Army's latest Panther and Tiger tanks possessed superior long-range firepower, the Soviet tanks raced forward to close the range between the opposing forces. In a chaotic series of armoured encounters that lasted throughout much of the day, the Soviet armoured counterstrikes halted the German advance north through Prokhorovka, preventing the Germans from capturing this formerly innocuous hamlet. At the end of the battle, the Soviets had lost 412 tanks and the Germans 149. Although this might seem to indicate that the encounter had been a German tactical victory, it was in fact a German strategic disaster. For the battle not only inflicted severe AFV losses that the Germans would struggle to replace, but it also wrested away what little operational initiative the Germans retained in the Citadel offensive. Recognizing this, Hitler decided to cancel Citadel on the 13th; this decision was also motivated, however, by Hitler's desire to transfer elite armoured forces from the Kursk region to Sicily to repel the Western Allied amphibious landings that had just occurred. During 13-24 July, Hoth and Kempf's forces had to conduct a stubborn fighting withdrawal back to their original positions in the face of fierce Soviet attacks. Citadel was a costly failure, with the German's suffering 883 AFV and 54,100 personnel casualties for a gain of just a few miles of insignificant terrain. In fact, the German Army in the East never recovered from this debacle.

Citadel crumbles

Yet even worse events were about to engulf the German Army. The Soviets had envisaged the halting of Citadel as the first part of a coordinated strategic counteroffensive. Consequently, on 12 July the Red Army initiated Operation Kutuzov, their own counteroffensive into the vulnerable northern flank of Model's forces then locked in combat along the salient's northern shoulder. The attack against the Second Panzer Army caught the Germans by surprise, as their attention still remained focused on Citadel. Consequently, the rapid Soviet advance west soon threatened to cut the German lines of supply. Even worse, on 3 August the Soviets initiated Operation Rumyantsev, an offensive by the Voronezh and Steppe Fronts against Fourth Panzer Army and Army Detachment Kempf, designed to eliminate the Kursk salient's southern shoulder. By 7 August the Soviet armour had advanced swiftly southwest to capture the key city of Kharkov. Subsequently, over the ensuing weeks the Red Army would widen these operations into a general strategic counteroffensive toward the Dnieper River.

UKRAINE AND THE CRIMEA 1944

On 1 January 1944 the German front along the Eastern Front's southern sector ran in an undulating fashion from the Pripyet marshes south through Zhitomir, Kirovograd and Nikopol, down to the Crimean isthmus. Over the ensuing five months the Red Army initiated a series of offensives, named collectively the Dnieper-Carpathian operation. On 5 January 1944 Konev's Second Ukrainian Front initiated an offensive in the Kirovograd region, while to the north the First Ukrainian Front continued to advance west through Tarnopol to close on Lvov. Next, on 24 January, Konev's southern flank launched another offensive. By the 28th these assaults had encircled 73,000 Axis troops in the Cherkassy-Korsun pocket; a German armoured relief attack launched from the west and a simultaneous break-out from the pocket was partially successful, but those left behind capitulated on 17 February. Meanwhile, forces from Third Ukrainian Front had assaulted the German positions in the salient that jutted west into enemy lines around Nikopol; by late February these attacks had eliminated the salient.

Red Army surge

During March the Soviets expanded their offensive operations. From 4 March, the First and Second Ukrainian Fronts resumed their offensive. Advancing swiftly west toward the Soviet border, by 27 March the Red Army's armoured spearheads had successfully encircled much of the First Panzer Army – some 180,000 troops – in the Kamenets-Podolsk pocket. The encircled forces successfully broke out of the encirclement and struck west; by 5 April they had linked up with the German relief attacks at Buczacz. Meanwhile, to the south, Konev's forces had again struck southwest, capturing Uman and pushing on toward Kishinev, inside the Romanian border. Further south still, the Third Ukrainian Front also advanced rapidly southwest to capture Odessa on the Black Sea coast on 10 April, before advancing into Romanian territory. In the process, Soviet armoured spearheads had encircled the German Sixth Army near Nikolayev, although a proportion of these forces successfully fought their way southwest to rejoin the main German line on the River Bug.

As these Soviet advances unfolded, in the extreme south on 8 April the Fourth Ukrainian front commenced its re-conquest of the Crimea, which was defended by the isolated German Seventeenth Army. By mid-April the Germans had been forced back to an enclave surrounding the key port-city of Sevastopol, which they intended to hold as a citadel as they had during the first battle for the Crimea in 1941-42. Continuing Soviet advances forced the surviving German forces to evacuate the city by sea during 7-9 May back to the Romanian port of Constanta. The battle as a whole cost the Axis some 85,000 casualties, with the Soviets suffering 95,000.

RED ARMY SURGE (OPPOSITE)
In the winter and spring of 1944, Red Army offensive operations in the Crimea isolated thousands of German troops in Sevastopol, while great numbers of German prisoners were taken during Soviet westward thrusts into the Ukraine.

HAIL VICTORS
Flush with victory, Soviet tankers push their T-34s forward to the cheers of a recently liberated village in the Ukraine. The versatile T-34 was available in great numbers and proved to be a war-winning weapon on the Eastern Front.

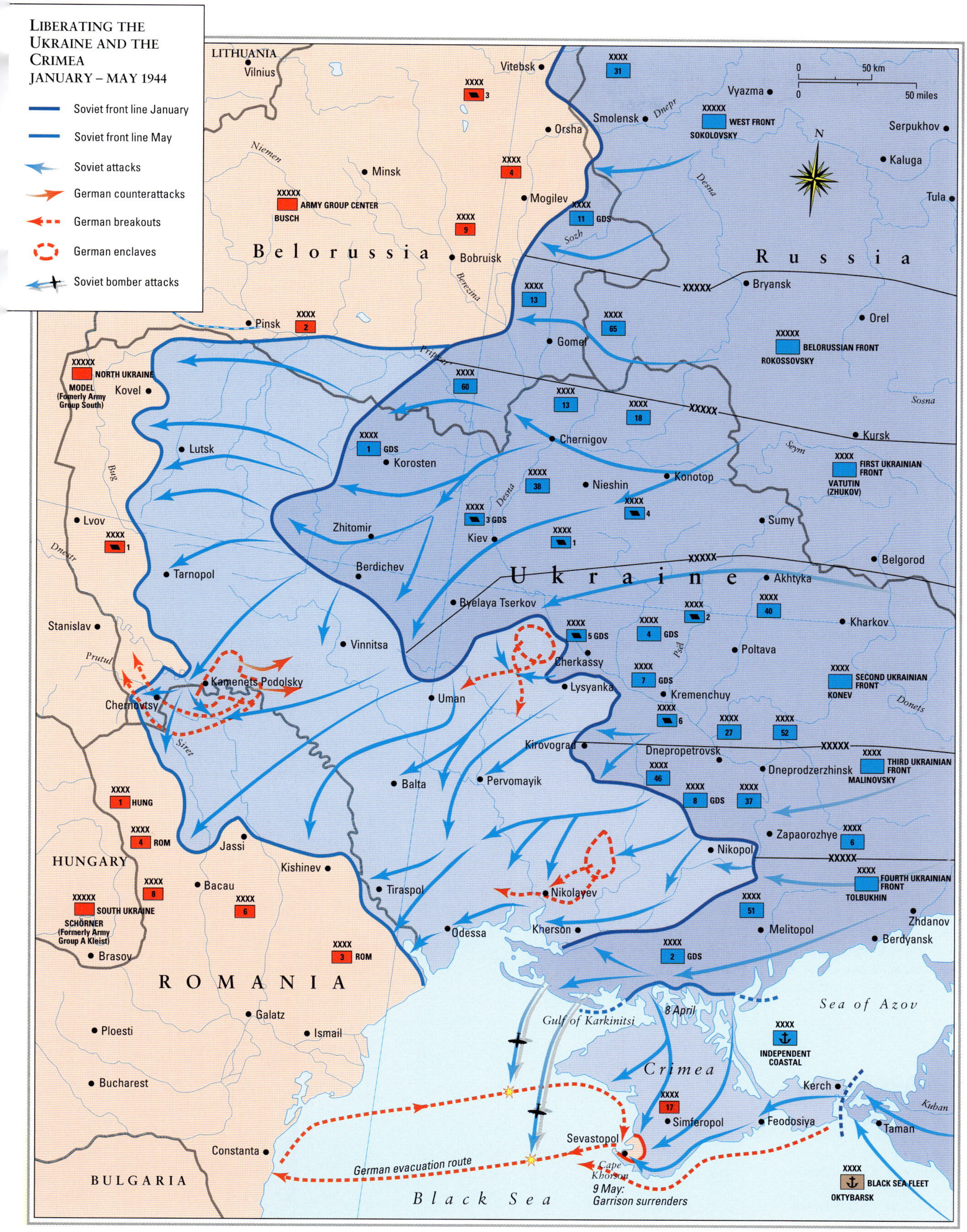

LIBERATING THE UKRAINE AND THE CRIMEA
JANUARY – MAY 1944
Soviet front line January
Soviet front line May
Soviet attacks
German counterattacks
German breakouts
German enclaves
Soviet bomber attacks
0 50 km
0 50 miles
N
LITHUANIA
Vilnius
Vitebsk
Orsha
Smolensk
Vyazma
Serpukhov
Kaluga
Tula
Minsk
Mogilev
Bobruisk
Bryansk
Orel
Gomel
Pinsk
Kovel
Lutsk
Korosten
Chernigov
Konotop
Nieshin
Kursk
Sumy
Belgorod
Lvov
Zhitomir
Kiev
Tarnopol
Berdichev
Byelaya Tserkov
Akhtyka
Kharkov
Stanislav
Vinnitsa
Poltava
Cherkassy
Lysyanka
Kamenets Podolsky
Chernovtsy
Uman
Kremenchuy
Kirovograd
Dnepropetrovsk
Dneprodzerzhinsk
Balta
Pervomayik
Zapaorozhye
Nikopol
Jassi
Kishinev
Bacau
Tiraspol
Nikolayev
Odessa
Kherson
Melitopol
Zhdanov
Berdyansk
Brasov
Galatz
Ismail
Ploesti
Bucharest
Constanta
Simferopol
Feodosiya
Kerch
Taman
Sevastopol
Cape Khersön
Belorussia
Russia
Ukraine
HUNGARY
ROMANIA
BULGARIA
Crimea
Sea of Azov
Black Sea
Gulf of Karkinitsi
Niemen
Dnepr
Desna
Sozh
Berezina
Pripyat
Bug
Dnestr
Prutul
Siret
Seym
Sosna
Psel
Donets
Kuban
XXXX 31
XXXXX WEST FRONT SOKOLOVSKY
XXXX 3
XXXX 4
XXXXX ARMY GROUP CENTER BUSCH
XXXX 9
XXXX 11 GDS
XXXX 13
XXXX 65
XXXX 2
XXXXX BELORUSSIAN FRONT ROKOSSOVSKY
XXXXX NORTH UKRAINE MODEL (Formerly Army Group South)
XXXX 60
XXXX 13
XXXX 18
XXXX 1 GDS
XXXX 38
XXXX FIRST UKRAINIAN FRONT VATUTIN (ZHUKOV)
XXXX 3 GDS
XXXX 4
XXXX 1
XXXX 1
XXXX 40
XXXX 2
XXXX 5 GDS
XXXX 4 GDS
XXXX 7 GDS
XXXX SECOND UKRAINIAN FRONT KONEV
XXXX 6
XXXX 27
XXXX 52
XXXX THIRD UKRAINIAN FRONT MALINOVSKY
XXXX 46
XXXX 8 GDS
XXXX 37
XXXX 1 HUNG
XXXX 4 ROM
XXXX 6
XXXX 8
XXXXX SOUTH UKRAINE SCHÖRNER (Formerly Army Group A Kleist)
XXXX 6
XXXX FOURTH UKRAINIAN FRONT TOLBUKHIN
XXXX 51
XXXX 3 ROM
XXXX 2 GDS
XXXX INDEPENDENT COASTAL
XXXX 17
XXXX BLACK SEA FLEET OKTYBARSK
8 April
German evacuation route
9 May: Garrison surrenders

THE SHERMAN

M4A3 SHERMAN
A Sherman tank rolls down an embattled street somewhere in Europe. Although German tanks were superior in armour and firepower, the Sherman used greater numbers to prevail on the battlefield.

One of the most easily recognized symbols of the Allied victory in World War II is the Medium Tank M4, popularly known as the Sherman. The product of rapid development in the United States, the M4 prototype, designated the T6, was completed and undergoing tests by September 1941, a scant three months prior to the U.S. entry into the war. The Sherman was designed for mass production and intended for a two-fold purpose: to correct the shortcomings of the Grant and Lee variants of the earlier M3 design that was something of a hybrid, both anachronistic and modern with a 75mm (3in) hull-mounted cannon and a 37mm (1½in) turret gun; and to perform in the dual role of infantry support with the ability to take on enemy tanks.

Production of the M4 neared 50,000 by the end of World War II, and several variants of the tank were introduced as the war progressed. When the Sherman made its combat debut as a Lend-Lease weapon with the British Eighth Army at El Alamein in October 1942, it became readily apparent that its 75mm (3in) main cannon was sufficient to fight it out with the early PzKpfw III and IV tanks of the Panzer Armee Afrika.

During the course of the war, several variants of the basic Sherman were produced, including the M4A1, M4A2, M4A3 and M4A4, primarily distinguished by their varying powerplants and some changes to hull and body configurations. The hull of the M4A1 was fully cast rather than a combination of cast and welded construction, and the vertical volute suspension system that equipped earlier variants was replaced in the M4A3 with a horizontal arrangement. With the M4A4, the hull was lengthened somewhat to make room for the Chrysler A57 multi-bank engine, which was actually the combination of five six-cylinder automobile engines performing in unison. Other engines included the Ford GAA III, Caterpillar nine-cylinder diesel, Continental R975, and the Wright Whirlwind.

Bigger bang

In combat, the early Sherman was soon hopelessly outclassed by a new generation of German tanks. The PzKpfw V Panther, armed with a high-velocity 75mm (3in) cannon, and the PzKpfw VI Tiger and Tiger II, with powerful 88mm (3½in) weapons, were capable of dispatching the Sherman from great distances, sometimes more than a mile, well before the Allied tank could fire a shot at anything approaching an effective range. Moreover, the low muzzle velocity of the 75mm (3in) gun failed to produce enough penetrating power to damage the hull armour of the Panther or Tiger.

Often, an entire platoon of four or five Sherman tanks were required to take on a single Tiger in open combat, and it was expected that two of the Shermans would fall victim to the Tiger's '88' before one Sherman was able to manœuvre into an advantageous firing position to the rear, or from a

SHERMANS DEPLOYED
Offloading from an LST (Landing Ship, Tank), an M4 Sherman rolls forward with other vehicles of its armoured unit, followed closely by a second tank just exiting the vessel's hold. The Sherman was easily recognized due to its high silhouette.

high or low angle into the areas where the Tiger's armour protection was most thin.

In response to the introduction of heavier tanks by the Germans, some notable attempts were made to improve the firepower of the Sherman, including the enlargement of the turrets on later models to accommodate the 76mm (3in) high-velocity gun, increasing the effective range of the main weapon substantially. In 1943, the British opted to install the Ordnance QF 17-pdr (76mm/3in) gun atop the chassis of the M4A4, providing an improvement in effective range from the minimal 180m (200 yards) of the old 75mm (3in) to 1,830m (2,000 yards), roughly equivalent to that of the German Tiger.

Sherman superlatives

Despite its shortcomings, the Sherman did possess two distinct advantages. Its top speed of 47km/h (29mph) provided excellent manoeuvrability, somewhat negating the lethal advantage in German firepower, the inviting target the high silhouette of the Sherman provided to enemy gunners, and its thin armour protection. Much more telling, however, was its sheer weight of numbers deployed on the battlefields of North Africa, Italy and Western Europe. American industrial capacity could easily make good the heavy losses suffered by Allied armoured units, while German engineers had opted for quality over quantity and found that production could not keep pace with battle losses or mechanical problems in the Panther and Tiger that could not be resolved prior to deployment due to wartime demands.

Among the more intriguing variants of the Sherman were specialized tanks that were equipped with such features as a flail system for detonating land mines, light bridging equipment, a heavy spigot mortar, a flamethrower and other innovations that were developed by British General Percy Hobart and tested through his 79th Armoured Division. The most famous of Hobart's 'Funnies' as they were called, was the Duplex Drive (DD) Sherman, equipped with a canvas flotation screen and intended for amphibious deployment to provide fire support to infantry landing in Normandy on D-Day, 6 June 1944.

While the Sherman eventually overwhelmed its adversaries in Western Europe, it was clearly superior to any tank the Japanese deployed in the Pacific. Although its operations were often restricted by inhospitable terrain, the Sherman proved invaluable during close-quarter fighting against fortifications on Japanese-held islands.

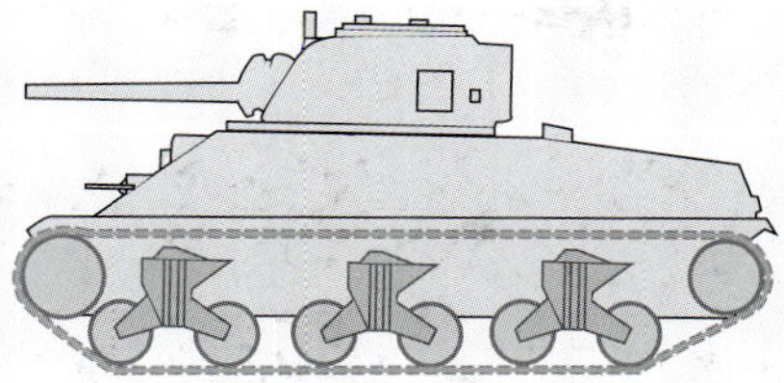

Sherman M4

Manufactured	1941
No. built	49,234
Weight	53.4 tonnes
Length	5.89 m
Crew	5
Armour	40–68 mm
Main Armament	75 mm M3 L/40
Secondary Armament	0.50 cal. HMG
Speed	30 mph (48 kph)

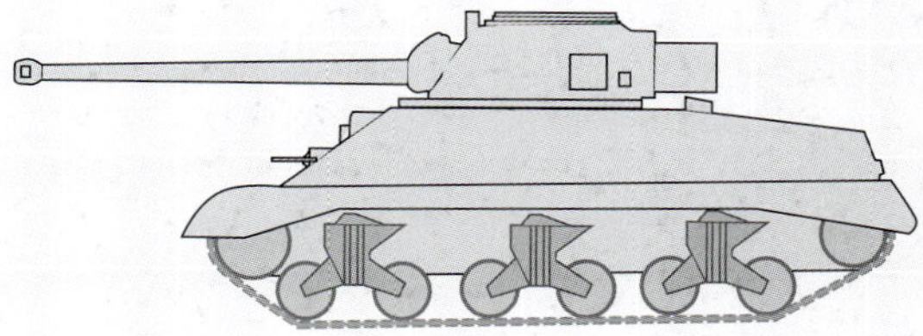

Sherman Firefly

Manufactured	1943
No. built	c. 2,400 (converted from standard M4)
Weight	53.4 tons
Length	5.89 m
Crew	5
Armour	89–40 mm
Main Armament	QQF 17-pounder
Secondary Armament	0.50 cal. HMG
Speed	30 mph (48 kph)

SHERMAN VARIANTS
The early versions of the M4 Sherman tank were armed with a rather ineffective 75mm (3in) cannon, generally unable to penetrate the armour of German tanks. The British Firefly introduced higher-velocity guns such as the 17-pounder.

RANGE IMPROVEMENTS
With 10 times the range of the earlier low-velocity 75mm (3in) gun, the Firefly's 17-pounder gave the Allied Sherman a fighting chance in tank-versus-tank combat with the German Tiger and Panther.

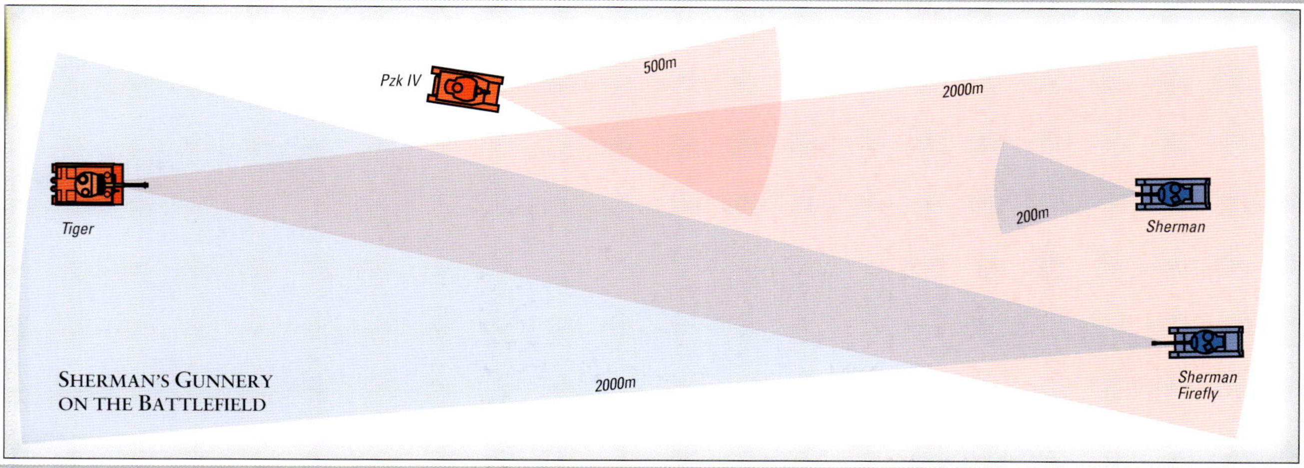

SHERMAN'S GUNNERY ON THE BATTLEFIELD

D-DAY, THE ASSAULT

DRESS REHEARSAL
A British Churchill tank lies abandoned at Dieppe, France, in 1942, as German troops mill about on the beach. Lessons learned during the disastrous raid on Dieppe were put into practice during the D-Day landings in Normandy two years later.

The Allied officers who planned the D-Day assault, the long-awaited invasion of Hitler's 'Fortress Europe', realized that the availability of mobile firepower on the beaches of Normandy and beyond would play a key role in the success of the invasion, codenamed Operation Overlord, on 6 June 1944. However, the difficulty inherent in the deployment of tanks on to a contested beachhead and their ability to operate in support of an infantry assault had been made abundantly clear during the disastrous Dieppe Raid of August 1942.

British General Percy Hobart met the problem head-on. Commanding the 79th Armoured Division, he led the development of numerous specialized armoured fighting vehicles, the most notable of which was the DD (Duplex Drive) Sherman tank, which employed a canvas flotation and water-borne propulsion system that was disengaged when the tank reached land. Once Allied troops had gained a satisfactory foothold in Normandy, conventional tanks could be offloaded directly on to the invasion beaches and later in the major ports which were liberated as the offensive progressed.

Seaborne Shermans

For Overlord, 10 battalions of DD Shermans were deployed. Theoretically, they would be released from landing craft roughly 3km (2 miles) offshore and proceed directly to the beaches to engage German

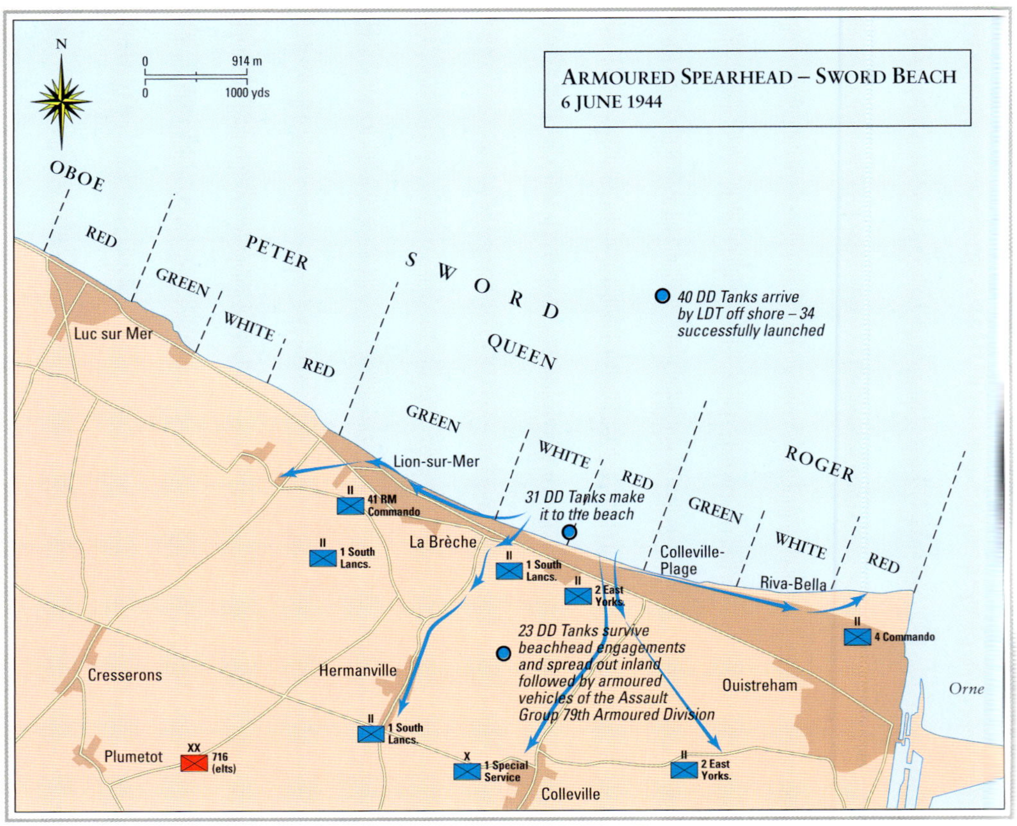

INVASION SUPPORT
Specially developed DD (Duplex Drive) Sherman tanks were deployed with varying degrees of success in support of the infantry landing on the beaches of Normandy on 6 June 1944. Heavy seas swamped a number of the tanks.

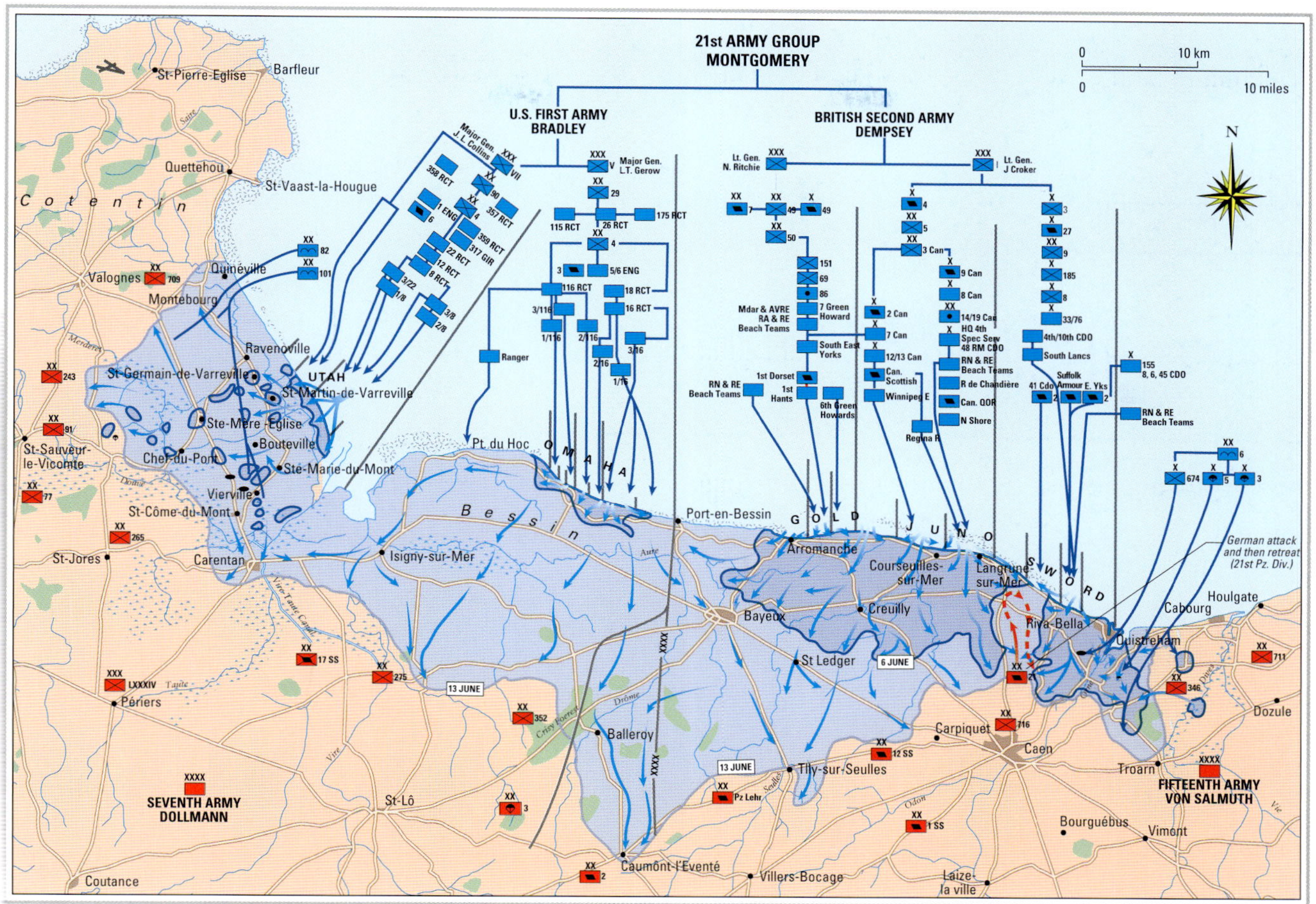

ALLIED PROGRESS
During the first week following the Normandy landings, Allied troops were able to move inland; however, weeks of hard fighting remained before the eventual breakout from the Norman hedgerows and into open country.

positions. In the event, the DD tanks at British Sword and Gold beaches performed reasonably well despite rough seas, with eight tanks swamped and sunk off Gold, and several others falling victim to German anti-tank guns. Nevertheless, the surviving tanks of the Sherwood Rangers Yeomanry and 7th Royal Dragoon Guards took on strongpoints and assisted the infantry in consolidating and then expanding the beachhead. On Juno Beach, Canadian DD tanks of the 1st Hussars supported the 7th Infantry Brigade as it moved inland, with 21 of a complement of 29 tanks successfully deploying from the beach. Rough seas prevented some tanks from landing.

The American forces at Utah Beach encountered little opposition, particularly due to the fact that the assault came ashore some distance from its designated landing position. Although Nazi gunfire sank one transport, along with its cargo of four DD tanks, 27 tanks did reach the beach against little opposition.

The most hotly contested of the invasion beaches was Omaha, and there elements of the 1st and 29th US Infantry Divisions clung desperately to a sliver of ground under heavy enemy fire. The situation became so tenuous that General Omar Bradley, commander of American forces for the invasion, considered withdrawing from Omaha. Compounding the problems for the embattled Americans was the ordeal of the 64 DD Shermans of the 741st and 743rd Tank Battalions. Twenty-nine DD Shermans of the 741st Battalion were launched a full 5km (3 miles) from shore; 27 of these foundered and sank to the bottom of the English Channel. Therefore, the remaining tanks of the 741st and the entire armoured complement of the 743rd were landed directly on the beach as the situation at Omaha improved.

Defensive disagreement

Allied operations on D-Day were actually aided by a disagreement among top German commanders as to the deployment of armoured reserves. Field Marshal Erwin Rommel, commanding the defences in Normandy, advocated the rapid deployment of armoured divisions against the invasion beaches to drive the enemy into the sea. Field Marshal Gerd von Rundstedt, overall commander in the West,

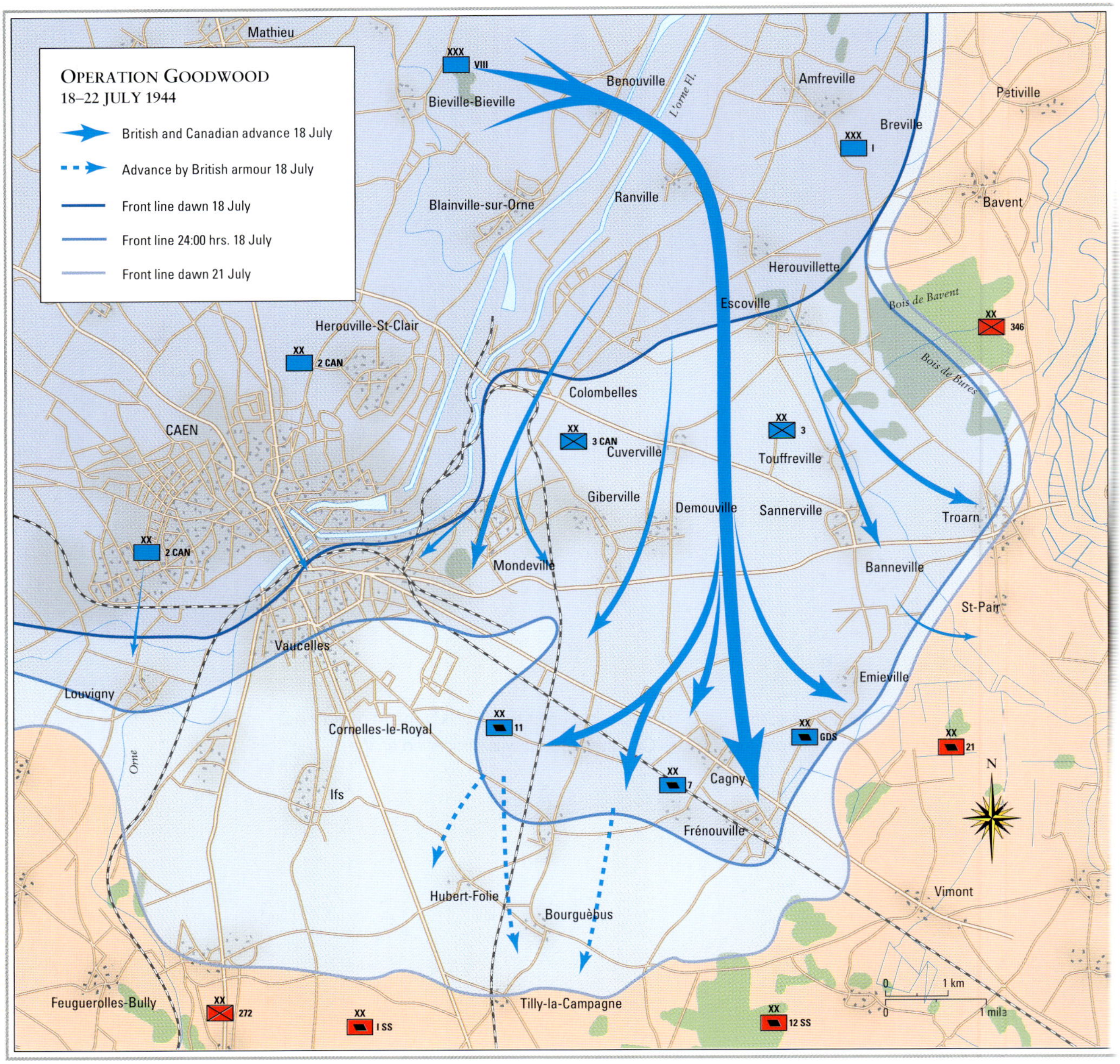

GOODWOOD GAINS
Although Allied planners had intended to capture Caen on D-Day, the Norman town remained in German hands for nearly a month. Several Allied attempts to capture Caen were repulsed. Operation Goodwood finally succeeded but at a high cost.

preferred to hold most Panzer formations in reserve to attack and annihilate the Allies once they reached the continent in large numbers. Furthermore, Hitler himself retained control of the majority of the German armoured divisions and declared that they were not to be committed without his direct order.

On 6 June Hitler refused to believe that Normandy, not the Pas de Calais, was the location for the 'real' invasion. The bulk of German armour was held in reserve. Only elements of the 21st Panzer Division were able to mount any significant armoured counterthrust against the Allies on D-Day, and this did not occur until late in the afternoon.

Around 16.00hrs, the attack commenced. German tanks reached the Channel coast at Lion-sur-Mer, but by that time daylight was fading, and no significant reserves were available to exploit the success. The best opportunity to defeat the Allies in the West had slipped away.

Capturing Caen

Perhaps overly ambitious, Allied commanders had expected to capture the communications and transport centre of Caen on D-Day. However, due to heavy German resistance and delays encountered at Sword Beach, British troops were unsuccessful in seizing this objective. Caen was to remain in German control for six weeks, and a series of bloody engagements, including Operations Epsom and Charnwood, occurred prior to the capture of the town during Operation Goodwood, 17-20 July.

In the vicinity of Caen, the British VIII Corps, under General Richard O'Connor, launched elements of the 7th, 11th and Guards Armoured Divisions, while I Corps, commanded by General J.T. Crocker, deployed the 3rd and 51st Infantry Divisions to battle stubborn German troops of the 1st SS Panzer Division Liebstandarte Adolf Hitler, 12th SS Panzer Division Hitler Youth, and the 21st Panzer Division.

Stunned by heavy Allied bombing, the Germans quickly recovered and vicious fighting erupted. The 11th Armoured Division alone had 126 tanks and armoured vehicles destroyed or damaged, while German forces lost a number of heavy Tiger tanks to air attack and massed gunnery from the smaller but highly manoeuvrable British Shermans.

British forces advanced approximately 11km (7 miles) to the vicinity of Bourguébus Ridge, but armoured formations of the VIII Corps lost some 200 tanks and more than 600 casualties during Operation Goodwood. Caen had been secured and the bridgehead across the River Orne expanded; however, controversy persists to this day as to the scope of Goodwood. Some historians insist that Goodwood was intended as a breakout from the *bocage*, or hedgerow, country of Normandy, while others believe that Goodwood succeeded in its stated objective – the capture of Caen.

SHERMAN'S STRUGGLE
A Sherman tank, its turret hatch open, lies abandoned on Omaha Beach shortly after D-Day. The DD Shermans intended to support the Normandy landings struggled in high surf, and several were swamped before they reached the beach.

Operation Cobra

While it is academic as to whether Operation Goodwood was completely successful, it must be acknowledged that British and Canadian forces confronted the bulk of German armour, more than six divisions, during the frustrating Allied effort to break out of the Norman hedgerow country. Only 11 understrength divisions, scarcely two of them armoured, faced the Americans to the south, prompting Allied commanders to shift the energy of the breakout effort to the U.S. zone of operations.

Operation Cobra was conceived as the sledgehammer blow that would break the potential stalemate in Normandy once and for all, possibly restoring the Allied timetable, which had fallen

AMPHIBIOUS ACTION
Loaded with British infantrymen, amphibious DUKW vehicles, commonly known as Ducks, operate in shallow water off the Normandy coast. The DUKW was instrumental in providing mobility to Allied troops across a variety of terrain.

OPERATION COBRA
25–31 JULY 1944

Allied front line 25 July
Allied front line 28 July
Allied front line 31 July

OPERATION COBRA
In July 1944, the Allies launched Operation Cobra as a ground offensive followed saturation bombing to break through German lines and clear the hedgerows of Normandy. The subsequent Allied dash across France liberated large areas of the country.

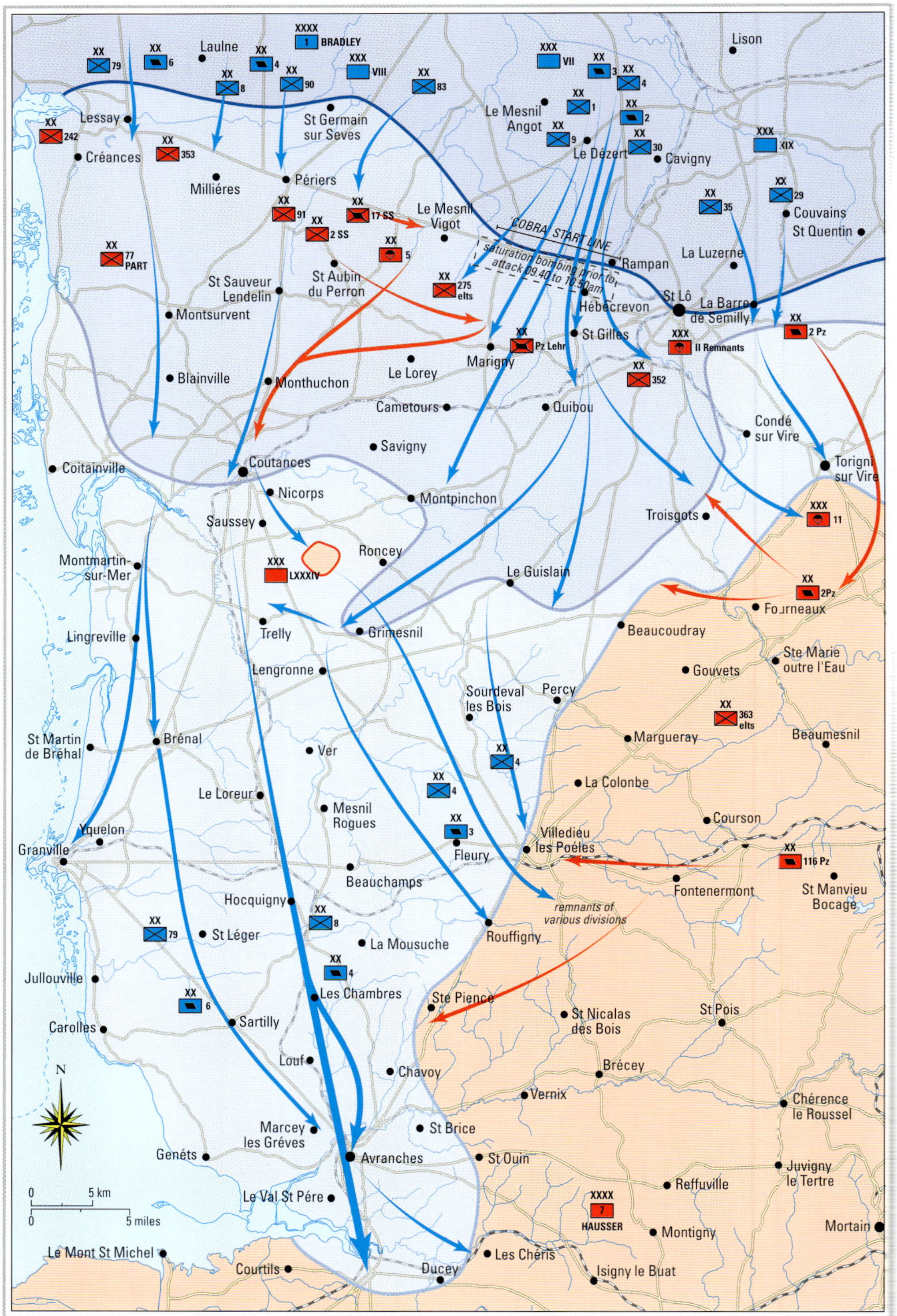

agonizingly behind. On the heels of Operation Goodwood, hundreds of Allied aircraft were to mount a saturation bombing effort along a 6km (4-mile) front. Subsequently, two corps of the U.S. First Army, the VII and XIX, would hurl their armour and infantry through the breach in German lines, racing into open country.

On 25 July 1944, more than 3,000 US bombers attacked German positions, decimating some formations of the Panzer Lehr Division. Remnants

Panzer Lehr and the 5th Parachute Division fought the advancing Americans, and on the following day the U.S. 1st Infantry and 2nd Armored Divisions entered the battle. Later, the 8th and 90th Infantry Divisions of VIII Corps joined in.

Within 72 hours, U.S. formations had cleared much of the organized German resistance and elements of three Army corps had begun to advance rapidly. German units faced potential encirclement, and the 2nd SS Panzer and 17th SS Panzergrenadier divisions launched counterattacks. In less than a week, the Americans had transformed the war in the West from stagnation to swift movement. On 1 August the U.S. Third Army, under the flamboyant General George S. Patton, was activated, and soon seven fresh divisions were rolling into southern Normandy and Brittany. British and Canadian forces moved efficiently southward, with the Canadians intent on trapping thousands of German troops in Normandy.

The crucible of Falaise

By the middle of August, Patton's Third Army had swept through Avranches and reached Argentan.

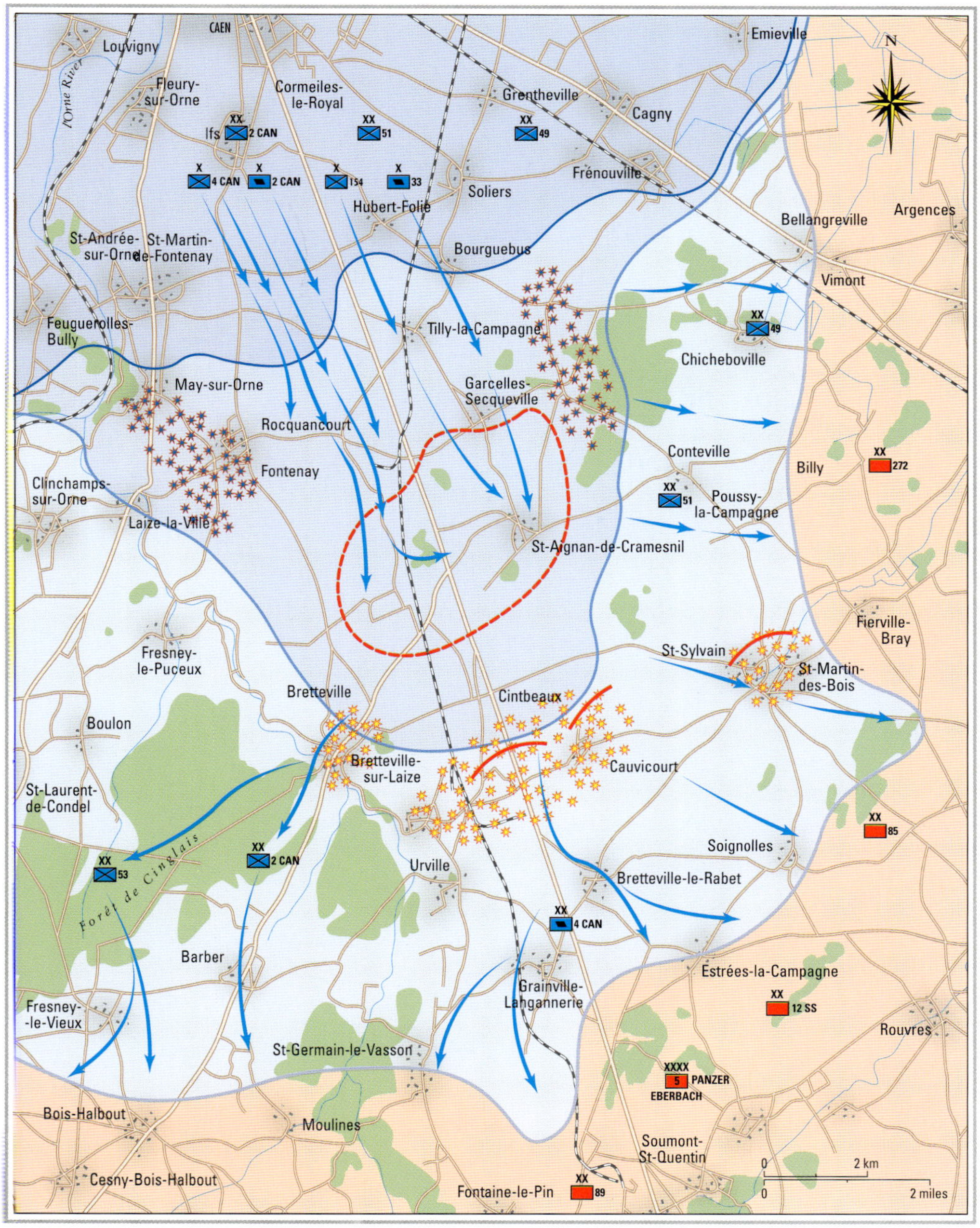

OPERATION TOTALIZE
7–11 AUGUST 1944

- Allied front 7 August
- Allied front 8 August
- Allied front 11 August
- German reserve positions
- German defence zone
- Night bomber targets 7 August
- Day bomber targets 8 August

OPERATION TOTALIZE
In August 1944, an Allied thrust led primarily by Canadian troops attempted to close the ring around thousands of German troops retreating from Normandy. The drive to Falaise resulted in the capture of at least 50,000 enemy soldiers.

WITTMANN'S DEATH CHARGE
8 AUGUST 1944

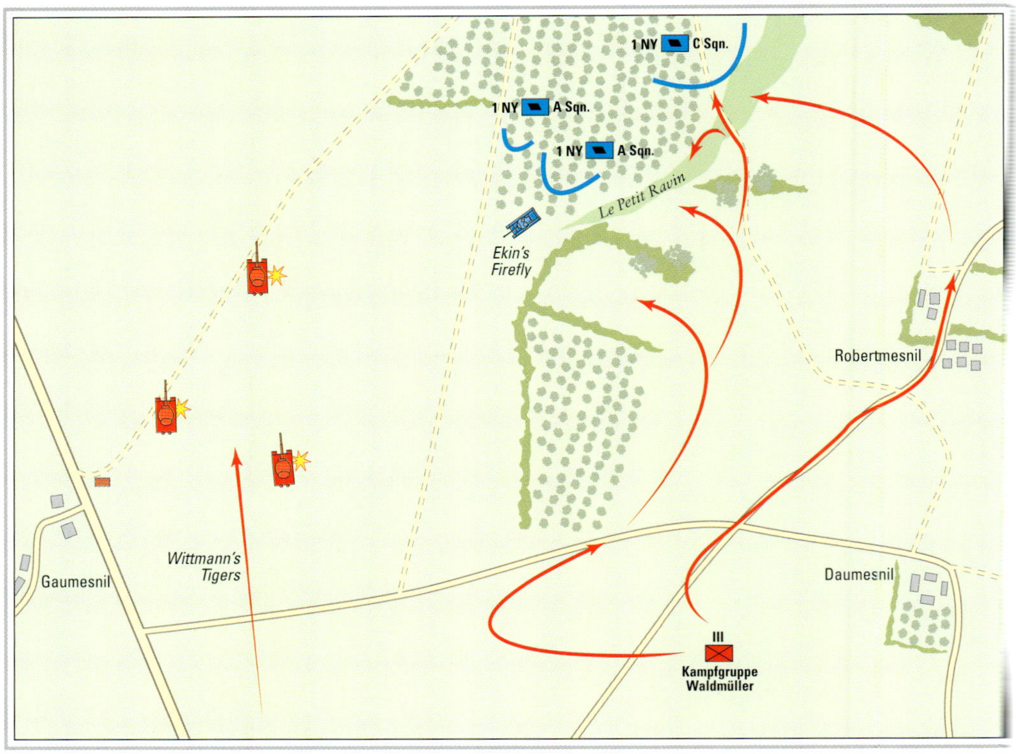

WITTMAN'S DEATH
Some controversy persists as to how German Tiger tank ace Michael Wittman met his death. He was killed in action near the town of Cintheaux, France, while fighting elements of the 1st Northamptonshire Yeomanry and Sherbrooke Fusiliers.

Troops of the U.S. First Army, particularly the 30th Infantry Division, had blunted the German counteroffensive known as Operation Lüttich at Mortain. Meanwhile, the British Second Army held firm and the advance of the Canadian First Army was painfully slow as the northern hinge of the trap closed around the village of Falaise. Commanded by General Henry Crerar, the First Canadian Army was a polyglot force, including the 2nd French, 1st Polish and 4th Canadian Armoured Divisions, along with the U.S. 90th and Canadian 2nd Infantry Divisions.

On 8 August Crerar launched Operation Totalize to seize high ground north of Falaise and shut off the escape route of the German Seventh and Fifth Panzer Armies. Preceded by heavy air bombardment, the Allied forces advanced slowly against the 12th SS Panzer Division and Tiger tanks of the 101st SS Heavy Tank Battalion. Canadian and Polish forces failed to reach Falaise, but took control of several commanding positions in the vicinity. When Totalize lost momentum, infantry units were moved forward to consolidate the gains. However, the gap remained unclosed.

Operation Tractable, the final push to slam the door at Falaise, began on 14 August, with assaults by the Canadian 4th and Polish 1st Armoured Divisions. Counterattacks by German Tiger tanks of Heavy SS Panzer Battalion 102 slowed progress. By 17 August, German Field Marshal Walther

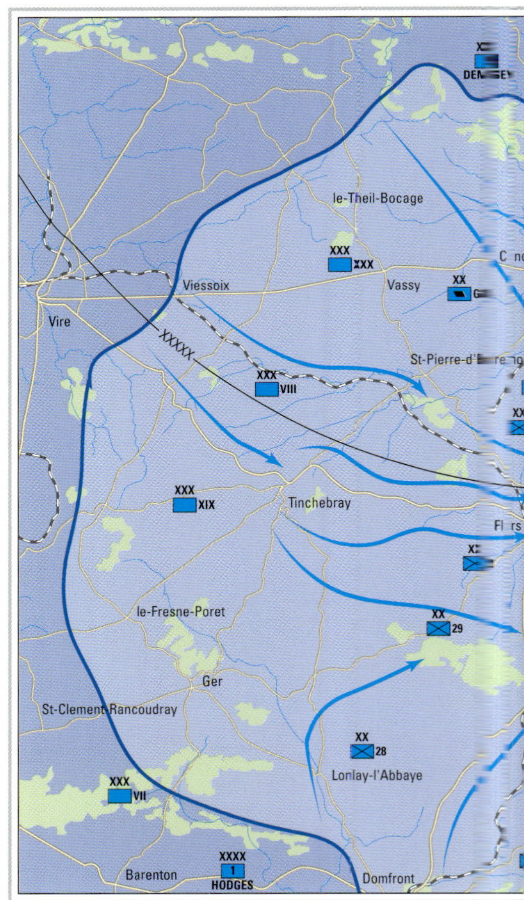

Model had ordered a general retreat from Falaise while elements of the II SS Panzer Corps sacrificed themselves to keep the narrow corridor open, and troops of the 2nd Canadian Infantry Division had occupied the town.

Three days later, Canadian troops captured the village of Trun and Allied forces had actually completed their encirclement of the Germans around Falaise as combat commands of the 1st Polish Division advanced ahead of the main First Army thrust. However, several avenues of escape remained open. Particularly troublesome was a corridor near Polish positions established on Mount Ormel Ridge. Units of the Polish 1st Armoured Division made tenuous contact with American forces, but were not of sufficient strength to prevent elements of the 2nd, 9th, 10th, and 12th SS Panzer Divisions and the 116th Infantry Division from forcing a road open for six crucial hours, allowing several thousand German soldiers to escape.

At long last, the Falaise Pocket was closed as daylight waned on 21 August. The Polish troops had withstood numerous counterattacks along Mount Ormel Ridge, raining artillery fire on the retreating Germans below, although they had been unable to completely cut off the line of retreat. The Polish position served as a base with which the remainder of the First Army could link up, and tanks of the Canadian 4th Armoured Division reached Polish troops at Coudehard to close the noose.

Fearful toll

While the attempt to close the Falaise Gap may have failed to realize its full potential, allowing thousands of German troops to escape eastward and fight another day, estimates of the number of enemy soldiers killed or captured at Falaise approach 100,000, with more than 15,000 dead. The Germans had been repeatedly pounded by Allied tactical air forces. Their combat effectiveness had been severely degraded, and their losses of nearly 350 tanks, 2,400 other vehicles and 250 artillery pieces were irreplaceable. Several combat divisions had been reduced to less than 70 percent of their standard strength in men and guns. On 25 August, Allied forces liberated Paris and Operation Overlord came to a close.

FALAISE ENCIRCLEMENT
Commonwealth attempts to close the northern shoulder of the trap at Falaise were frustrating slow, while the southern shoulder held firm near Argentan. Pounded by artillery and air attacks, thousands of German troops managed to escape capture.

CLOSING THE FALAISE POCKET
16 AUGUST 1944

Allied advance
Front line 13 August
Front line 16 August
Front line 19 August
German retreat

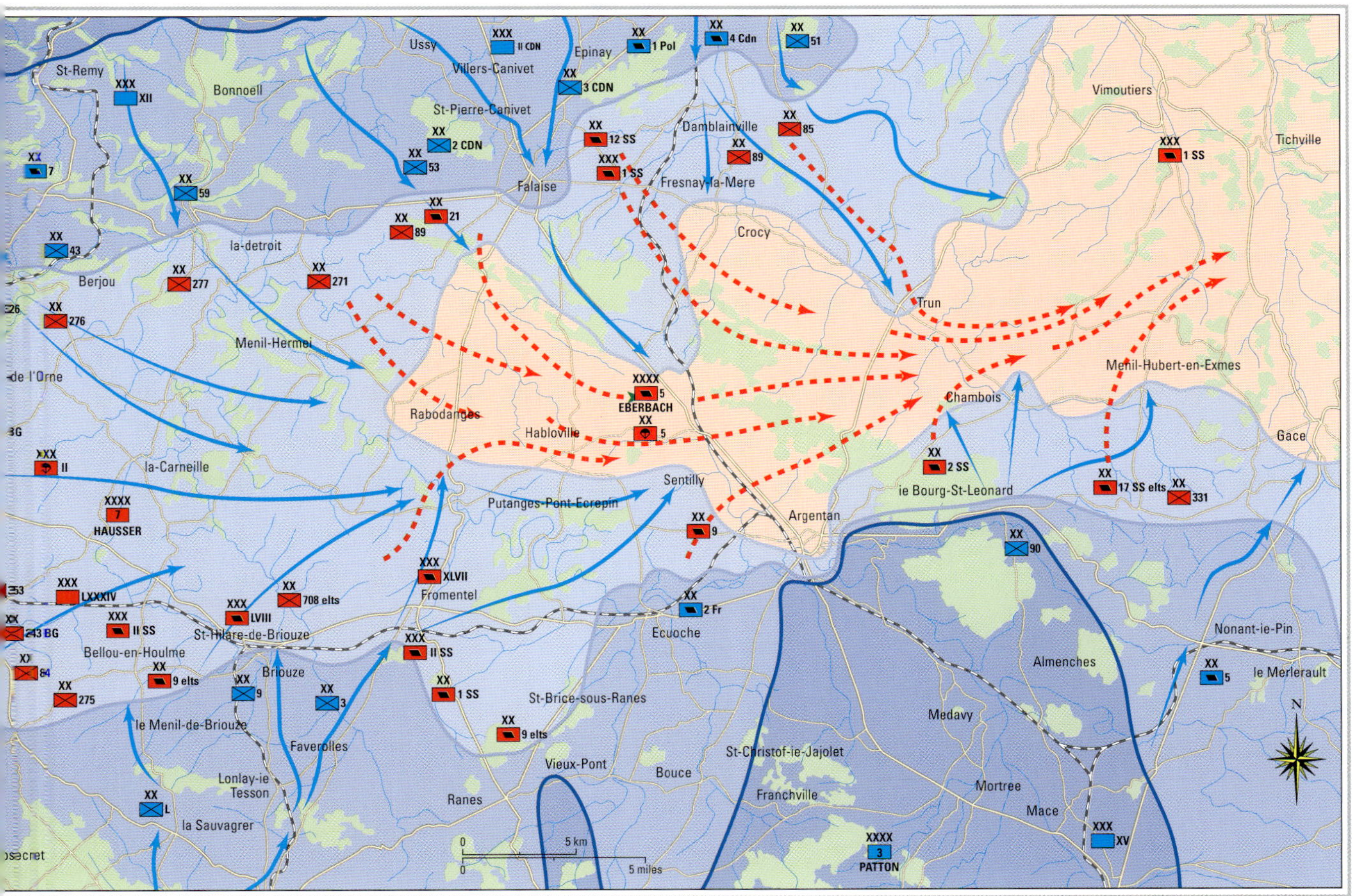

MARKET GARDEN 1944

CAUTIOUS APPROACH
Utilising a Sherman tank for cover, Allied infantrymen advance cautiously into the rubble of a town somewhere in Western Europe. The advance of XXX Corps during Operation Market Garden was slowed by a single road and strong resistance.

As the Allied broad front strategy breached the frontier of the Third Reich in the autumn of 1944, British Field Marshal Bernard Montgomery conceived a combined airborne and ground operation that could potentially end World War II by Christmas.

Not known for risk-taking, Montgomery nevertheless advocated Operation Market Garden, an ambitious offensive involving three Allied airborne divisions, the British 1st and the American 82nd and 101st, to seize key bridges across the Maas, Waal and Neder Rhine rivers in the Netherlands, holding the spans until relieved by armour and troops of British XXX Corps, theoretically racing hell-for-leather to consolidate these positions. Allied troops could then flood into the Ruhr, the industrial heart of Germany, and force a Nazi surrender.

Painful progress

Air and ground operations commenced on 17 September 1944, and initially made good progress. Key bridges at Eindhoven and Nijmegen were captured. However, the grand prize was the bridge across the Neder Rhine at Arnhem, 100km (63 miles) from the Allied front. U.S. paratroopers fought to keep the XXX Corps route, dubbed 'Hell's Highway' open, but the ground advance proved painfully slow. Tanks and half-tracks of XXX Corps were confined to the narrow road, elevated above marshy terrain on either side.

As German troops roused to block the armoured advance, the high profiles of the British Sherman tanks were silhouetted against the sun, perfect targets for German anti-tank weapons. The disabling of a single tank required a deployment of infantry to clear the way and the removal of a wrecked vehicle that blocked an entire column. As the timetable lagged, the 6th Airborne situation grew more tenuous, particularly at Arnhem.

Arnhem

Soon after reaching its objective, the 2nd Battalion, Parachute Regiment, under Lieutenant Colonel John Frost, took up positions at the north end of Arnhem bridge. Prior to Market Garden, Montgomery and other Allied commanders discounted aerial reconnaissance photos which indicated a German armoured presence in the Arnhem area. Actually, the 6th Airborne had landed virtually on top of the IX and X SS Panzer Divisions sent for refitting after being mauled in Normandy.

Frost's lightly armed paras, numbering some 740 men, held out for nearly four days against German armoured counterattacks, sustaining heavy casualties before being compelled to surrender. The 1st Airborne Division began Market Garden with approximately 10,000 troops, and roughly 80 percent of these became casualties. The costly failure of Operation Market Garden is popularly remembered in the film, *A Bridge Too Far*.

CHAFFEE CHALLENGE
The M24 Chaffee light tank was introduced late in World War II with the U.S. Army, replacing the earlier Stuart models. The Chaffee's firepower was substantial by earlier standards, with a 75mm (3in) main gun.

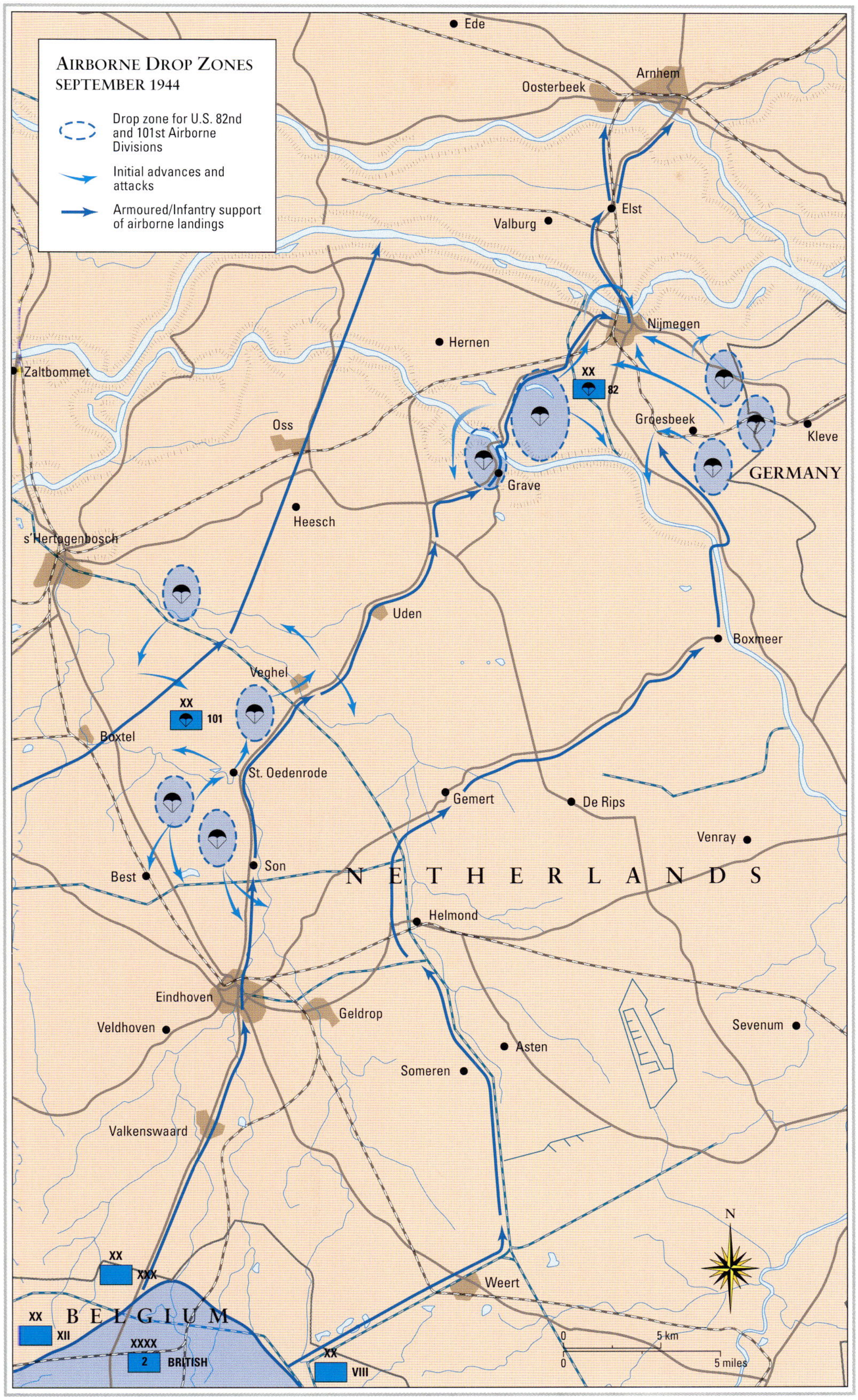

MARKET GARDEN AIRDROPS
The initial airborne phase of Operation Market Garden proceeded fairly well; however, the situation rapidly deteriorated at Arnhem, where elements of the British 1st Airborne ran into two German SS Panzer divisions.

CHINA 1937-45

With the inception of the Second Sino-Japanese War in 1937, Chinese forces fielded only the equivalent of three armoured battalions, and these consisted of obsolete Vickers tanks constructed in Britain, vulnerable Italian CV33 tankettes, and the German Panzerkampfwagen I, armed only with machine guns.

Although Japanese armoured forces were not substantial, the 37mm (1½in) main weapon of the Type 95 Ha-Go light tank and the 57mm (2¼in) cannon of the Type 97 Chi-Ha medium tank were far superior to the armament of the light Chinese armoured vehicles. Tank-versus-tank encounters between Japanese and Chinese tanks were rare; therefore, the Japanese armour followed traditional military doctrine and functioned primarily in an infantry support role.

Soviet surprise

One of the most significant armoured battles on the Asian mainland took place in the summer of 1939 at Nomonhan, also known as Khalkin Gol, in Manchuria. During an undeclared border war with the Soviet Union, senior commanders of the Japanese Sixth Army authorized an operation in early July to remove the threat of Soviet intervention against Japanese expansion in China and Manchuria.

The Japanese committed large numbers of infantry and the armoured vehicles of their 3rd and 4th Tank Regiments. These units included roughly 85 tanks and tankettes of various fighting quality. Among them was the elderly Type 89 medium tank, developed in the 1920s and already considered obsolete at the outbreak of war with China.

Launching a two-pronged offensive, the Japanese concentrated most of their armour on the southern front, and the Red Army contingent commanded by General Georgi Zhukov, launched a counterattack with more than 400 tanks, nearly surrounding the bulk of Japanese forces in the north. Moving at night, a force of Japanese tanks counterattacked at daylight on 2 July. The Japanese armoured force was devastated, losing more than half its tanks. A week later, the unit had been so badly mauled that it was disbanded.

In August, Zhukov retaliated. Supported by heavy artillery bombardment and air cover, three divisions of Red Army troops and a battalion of tanks struck the Japanese positions. Rapidly moving Soviet armour swept around both Japanese flanks, executing a magnificent double envelopment. Japanese forces that had occupied the Soviet side of the disputed Manchurian border were virtually annihilated.

Having deployed only two armoured regiments at Nomonhan, the Japanese armoured formations were hopelessly outnumbered by the Soviets. Compounding their error, senior Japanese war planners failed to heed this lesson in armoured warfare and seldom deployed significant tank forces during the subsequent war with the United States.

Soviet armoured assets included the BT-7 light tank, of which more than 5,000 were produced from 1935 to 1940. The BT-7 mounted a 47mm (1¾in) main cannon in a traversing turret, along with a pair of 7.62mm (⅓in) machine guns. Although it had been rendered obsolete by Western standards by the outbreak of World War II, the BT-7 continued to serve with the Red Army on the Eastern Front and against the Japanese in 1945.

JAPANESE ARMOUR
Soldiers of the Chinese People's Liberation Army occupy captured Japanese Type 97 tanks. Following the end of World War II, both communist and nationalist forces utilized equipment abandoned by the defeated Japanese army.

War in China July 1937 – December 1944

- Territory seized to July 1937
- Territory held by end of 1941
- Territory held by December 1944 after Ichi-go Operations
- Territory controlled by Communists from 1935
- Major Chinese advances 1937–41
- Operation Ichi-go lines of advance
- Japanese conflicts with USSR 1939

JAPANESE EXPANSION
The Japanese began their expansion into China in 1931 and by 1944 had captured large coastal areas and pushed into the country's interior with limited success. The vastness of China frustrated the Japanese.

BAGRATION 1944

STUG STALWARTS
A column of German Sturmgeschütz III self-propelled assault guns advances along a dirt road in Russia. The Sturmgeschütz was effective in the infantry support and anti-tank roles and cheaper to build than a standard tank.

OPERATION BAGRATION
(Opposite) In the summer of 1944, the Soviet Union launched Operation Bagration, a massive offensive that cleared the Germans from Belorussia and Eastern Poland. Estimates of German casualties during two months of heavy fighting run as high as half a million.

PANZERFAUST POWER
The shoulder-fired anti-tank Panzerfaust was a standard-issue weapon of the German Army during World War II. Its hollow-charge warhead was innovative and effective against heavy Allied armoured vehicles.

On 22 June 1944 (the third anniversary of the German Operation Barbarossa invasion of the Soviet Union), four powerful Soviet Front commands initiated the Bagration offensive against the German Army Group Centre, which held a defensive line that ran roughly north-south through eastern White Russia. Within the first two days the Red Army had achieved several decisive ruptures of the German line. Subsequently, Soviet armoured spearheads raced west deep into the German rear areas to capture the city of Minsk on 4 July. This accomplishment encircled most of the German Fourth Army, plus elements of the Ninth Army, in the Bobruisk-Rogachev region. Further north, meanwhile, forces from the First Baltic and Third White Russian Fronts (or armies) had captured the important towns of Vitebsk and Polotsk by 4 July. During the rest of July, amid the offensive's next main phase, Soviet armour drove rapidly west against weak German resistance to capture the cities of Brest-Litovsk, Grodno, Vilnius and Dvinsk.

Reclaiming Mother Russia

Encouraged by these successes, Soviet flanking forces had initiated powerful supporting operations that significantly widened the frontage of attack. In the north, Soviet forces drove back the German Army Group North from the areas around Pskov and Daugapils. By 21 August, these Soviet advances had left Army Group North, still defending Estonia, only precariously connected to the rest of the German front by a narrow land corridor located southwest of Riga in Latvia. Meanwhile in the south, the First White Russian Front attacked west from the region south of Pinsk to capture Lublin in the General-Gouvernement of Poland by advancing to

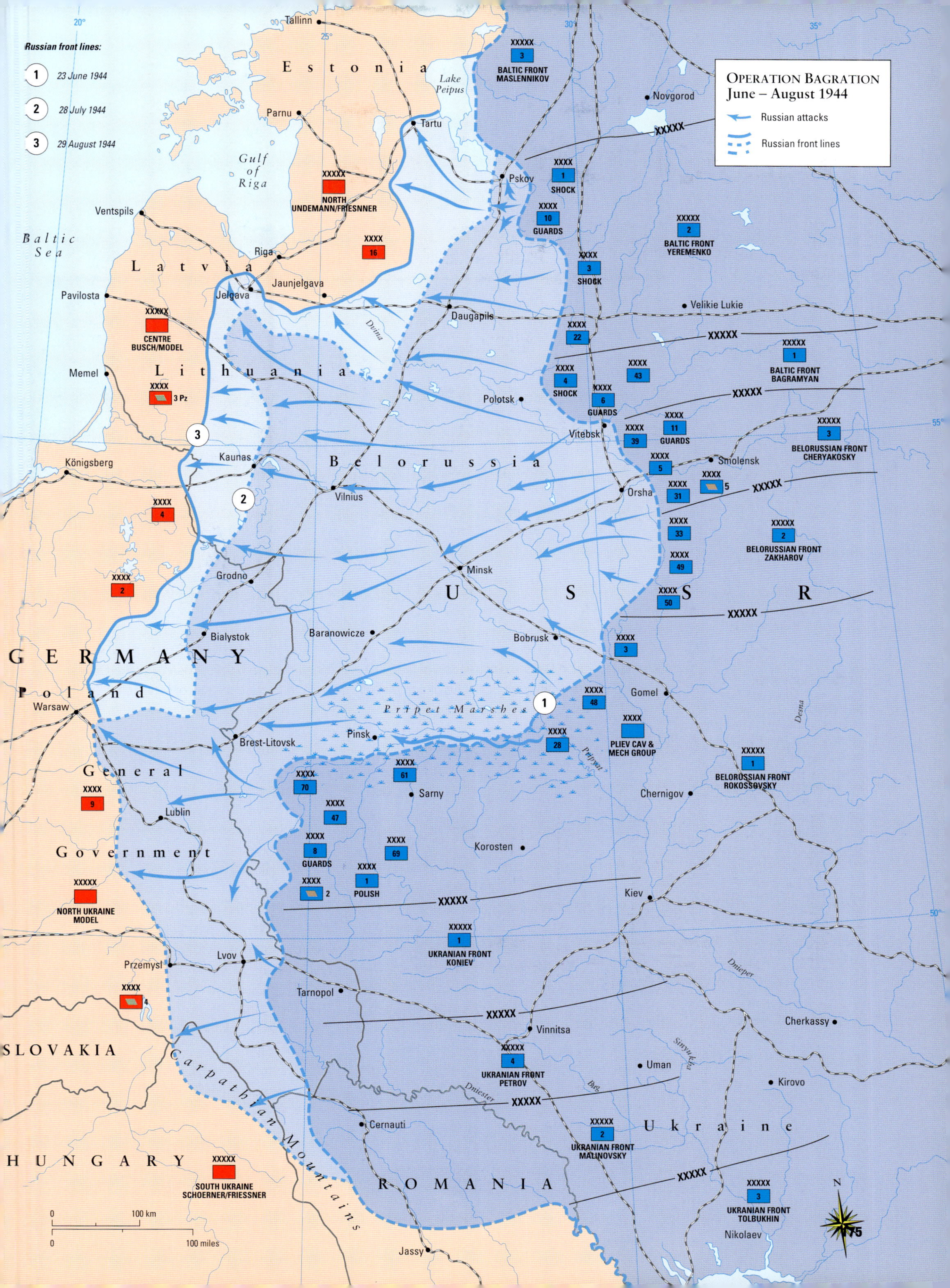

Operation Bagration
June – August 1944
Russian attacks
Russian front lines
Russian front lines:
1 23 June 1944
2 28 July 1944
3 29 August 1944
Estonia
Latvia
Lithuania
Belorussia
U S S R
GERMANY
Poland
General Government
SLOVAKIA
HUNGARY
ROMANIA
Ukraine
Baltic Sea
Gulf of Riga
Lake Peipus
Pripet Marshes
Carpathian Mountains
Dvina
Pripyat
Desna
Dnieper
Dniester
Bug
Sinyukha
Tallinn
Parnu
Tartu
Pskov
Novgorod
Ventspils
Riga
Jaunjelgava
Jelgava
Pavilosta
Daugapils
Velikie Lukie
Memel
Polotsk
Vitebsk
Smolensk
Königsberg
Kaunas
Vilnius
Orsha
Minsk
Grodno
Bialystok
Baranowicze
Bobrusk
Gomel
Warsaw
Brest-Litovsk
Pinsk
Sarny
Chernigov
Lublin
Korosten
Kiev
Przemysl
Lvov
Tarnopol
Vinnitsa
Uman
Cherkassy
Kirovo
Cernauti
Jassy
Nikolaev
XXXXX 3 BALTIC FRONT MASLENNIKOV
XXXXX NORTH UNDEMANN/FRIESNER
XXXX 16
XXXX 1 SHOCK
XXXX 10 GUARDS
XXXXX 2 BALTIC FRONT YEREMENKO
XXXX 3 SHOCK
XXXXX CENTRE BUSCH/MODEL
XXXX 22
XXXX 3 Pz
XXXX 4 SHOCK
XXXX 43
XXXXX 1 BALTIC FRONT BAGRAMYAN
XXXX 6 GUARDS
XXXX 39
XXXX 11 GUARDS
XXXXX 3 BELORUSSIAN FRONT CHERYAKOSKY
XXXX 5
XXXX 5
XXXX 31
XXXX 4
XXXX 33
XXXXX 2 BELORUSSIAN FRONT ZAKHAROV
XXXX 49
XXXX 2
XXXX 50
XXXX 3
XXXX 48
XXXX PLIEV CAV & MECH GROUP
XXXX 28
XXXXX 1 BELORUSSIAN FRONT ROKOSSOVSKY
XXXX 70
XXXX 61
XXXX 9
XXXX 47
XXXX 8 GUARDS
XXXX 69
XXXX 1 POLISH
XXXX 2
XXXXX NORTH UKRAINE MODEL
XXXXX 1 UKRANIAN FRONT KONIEV
XXXX 4
XXXXX 4 UKRANIAN FRONT PETROV
XXXXX 2 UKRANIAN FRONT MALINOVSKY
XXXXX SOUTH UKRAINE SCHOERNER/FRIESSNER
XXXXX 3 UKRANIAN FRONT TOLBUKHIN
0 100 km
0 100 miles
N

BATTLES FOR BUDAPEST AND VIENNA
DECEMBER 1944–MAY 1945

Soviet frontline 15 December
Soviet frontline 9 May
Soviet advance
German counterattack
German withdrawal
Area of Slovak uprising

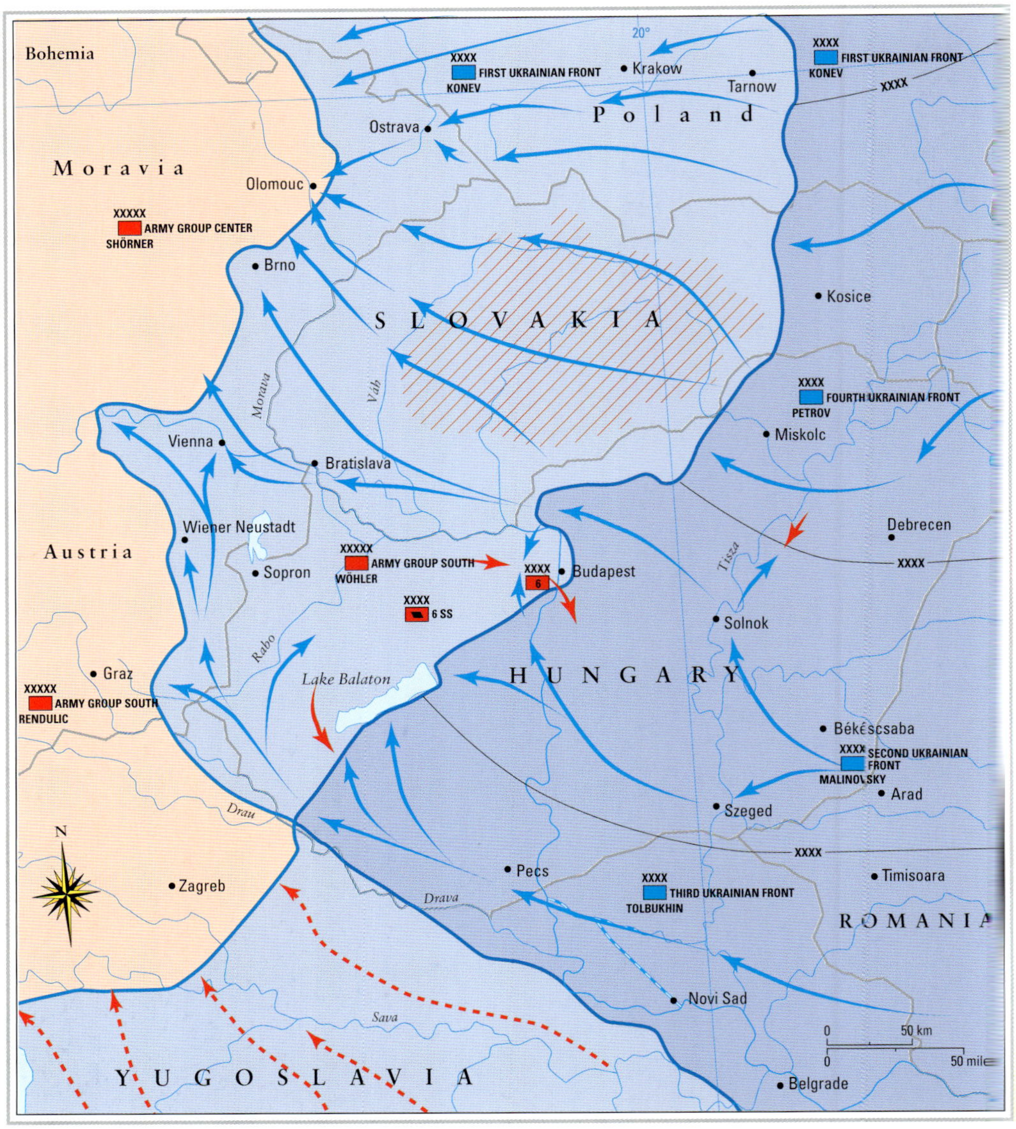

SOVIET SURGE
In the winter of 1944 and spring of 1945, the Soviet Red Army executed deep penetration offensives, encircling large numbers of German troops and capturing Vienna and Budapest during the westward advance toward the Reich.

the River Vistula east of Radom. By the time the Bagration offensive, and its flanking subsidiaries, had petered out in late August, the Germans had been driven out of the Soviet Union back to the border of East Prussia and to the River Vistula south of Warsaw in German-occupied Poland. In the process, the Soviets had destroyed 26 entire German divisions, inflicted 398,000 casualties, and captured 170,000 prisoners. Crucially, the Soviets also secured three threatening bridgeheads over the Vistula south of Warsaw, from which they would launched their next offensive in January 1945. In just nine weeks of bitter fighting the Bagration offensive had inflicted the most crushing military defeat experienced by the German Army during the entirety of World War II, a débâcle from which it never recovered.

Advance west

In the wake of the disaster experienced by the German Army Group Centre in White Russia during summer 1944, the Soviets continued their offensive actions across the entire central and southern sectors of the Eastern Front over the ensuing months. Marshal Ivan Konev's First Ukrainian Front, operating in January 1945 on the southern flank of the Soviet Vistula-Oder offensive that raced across Poland, successfully advanced west through Tarnow and Kraków. By March its forces had captured the Ostrava region of eastern Moravia. Further south, during spring 1945, the Fourth Ukrainian Front advanced west through Slovakia in the process linking up with the remaining surviving guerrilla units that had earlier launched the Slovak Uprising against the pro-German Axis Slovak state during the second half of 1944. By March 1945, these Soviet forces had advanced further west against the opposition of Army Group Centre to reach the Brno-Olomouc region in south-eastern Moravia.

Further south still, during December 1944 Marshal Rodion Malinovsky's Second Ukrainian Front continued its drive west, forcing back Infantry General Otto Wöhler's German Army Group South. In the process the Red Army encircled the German IX SS Mountain Corps in Budapest, the Hungarian capital. During 1–27 January the German forces launched three abortive counterattacks to relieve the encircled Budapest garrison, which eventually succumbed to the Soviet assaults. Subsequently, during 6-16 March, the Germans launched their 'Spring Awakening' offensive in the Lake Balaton region to protect the sole remaining oilfield in Axis hands, but this operation was soon halted by Soviet counterattacks. The Soviet forces then rapidly drove the beleaguered German forces west over the western Hungarian border into Austria. On 13 April, the Soviets captured Vienna, the Austrian capital, before pushing further northwest along the valley of the River Danube and west toward the city of Graz during the war's final days.

Balkan guerrilla warfare

Further south still, during late 1944, Soviet forces rapidly advanced through Romania to link up with the Yugoslav partisan armies that held positions along the Drava River valley. During January 1945, Soviet, Bulgarian, and Yugoslavian units attacked the positions held by Field Marshal Maximilian Freiherr von Weichs' Supreme Command South-East. This command controlled eastern Bosnia-Herzegovina to cover the key cities of Mostar, Sarajevo and Višegrad; in these areas the Croat-German forces also remained locked into a brutal guerrilla warfare struggle with Yugoslavian partisans that had raged for several years. Finally, in early May 1945, during the war's last hours, the Axis forces rapidly withdrew west in order to avoid falling into Communist captivity.

TIGERS FORWARD
A column of formidable Tiger tanks rolls forward in Russia along a dirt road. The German Army made remarkable progress during favorable weather conditions; however, the harsh winter and difficulties in maintaining heavy armour during cross-country operations slowed their advance.

BATTLE OF THE BULGE 1944

Hitler's last desperate gamble in the West, Operation Watch on the Rhine erupted in the lightly defended Ardennes sector of the Allied front on 16 December 1944. Marshalling his forces for weeks, the Führer personally directed three armies, the Sixth Panzer under SS General Josef 'Sepp' Dietrich in the north; the Fifth Panzer under General Hasso von Manteuffel in the centre; and the Seventh Army under General Erich Brandenberger to the south. They were to attack along a 95km (60-mile) front into Belgium, France and Luxembourg, capture the major Belgian seaport of Antwerp, and drive a wedge between the Allied 12th and 21st Army Groups.

Panzer progress

A successful winter offensive might compel the Allies to negotiate. Germany could then concentrate forces against the Soviets along the River Vistula in the East. With 275,000 troops, 1,800 tanks and armoured vehicles, and 2,000 artillery pieces, the Germans struck in bad weather that prevented Allied tactical air sorties.

Initially successful efforts in the north were slowed by stubborn resistance from the U.S. 99th Division along Elsenborn Ridge, although two entire regiments of the 106th Division surrendered to Dietrich in the heavily wooded Schnee Eifel. At St Vith, elements of the 7th Armored Division held out for six days.

In the south, the U.S. 9th Armored and 4th Infantry Divisions stopped Brandenberger cold. The shoulders of a great pocket developed, which gave the offensive its name, the Battle of the Bulge. At the crossroads town of Bastogne, surrounded troops of the 101st Airborne and 9th and 10th Armored Divisions held out until relieved by General George Patton's Third Army, which had disengaged from its eastward advance, pivoting 90 degrees north in a spectacular combat manoeuvre to relieve the beleaguered comrades.

ARMOURED SCOUT
U.S. soldiers halt their M8 Greyhound armoured car amid other vehicles while proceeding on a reconnaissance mission somewhere in France. The wheeled M8 entered Ford Motor Company production in the spring of 1943.

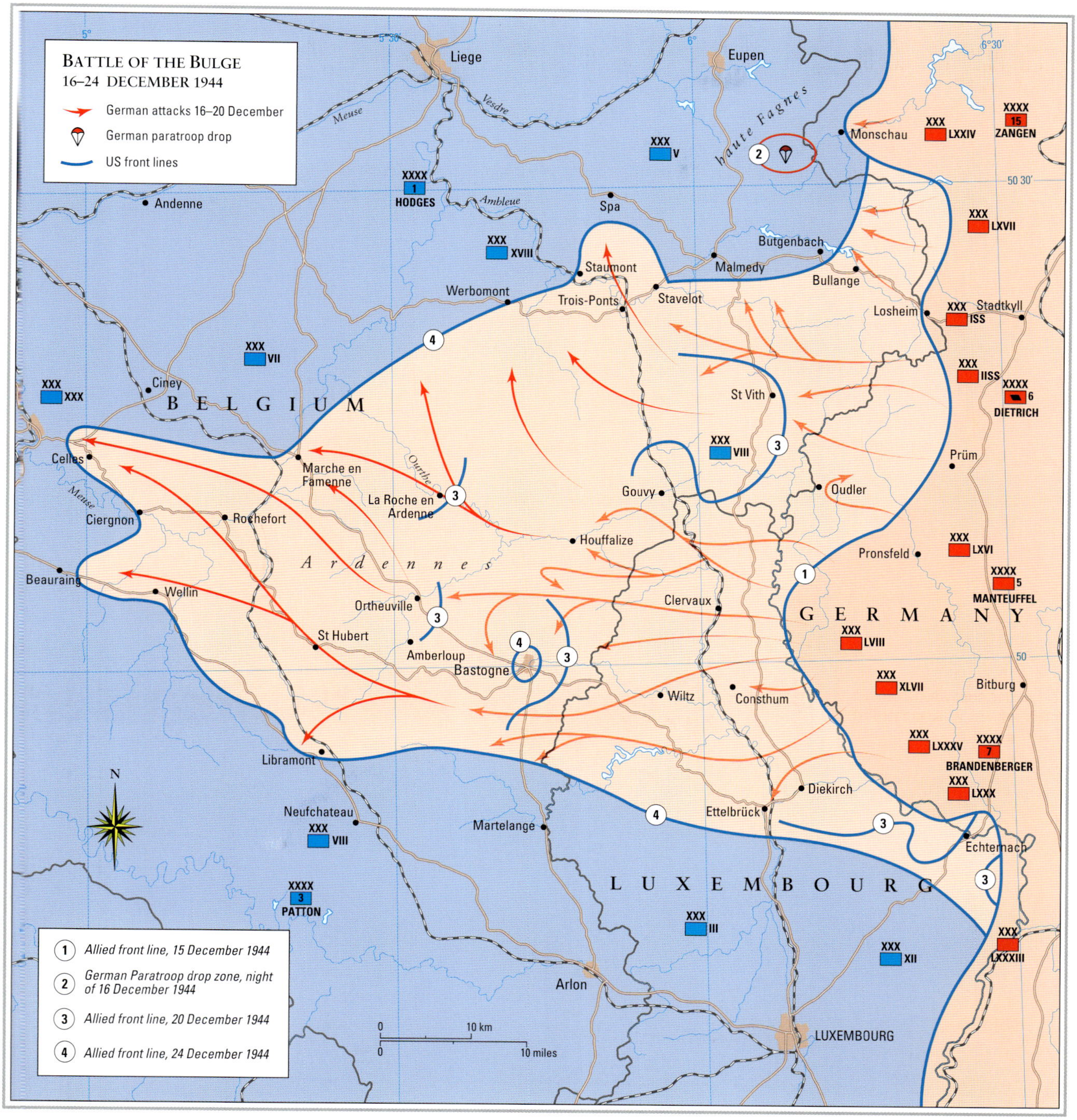

Halting the Germans

Spearheading the German advance in the centre, elements of the 2nd SS Panzer Division under Colonel Joachim Peiper rushed toward crucial bridges across the River Meuse, but were stymied by combat engineers who blew up bridges and by continuing shortages of precious fuel. Peiper, later convicted of war crimes for the murder of American prisoners at Malmédy, neared the Meuse at Dinant on 24 December but could advance no further.

As Allied resistance stiffened, many German tanks ran out of fuel and were abandoned. Eventually, the weather cleared, and Allied aircraft hit ground targets. By 15 January 1945, the German salient had been erased. Hitler's losses were staggering: there were at least 120,000 casualties and 600 armoured vehicles destroyed. U.S. losses were nearly 80,000 killed, wounded and captured.

WATCH ON THE RHINE
The German Ardennes offensive fell short of its ultimate goal, the capture of the great port of Antwerp, Belgium, and the splitting of the Allied front in two. Hitler's desperate gamble resulted in irreplaceable losses in troops and armoured vehicles.

THE PACIFIC 1944–45

IWO JIMA ASSAULT
Three divisions of U.S. Marines assaulted Iwo Jima in February 1945, wresting control of the island from the Japanese after more than a month of fighting. The island's airfields were in use before Iwo Jima was declared secure.

LANDINGS ON IWO JIMA
19 FEBRUARY – 26 MARCH 1945

U.S. advances
U.S. front lines with dates
Japanese line

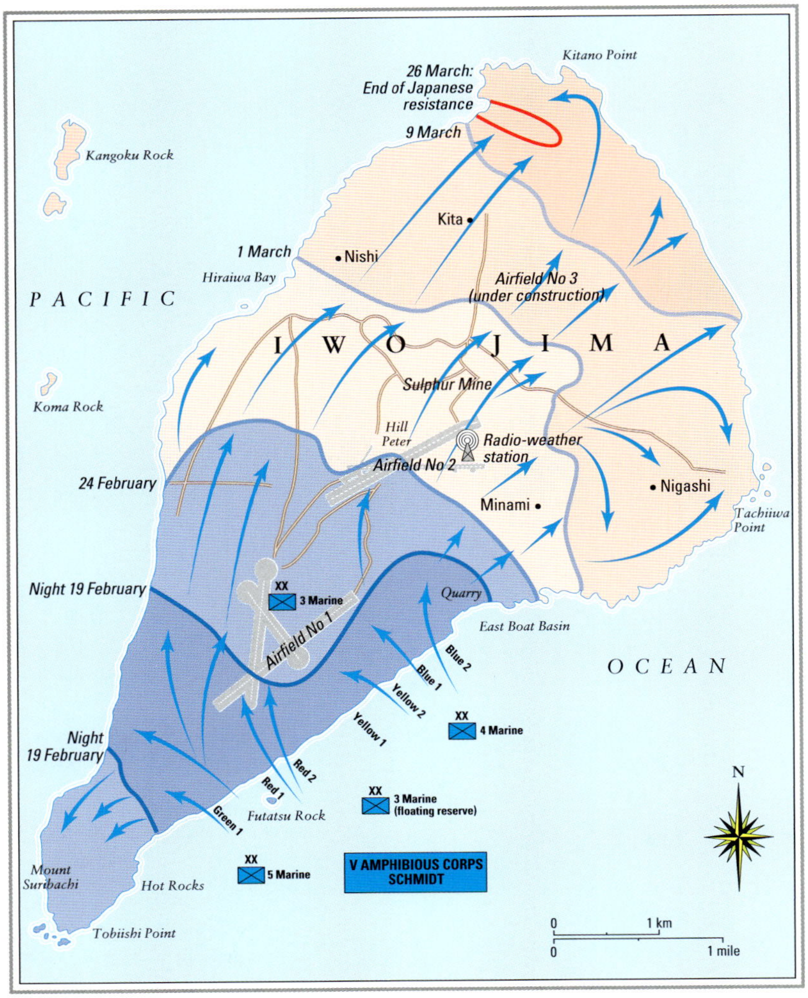

During World War II, armoured action in the Pacific Campaign was limited due to several circumstances. Thick jungle, difficult terrain, and lengthy periods of inclement weather made armoured operations difficult, while the small land masses of many Pacific islands made the deployment of numerous tank formations impractical. Reclaiming the scattered islands of the Pacific captured by the Japanese was a slow process of island-hopping that began in mid-1942.

As Japan moved toward war in the 1920s and 1930s, the nation's military establishment paid little attention to the development of tanks and armoured fighting vehicles in comparison to the Western powers. Most Japanese tanks were lightly armed and armoured. Typical of these was the Type 97 Chi-Ha medium tank, which mounted a 57mm (2¼in) cannon in a turret that could not be elevated. Intended for infantry support, as were most of the Japanese designs deployed during World War II, the Type 97 was the most widely produced medium tank in the Japanese arsenal. Even so, only about 3,000 of the Type 97 and an improved version, the Kai Chi-Ha, were built. The deficiencies of these tanks were largely hidden in the early stages of the Pacific war, by Japan's superiority of force and seizure of the initiative in individual battles.

Guadalcanal and Tarawa

The United States began to deploy tanks to the Pacific to support Army and Marine Corps landings on numerous enemy-held islands, and these armoured vehicles provided essential firepower as early as August 1942 during operations on Guadalcanal in the Solomon Islands. During the decisive Battle of the Tenaru (misnamed and actually fought at the mouth of the Ilu River), American M3 Stuart light tanks inflicted significant casualties on attacking Japanese troops with their 37mm (1½in) cannon and 7.62mm (⅓in) machine guns. Although the Stuart was too light for tank-versus-tank combat in Europe, it proved more than capable of taking on Japanese machine gun nests and troop concentrations at Guadalcanal and elsewhere.

At Tarawa in the Gilbert Islands in November 1943, a single M4 Sherman medium tank, nicknamed 'Colorado', of the Marines' Company C, 1st Corps Medium Tank Battalion, destroyed numerous Japanese pillboxes and strongpoints, assisting in the capture of the island during a 76-hour assault. Fire from Japanese anti-tank guns failed to penetrate the Sherman's armour, turning the tank's exterior lemon yellow where hits had glanced off.

Iwo Jima and Okinawa

The bloodiest island fighting of the Pacific War took place at Iwo Jima and Okinawa during the winter and spring of 1945. The recapture of these islands was critical, and in both cases, the Japanese commanders employed the technique of defence in depth. The Japanese defenders had heavily fortified the islands, placing artillery pieces in caves and manning concrete reinforced bunkers and pillboxes. Although the confined spaces in which these U.S.

tanks were often required to operate made them susceptible to mines, anti-tank guns and tank-killer squads of Japanese infantry, they were instrumental in gaining the upper hand against fortified enemy positions. Mounting 75mm (3in) cannon, the M4 Shermans were used as mobile artillery and took on Japanese strongpoints at close range, blasting them into submission. Specialized Shermans mounting flamethrowers silenced bunkers and crushed machine gun emplacements beneath their treads.

AMERICAN TRIBUTE
American troops and tanks form ranks for ceremonies to honour the fallen of Okinawa and Iwo Jima. The cost in lives during the capture of these islands was tremendous; however, the Americans were positioned to invade Japan.

OKINAWA TAKEN
The fighting for Okinawa was bitter as U.S. Marines and Army troops battled for every yard of ground. Only 550km (340 miles) from the home islands, Okinawa was literally on the doorstep of Japan.

THE FALL OF THE REICH
MAY 1945

By the spring of 1945, the fall of the Third Reich was inevitable. Two Soviet army groups, or fronts, bore down on the German capital of Berlin, Marshal Georgi K. Zhukov's 1st Belorussian and the 1st Ukrainian under Marshal Ivan Koniev. Seeking a rapid victory and propaganda coup, Soviet Premier Josef Stalin pitted the two army groups against one another in a race to final victory.

The advancing Soviets were intent on swallowing Berlin in a giant pincer movement. The defending Germans massed a million troops, many of them walking wounded, Hitler Youth and old men of the

CROSSING THE RHINE MARCH 1945

- Allied Rhine crossings
- Allied attacks
- Allied front line
- German pockets
- Allied paratroop drop

Allied front lines:

1. *22–28 March 1945*
2. *4 April 1945*
3. *18 April 1945*

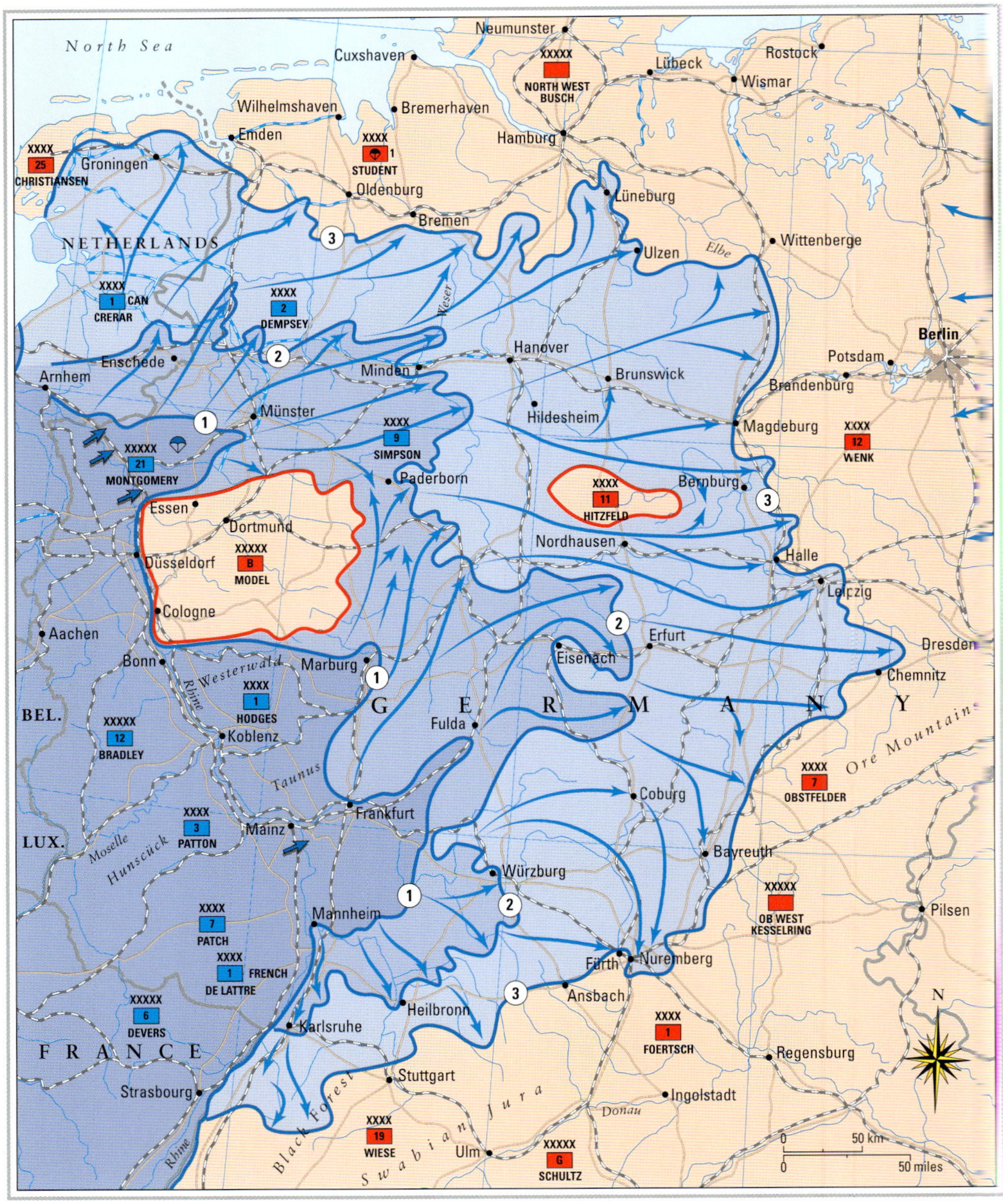

RHINE CROSSING
In March 1945, elements of the U.S. First Army captured the Ludendorff bridge across the Rhine near Remagen, Germany. Before the bridge collapsed, Allied troops and tanks were able to force their first bridgehead across the river.

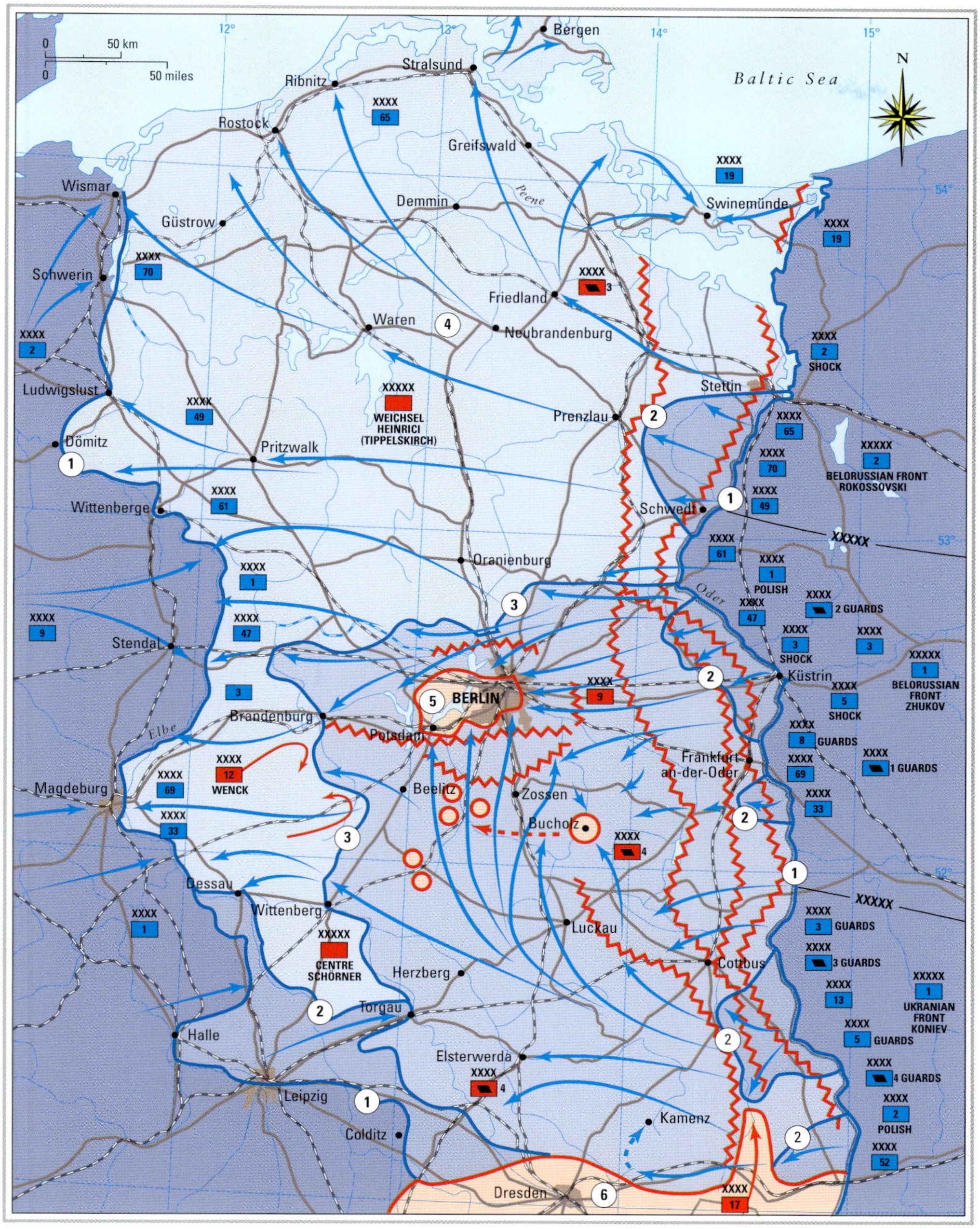

EMBATTLED BERLIN
Two Soviet army groups assaulted Berlin in the spring of 1945. Before its capture, severe fighting occurred eastward at Seelow Heights. Rival Red Army marshals Georgi Zhukov and Ivan Koniev vied for the honour of capturing the capital.

Volkssturm (or home guard). Mechanized Soviet formations struck at Seelow Heights to the east, encountering tank-killer groups armed with the lethal hollow-charge, shoulder-fired Panzerfaust.

Although hundreds of T-34 medium and Josef Stalin heavy tanks were lost, the Soviets pressed on to the streets of Berlin, reducing fortified positions in buildings that had once been occupied by the Nazi government. Fighting raged in the suburbs of the capital city as the Soviets were compelled to clear defenders house-by-house. Often, Soviet tanks were limited in their ability to operate in the confines the urban area. German armoured formations were virtually annihilated, and in the final days, a delusional Hitler issued orders to formations long since destroyed.

As the Red Army grip on Berlin tightened, thousands of German soldiers and refugees fled westward to avoid capture or detention at the hands of the vengeful Soviets. Western Allied forces crossed the Rhine and linked up with the Soviets at Torgau on the Elbe River. By 1 May, the Nazi Führer was dead and the banner of the hammer and sickle fluttered from the Reichstag.

MANCHURIA 1945

SOVIET SPEARHEADS
When the Red Army launched its invasion of Manchuria on 8 August 1945, more than 5,000 tanks were committed to the offensive. Here, a formation of Soviet tanks, some carrying infantrymen, roll forward across a desolate Manchurian plain.

On 8 August 1945, the Soviet Union declared war on Japan. The following day, more than a million troops invaded the long-disputed region of Manchuria. In some locations, Japanese troops were overwhelmed and virtually annihilated during the opening hours of the attack. During the Soviet offensive, which lasted little more than two weeks, Red Army forces eventually numbered more than 1.5 million men, 26,000 artillery pieces, and 5,600 tanks and armoured vehicles.

Attacking on three fronts, Soviet armoured spearheads struck deep into Manchuria. Difficult terrain initially impeded the progress of the Soviet tank formations, caused many of them to consume greater quantities of fuel than logistical efforts could easily support, and compelled them to halt for lengthy periods while fuel was brought forward. Nevertheless, in fewer than five days, elements of the 6th Guards Tank Army had advanced 220km (350 miles).

The Japanese defenders were unable to cope with the speed of the Soviet armoured advance. Facing the Red Army onslaught were approximately 600,000 troops of the Kwantung Army, including only two tank divisions with an entire complement of about 1,100 armoured vehicles. Many of these were obsolete light tanks and armoured cars.

Utilizing giant pincer movement tactics that had proved successful against the Germans on the Eastern Front, the Soviets enveloped large contingents of Japanese troops in an expansive area of operations. Although units of the Kwantung Army resisted stubbornly at times, particularly at the town of Hailar, where a Red Army tank brigade raced more than 100km (62 miles) beyond the village and trapped a significant number of retreating Japanese troops, the outcome of the offensive against the exhausted Japanese was inevitable.

Accompanied by the landings of Soviet amphibious forces in northern Korea, the Kurile Islands, and on Sakhalin Island, the Soviet juggernaut rolled forward. On Shumushu Island in the northern Kuriles, a tank regiment attached to the Japanese 91st Infantry Regiment was ordered to advance against Soviet landing beaches and managed to contain the invading force for a short time before being overwhelmed.

By September, the Red Army had achieved virtually all of its territorial aims, although complete occupation of the Korean peninsula was denied when U.S. troops came ashore at Inchon on 8 September.

Soviet superiority

Those Japanese tanks that offered battle against the Red Army mechanized units were often hopelessly outclassed, and the participation of Japanese armour in the Manchurian fighting was largely

CAUTIOUS ADVANCE
Soviet infantrymen peer from a covered position during the advance in Manchuria, while armoured vehicles proceed cautiously through territory where Japanese troops are known to be operating. The rapid Soviet advance in August 1945, routed Japanese defenders in Manchuria.

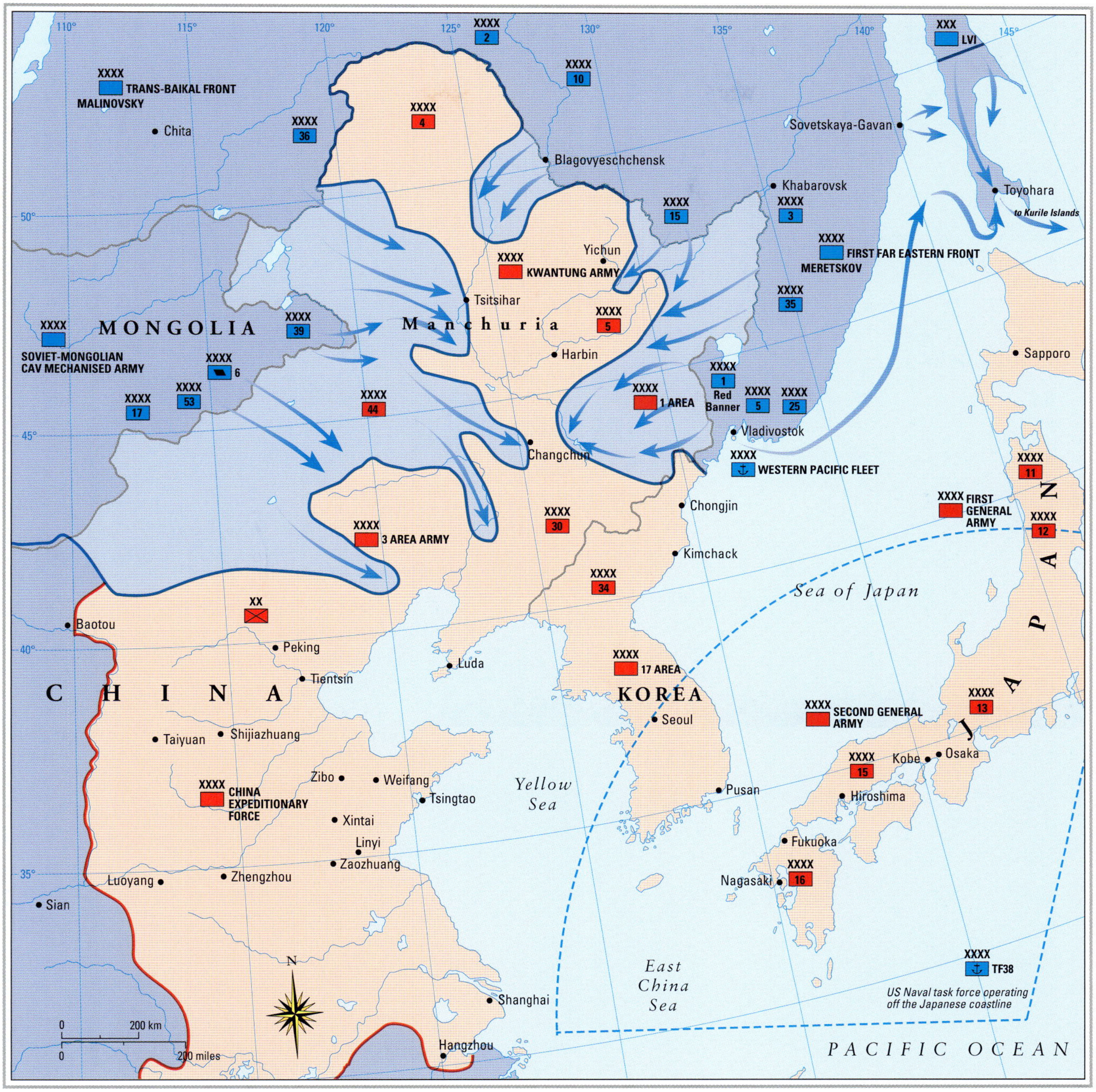

inconsequential. The 57mm (2¼in) guns of the Type 97 Chi-Ha medium tank were ineffective against Soviet armour.

In addition to the older BT-7 light tanks deployed on the Manchurian frontier, the Soviets reportedly engaged a number of the superb T-34 medium tank, arguably the finest all-around tank fielded during World War II. The basic T-34 combined superior firepower with a 76.2mm (3in) and later an 85mm (3¼in) main gun. Secondary armament included a pair of 7.62mm (3in) machine guns. Production of the T-34, with armoured protection of up to 60mm (2¼in), had begun in the mid-1930s. Intended as a replacement for the BT series and the earlier T-26 model, the T-34 entered service with the Red Army in 1940.

Unconfirmed reports exist that the Soviets may have deployed the heavy IS-2 and IS-3 Josef Stalin tank to Manchuria. Mounting a 122mm (4¾in) cannon and weighing in excess of 46 tons, these mammoth armoured vehicles would have been by far the heaviest of their kind in the Far East.

SOVIET INVASION OF MANCHURIA
SEPTEMBER 1945

- Soviet frontline on 1 September
- Japanese frontline 1 September
- Soviet advances

The Red Army threatened to close a massive encirclement of the Japanese Kwantung Army near the town of Changchon.

KOREA 1950–53

At the very end of World War II, the Soviet Union moved forces into territories still claimed or occupied by Japanese forces, including northern Korea. At the Japanese surrender the Soviets took control of Korea north of the 38th Parallel, while the Western allies were responsible for the south of the country.

The result was that the country became divided along an arbitrary line while attempts were made to create a national government. Finally, the Soviet Union installed a Communist government in what became the People's Republic of North Korea. Elections in the South produced an ostensibly democratic and pro-Western Republic of Korea.

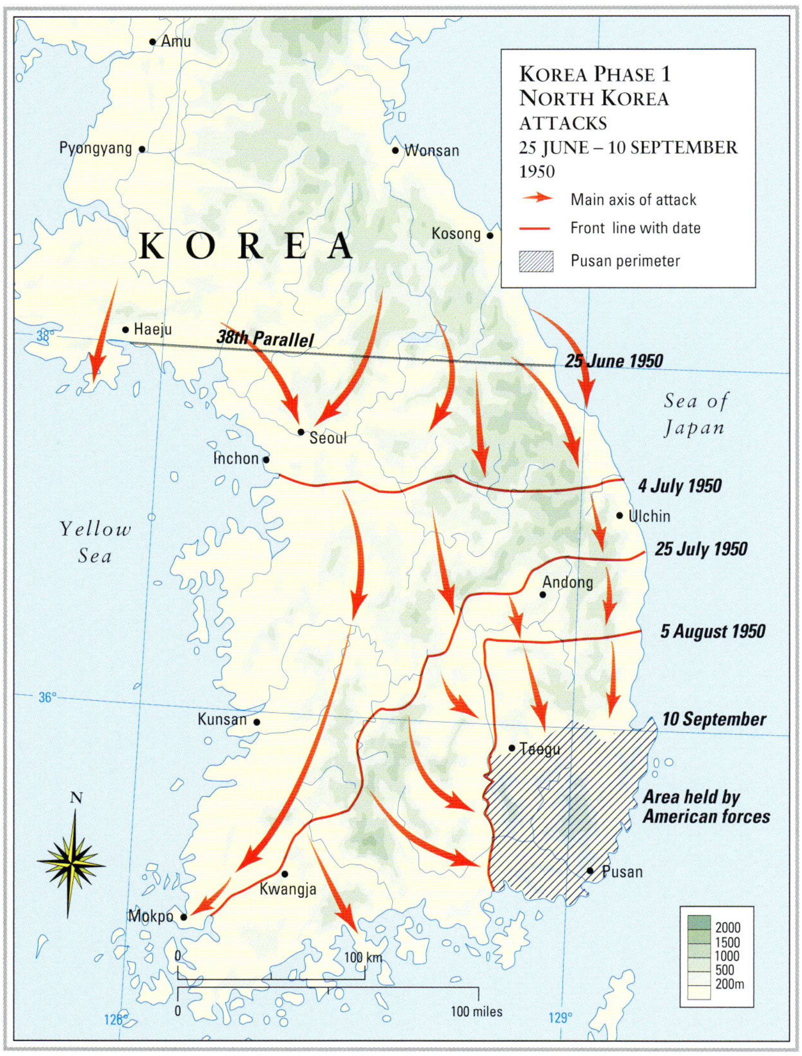

NORTH KOREAN ATTACKS
Although terrain channelled the North Korean offensive into predictable routes, the unprepared South Korean forces gained no advantage from this factor. Harsh terrain and lengthening supply lines slowed the North Korean advance as much as South Korean resistance.

Both governments wanted to reunify the country under their own ideology, which created a volatile situation. By the middle of 1949 both the U.S. and the U.S.S.R. had withdrawn their troops and were expecting the Koreans to deal with their own problems. Propaganda campaigns and border skirmishes became increasingly common, along with raids and terrorist attacks deeper within North and South Korea.

The North Korean invasion

Before departing, U.S. forces trained the South Korean Army. However, it was more of a security arm than a fully capable military force. Four divisions, at full strength, were deployed along the border with North Korea, while four more, which were at a much lower level of readiness, were deployed deeper in the country.

The South Korean Army was equipped with World War II U.S. equipment, and notably lacked anti-tank weapons. Those that did exist were primarily early-model bazookas and light anti-tank guns, both of which were ineffective against the later generations of tanks that emerged during the last years of the war. Air support was non-existent and South Korea lacked armoured forces with which to mount a mobile defence or a counterattack.

The North Korean People's Army, conversely, was well equipped with Chinese and Soviet equipment. It outnumbered the South Korean forces by around 135,000 to 95,000 personnel, and more importantly it possessed an armoured striking force. Eight full-strength North Korean divisions were available for offensive operations, with two weaker divisions held in reserve as security forces. A motorcycle regiment was available for reconnaissance, and the assault was led by an armoured brigade.

The North Korean People's Army was equipped with T-34 tanks supplied by the Soviet Union. Although these were a little dated by 1950, they were still formidable in action. Perhaps more importantly they were easy to maintain in the field, allowing the North Korean armoured force to remain effective throughout a long advance. Both tank and infantry formations benefited from the support of self-propelled artillery.

The North Korean Army was also better trained and more experienced than its Southern counterpart. Many personnel had fought in World War II and the

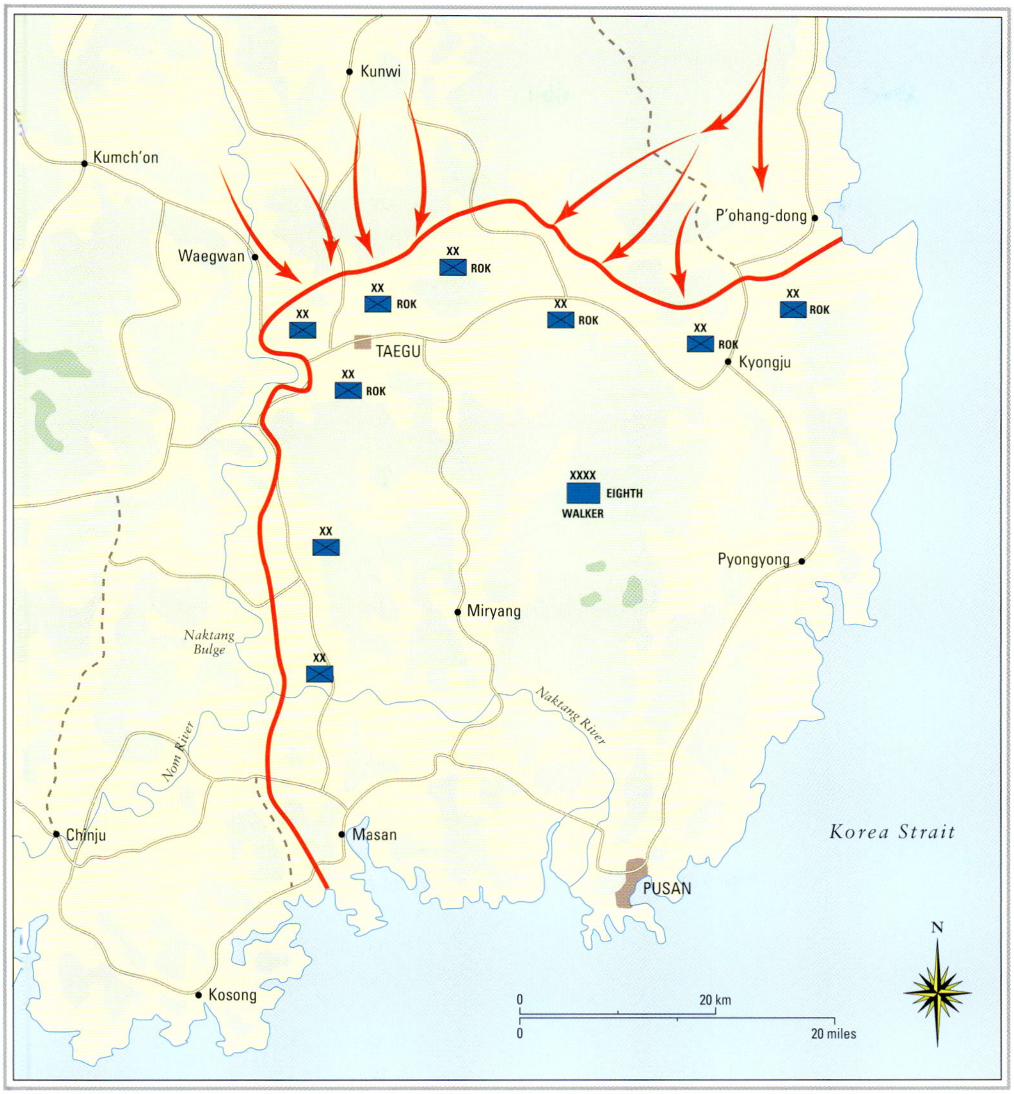

THE PUSAN PERIMETER
Only the arrival of reinforcements from overseas allowed the South Koreans to avoid total defeat. United Nations forces clung precariously to their positions around Pusan, under constant attack, before their increasing strength made a counteroffensive possible.

THE BATTLES OF THE PUSAN PERIMETER
1–15 SEPTEMBER 1950

UN Perimeter
Communist attack

Chinese Civil War, and had received training from the Soviets, who were familiar with infantry-armour cooperation techniques.

The invasion began in June 1950. Despite the fact that terrain channelled the North Koreans into predictable avenues of advance, there was little that the unprepared South Korean divisions could do and the capital, Seoul, was soon overrun. Facing few anti-tank weapons, the North Korean tank forces could punch straight through a South Korean position in a headlong advance, regrouping on the far side to attack the rear of the enemy while infantry forces pushed into the breach so created.

United Nations intervention

Unable to form a stable defensive front, the South Korean Army was driven southwards with the North Koreans in pursuit. U.S. troops were rushed from Japan and air attacks were made by American warplanes in the hope of slowing the advance. U.S. forces landing at the port of Pusan attempted to counterattack but were driven back, finally establishing a shaky perimeter around Pusan in early August 1950.

Backed by a United Nations' resolution, troops arrived to strengthen the Pusan Perimeter, notably from Britain and the Commonwealth, as well as additional U.S. forces. With tanks of their own and no shortage of effective anti-tank weapons, the United Nations' forces were able to defeat repeated North Korean attacks, weakening their army even as U.N. strength increased. However, a frontal assault was still a costly option, so an alternative was sought. This took the form of an audacious amphibious landing at Inchon and a drive on Seoul which not only recaptured the South Korean capital

KOREA PHASE 2
UNITED NATIONS COUNTER-ATTACK
15 SEPTEMBER – 25 NOVEMBER 1950

UN counter-attack
Front line with date
Pusan perimeter

THE U.N. COUNTEROFFENSIVE
Weakened by a disrupted supply chain, North Korean forces were driven before the U.N. counteroffensive. As the ground forces pushed northwards, amphibious operations took control of key coastal targets. A total North Korean collapse was prevented only by Chinese intervention.

but also cut the main logistics route supporting the North Korean forces around Pusan.

With the North Korean position around Pusan seriously undermined, the U.N. breakout began. Supported by heavy air attacks that depleted North Korean artillery and armoured strength, U.N. forces were led by armoured elements that broke the North Korean positions. The North Korean Army then began retreating northward in some confusion, pursued by U.N. forces.

The U.S. troops who had first arrived in Korea were not combat-ready, having come from occupation duty in Japan. By the time of the Pusan breakout, however, the U.N. forces were experienced and psychologically prepared for full-scale combat operations. Techniques for massed operations learned in World War II were implemented, with good cooperation between infantry, armour and artillery forces. The North Korean Army, which had been more than a match for the unready South Koreans, now faced the forces that had won World War II.

A new generation of equipment was being fielded by the U.N. forces. Some armoured vehicles had arrived too late to make much of an impact in 1945, but were vastly superior to the T-34s possessed by the North Koreans. Indeed, some of the hardware had been specifically developed to deal with the threat posed by late-war Soviet tanks. Excellent as it had been in its day, the T-34 was out-matched by the new vehicles pitted against it.

Just as significantly, the U.N. forces were supported by a good logistics 'tail' and a wealth of operational experience was now available. Formations functioned at a high level of efficiency and maintained their combat effectiveness as they advanced. The North Korean Army could do little more than fight a ragged rearguard action as it fell back across the 38th Parallel.

Chinese intervention

With Pyongyang, capital of North Korea, under threat, China made the decision to intervene in support of the fellow-Communist North Koreans. Chinese forces had gained recent combat experience in their civil war, but lacked armour or anti-tank weapons. Improvised measures included satchel charges which could immobilize a tank by blowing the tracks off, but were rarely effective against the hull. However, the Chinese troops were highly mobile due to their light equipment, and were capable of moving quickly cross-country with no reliance on predictable lines of supply.

The Chinese intervention reversed the situation inflicting sharp defeats on the U.N. forces which fell rapidly back to a defensive position along the 38th Parallel. A series of offensives and counter-offensives ensued throughout the first half of 1951 after which the war gradually settled into a static battle of attrition. Heavily fortified bunkers and trench complexes defended strategic high ground, resulting in bloody infantry battles which rarely achieved any lasting result.

Armoured forces were of limited use in this static stage of the war, and could not be of much assistance to an assault because of the terrain. The highly motivated infantry-based Chinese forces had numerous advantages in this kind of fighting, although these were offset by superior U.N. firepower. Armoured vehicles were useful as mobile fire-support platforms and could sometimes engage enemy positions using direct fire from nearby high ground. Tanks were used as bunkers at times, emplaced in positions where they received additional protection from sandbags or earthworks. They were also effective as a mobile reserve to contain an attack or launch a counterthrust.

The availability of mobile fire support was important, as the North Korean and Chinese forces employed careful reconnaissance to find the weakest points of the U.N. defences. This was usually the under-equipped South Korean units, whose relatively low firepower made them vulnerable to massed infantry assaults. Chinese troops were issued with large numbers of sub-machine guns rather than the more usual rifles, giving them an additional advantage in close combat.

The stalemate phase of the war dragged on for two more years, until July 1953 when armistice negotiations began. The country was partitioned along more or less the same line as before the conflict and a demilitarized zone created. Tensions were never really resolved, however.

The role of tanks in Korea

The terrain in Korea would have made large-scale armoured operations problematical even if both sides had possessed large numbers of tanks. Thus there were no large armoured battles and few tank-versus-tank engagements. Most North Korean tanks were dealt with by aircraft, anti-tank guns or infantry anti-tank weapons. Artillery weapons were at times used for direct fire against tanks, usually for lack of anything better. U.S.-supplied 105mm (4in) guns were successful in disabling some North Korean tanks early in the invasion, but were not mobile enough to avoid being overrun; most were lost in their first engagement.

By the time U.S. and British tanks arrived, North Korean tank strength was already waning. Thus armoured vehicles were more useful as infantry support platforms than anti-tank weapons. Post-war tanks were well suited to this role, as their guns were of large enough calibre to fire an effective high-explosive round. Early World War II tanks tended to have very small-calibre guns whose high-velocity projectile could punch through the armour of many contemporaries, but which relied on their secondary armament of machine guns for infantry support. The increase in calibre needed to defeat the improved protection of late-war vehicles also permitted tanks to function as assault guns if needed.

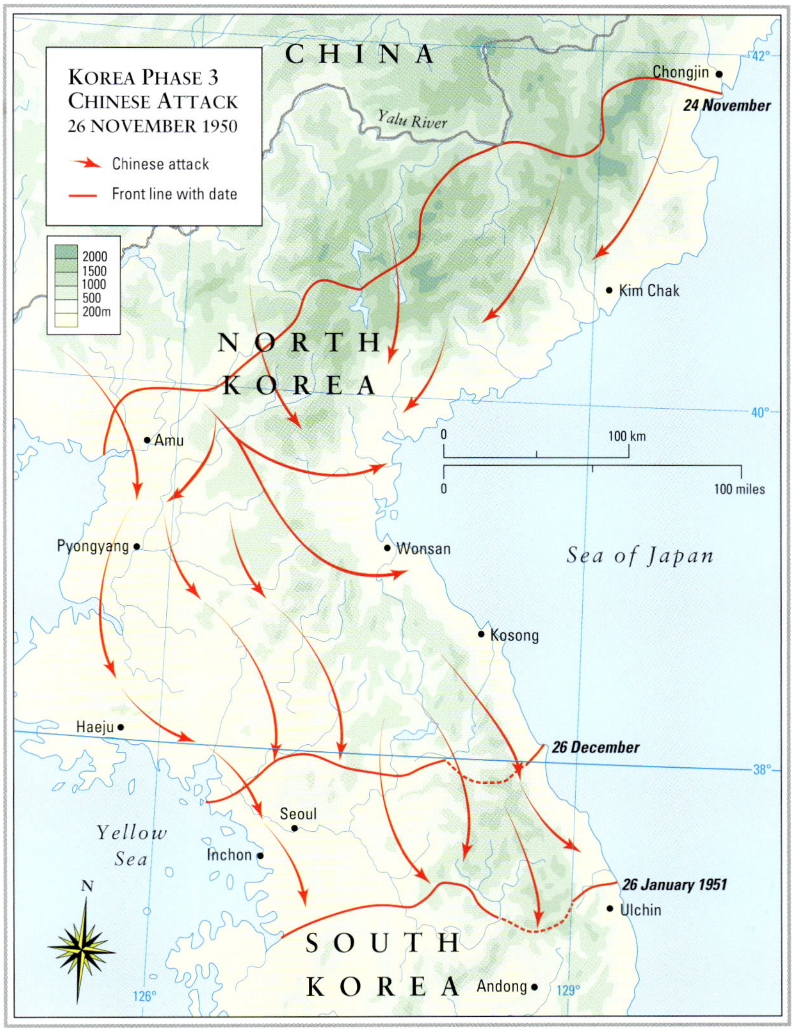

U.S. ARMOUR IN KOREA
The arrival of U.S. armoured forces eliminated the main advantage held by the North Koreans. Although there were no major tank actions, armour supported infantry in numerous engagements. The primary U.S. tanks were the M26 Pershing and the M46 Patton.

CHINESE OFFENSIVES
Chinese forces in Korea relied primarily upon huge numbers of infantry to overwhelm U.N. formations.

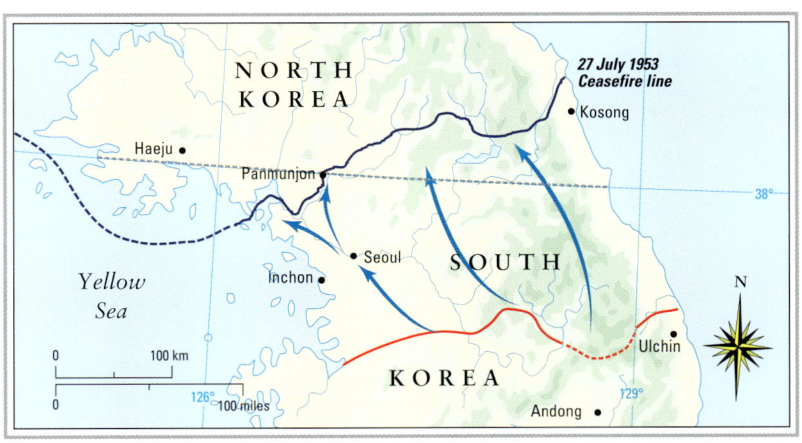

KOREA PHASE 4 – UN COUNTER-ATTACK
JANUARY 1951 – 27 JULY 1953

UN campaign
Ceasefire line with date
Maximum Chinese advance with date

ARAB-ISRAELI WARS 1948–82

On 15 May 1948, the day after the end of the British mandate in Palestine and Israel's unilateral declaration of itself as a state, five neighbouring Arab states invaded – Lebanon from the north, Syria, Transjordan and Iraq from the east, and Egypt from the south. The geographically ragged embryonic Israeli state was compelled to fight fanatically for its survival. The Israelis rapidly transformed their guerrilla forces, which had been used during the Palestine insurgency against the British (1939-48), into conventional armed forces, the Israeli Defence Forces (IDF). The combination of fanatical Israeli will to survive, coupled with internecine Arab squabbling and poor cooperation eventually enabled the Israelis to gain the upper hand, and during February-July 1949 a series of bilateral ceasefires were agreed.

Suez and the Sinai, 1956

During 1956 Israel's Arabs neighbours increased their level of overt hostility, largely due to domestic political reasons. Colonel Nasser, the new Egyptian champion of pan-Arab/ anti-Western imperialist aspirations, drove forward this increasing hostility. Egypt closed the Straights of Tiran to blockade the southern Israeli port of Eilat and then formed a Syrian-Jordanian-Egyptian integrated command structure. These actions were extremely provocative to Israel, which decided to strike first before being invaded. In initiating this conflict, the Israelis were cynically helped by Britain and France who wanted to reverse Nasser's nationalization of the former Anglo-French-controlled Suez Canal Company. Between 29 October and 7 November, Israeli light mechanized forces successfully conquered the Egyptian Sinai peninsula, exploiting the fact that Egyptian forces were concentrated around the Suez Canal to neutralize the Anglo-French forces that had landed at Port Said. After a U.N.-negotiated ceasefire, the two superpowers coerced Israel into handing back the captured territory which would remain demilitarized under U.N. peacekeepers.

The Six-Day War, 1967

During the spring of 1967 the Soviets attempted to take advantage of America's entanglement in

Suez and Sinai Campaign October – November 1956

- Israeli border 1948–1967
- Israeli airborne assaults
- Principal Israeli lines of advance, 29 October – 5 November 1956

SUEZ ASSAULT
During a lightning preemptive attack backed by Great Britain and France, the Israelis occupied large amounts of Arab territory; however, political pressure forced them to withdraw from the Sinai, which then came under United Nations jurisdiction.

ISRAEL ASSAILED (OPPOSITE)
With the birth of the nation of Israel, the new country was attacked by several Arab nations simultaneously. Israeli militia and military units managed to fend off the massive assaults from three directions.

Vietnam to stir up trouble in the Middle East. They announced that they had obtained intelligence that Israel was preparing to attack Syria. Nasser whipped up public opinion against Israel, partly for domestic political reasons. During 14-19 May, Egyptian ground forces moved into the demilitarized Sinai, sweeping aside the U.N. peacekeepers. On 22 May the Egyptians closed the Straights of Tiran to shipping, blockading the entry of goods into the southern Israeli port of Eilat, despite knowing that Israel would view this as an act of war. On 30 May, Egypt, Syria and Jordan formed a multinational Arab command structure, while the next day Iraqi forces moved onto Jordanian soil. By early June, 182,000 Lebanese, Syrian, Iraqi and Jordanian troops had concentrated on Israel's borders.

Israel responded by creating a Government of National Unity on 1 June and began covertly mobilizing on the 4th. Recognizing how important it was to main the initiative in any future war, given Israel's precarious geographical shape and lack of depth, the Israelis decided to strike first by launching a surprise pre-emptive air strike to gain air superiority over the region. This would then free up the Israeli Air Force (IAF) to support the ensuing ground offensive. In the latter Israeli ground forces would then strike into the Sinai to defeat the Egyptian Army; at the same time the Israeli Army would strike to secure the strategic Golan Heights on the northeastern front, as well as Jordan's West Bank enclave. The weaker Israeli forces would have to rely on surprise, speed, and professionalism to overcome the numerically superior enemy forces.

Early on 5 June, the IAF launched a surprise air strike against the Egyptian Air Force (EAF). Despite the build-up for war, the EAF remained on its standard peacetime routines. The IAF thus timed its attacks to coincide with the end of the EAF's regular dawn patrols, when their aircraft would be refuelling. Flying low to avoid radar detection and attacking from the rear (west), having flown out across the Mediterranean, the Israeli aircraft thus avoided most of the Egyptian surface-to-air missile (SAM) defences. In daring attacks on 17 airbases, the IAF destroyed 237 out of 482 Egyptian aircraft in those very first minutes. The Israelis then turned their attacks against the Syrians and Jordanians. By the end of 5 June, the IAF had destroyed over 370 enemy aircraft, giving air supremacy to the Israelis.

Following these air strikes, the IDF ground forces deployed along the River Jordan on the northeastern Israeli border, halting a limited Syrian offensive during 5-6 June. Next, on 9 June the IDF attacked the Syrian defences established along the strategically vital Golan Heights plateau, which overlooked northern Israel. During 9-10 June, four IDF brigades fought themselves on to the plateau. During these operations IDF units won a number of fiercely contested armoured encounters in the area around Kaffar Nafak and Merom Golan. Gradually, however, the cohesion of the Syrian troops began to collapse; as a result, by the 10th the Israelis had captured the entire plateau including the key town of Kuneitra. The next day both sides agreed to a ceasefire. The possession of the heights significant improved the defensive strength of the northern

THE ARAB ADVANCE TO ISRAEL'S BORDERS 14–30 MAY 1967

Arab advance

Military strength:

Troops

Combat aircraft

Tanks

ISRAEL STRIKES (FAR RIGHT)
Following devastating air attacks that destroyed most of the Egyptian Air Force on the ground, Israeli ground forces launched an offensive against the Egyptians in the southwest and the Syrians and Jordanians in the north.

ARAB BUILD-UP (RIGHT)
Spurred by false Soviet intelligence reports of a pending attack, the forces of Egypt, Syria and Jordan massed along the Israeli frontier in preparation for an attack. The Israelis, however, struck the first blow.

PATTON PUNCH (FAR RIGHT)
The Israeli Defence Forces deployed large numbers of American tanks during the Six-Day War. These modified M48 Patton tanks helped to rout the Syrian forces on the Golan Heights and capture the strategic plateau on the Israeli frontier.

Israeli border. Meanwhile, Israeli forces attacked into the West Bank, the enclave of Jordan that extended west across the River Jordan. Within several days the Israelis had captured the entire enclave.

Meanwhile at 08.15hrs on 5 June, General Tal's division attacked the northern Egyptian 'shield' defences at Gaza, Rafah and El Arish, while General Sharon's division attacked the central enemy 'shield' at Abu Ageila and Um Qatef. In a surprise move, the Centurion tanks of General Yoffe's division infiltrated across supposedly impassable sand dunes between these two enemy 'shields'. Yoffe's tanks then assumed hull-down ambush positions at the Bir Lafhan crossroads and waited for elements of the Egyptian 4th Armoured Division to move forward to assist the beleaguered shields, as per the Egyptian 'sword and shield' doctrine. These Egyptian armoured reserves unknowingly moved forward to aid the beleaguered 'shields' and blundered into the Israeli trap. With the benefit of surprise, the Israelis beat the Egyptian forces, while inflicting heavy losses upon them; the enemy front-line defensives 'shields' had thus been isolated from any friendly-force assistance. Meanwhile, the Israeli assault on the Egyptian front-line defences had continued to unfold. By using a combination of all-arms tactics, amphibious assaults, airborne drops and outflanking moves, the divisions of Generals Tal and Sharon quickly overran the two Egyptian shields during the war's first 36 hours, despite encountering fierce resistance in many locations.

With the enemy defensive positions smashed, the second phase of the Israeli offensive in the Sinai now unfolded during 7-8 June in a battle-space that had now become more fluid. The Israelis audaciously unleashed the armoured units from Sharon's and Yoffe's divisions in a mad charge west to capture the key Mitla and Gidi passes, which

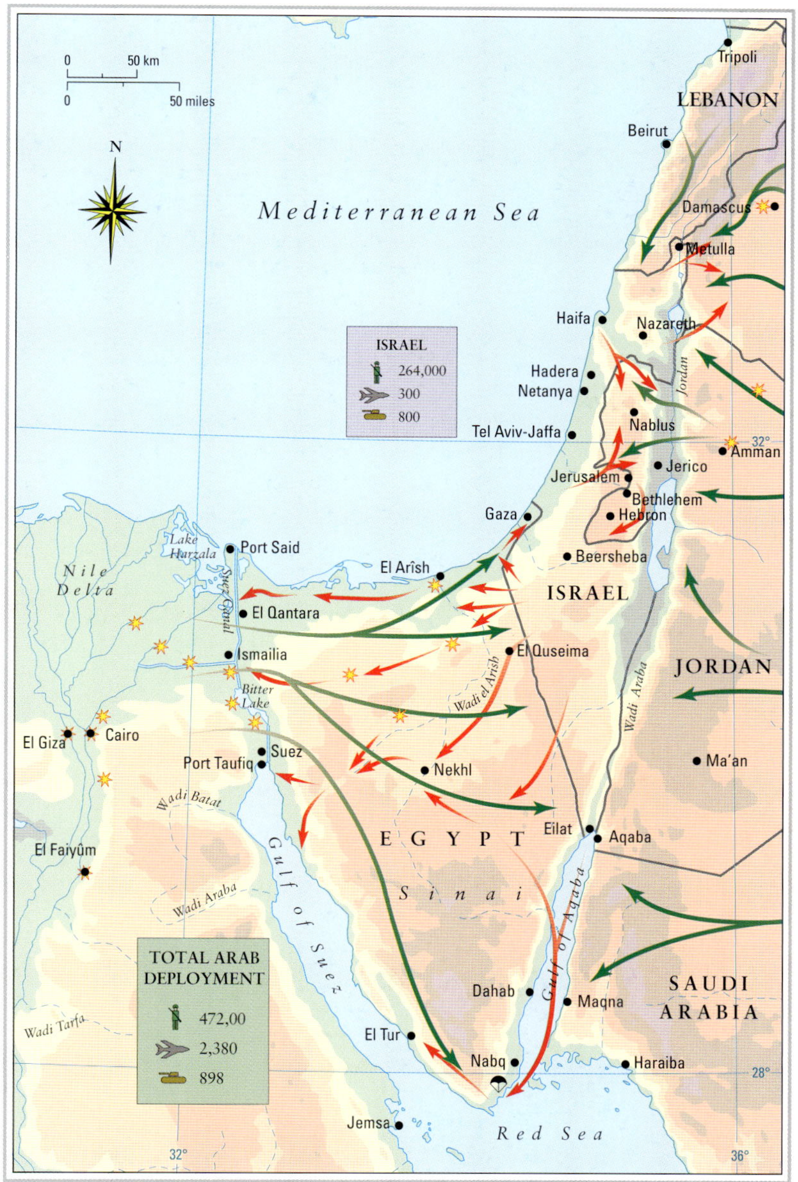

THE SIX-DAY WAR
ISRAELI COUNTER-ATTACK
14–30 MAY 1967

Main Israeli attacks
Israeli air strikes
Airborne landing
Arab advance

Military strength:

Troops
Combat aircraft
Tanks

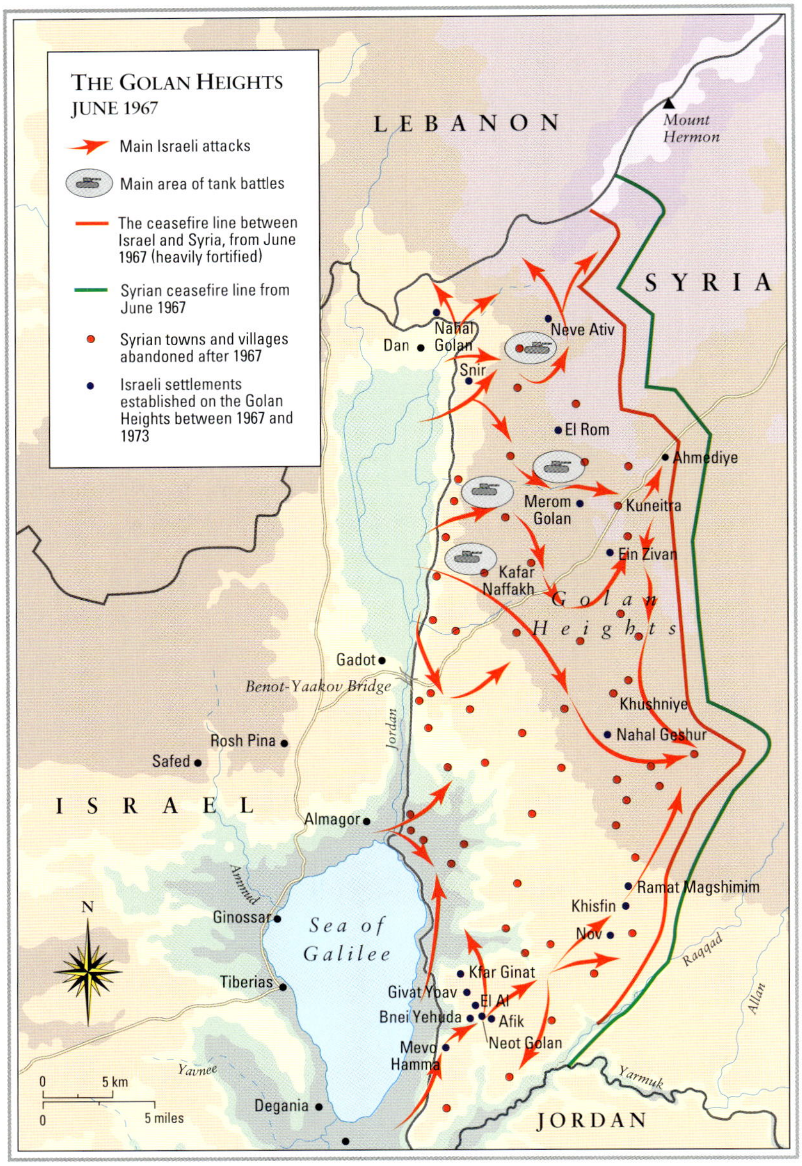

GOLAN BATTLES
Israeli and Syrian armour clashed repeatedly on the disputed Golan Heights. Within 48 hours, Syrian unit cohesion had eroded and the Israelis had captured the entire plateau. Possession of the Golan Heights improved Israeli border security.

controlled east-west movement through the Sinai. Simultaneously, other mobile elements of Yoffe's command raced west to seize the vital Bir Gafgafa crossroads. Between them, these Israeli advances cut off the escape routes of the Egyptian forces still located south of Abu Agheila in the Sinai, which had not yet been engaged by the Israelis. By 9 June, these armoured spearheads had reached, and then blocked, the passes. As enemy forces in the Sinai tried to withdraw west toward Egypt proper, on the Suez Canal's western bank, they found these escape routes denied to them by the well dug-in Israeli armour. Subsequently, the IAF subjected the increasingly demoralized Egyptian forces caught east of the passes to continuous attack. With the Egyptian forces trapped in the Sinai now beginning to surrender en masse, during 10-11 June Israeli forces raced along the Sinai peninsula's south-western perimeter and closed up on the Suez Canal between El Qantara and Suez. Finally, intense superpower pressure now forced both sides to agree a ceasefire on the 12th, the sixth day of the war. By then the Israelis had inflicted a swift and crushing defeat on the Egyptians and captured the Sinai. From the Israeli perspective the 1967 war had been an amazing triumph that seemed to vindicate its decision to build up the combat power of its air force and armoured units to wage a high-tempo air-land war of mobility – a campaign that in effect had been a modern variant of the German 'Blitzkrieg' of World War II. The Israelis gained vital additional territory to provide the geo-strategic depth they craved. This extra buffer territory made Israeli feel less insecure, but in reality, the war only deepened the hostility felt in the region toward the existence of the Jewish state. The defeat in 1967 had been a stain on the honour of the Arab states involved – one that needed to be avenged. This desire triggered a series of events that led to the 1973 Yom Kippur War, also known as the Ramadan War.

The Yom Kippur War, 1973

During the late 1960s, Syria and Egypt tried to compel Israel to withdraw from the occupied Sinai and Golan Heights by using political means, by waging a low-level military campaign of harassment and by supporting terrorism within the Jewish state. When these efforts failed to deliver the desired result, the two countries decided to employ military force. Massive spending helped rebuild their shattered armed forces, and this was coupled with a sophisticated strategy for attacking Israel. Both sides aimed to mount simultaneous strikes against Israel and then dig-in, defeating the inevitable

WALID FORWARD
Arab troops parade while riding aboard several Al-Walid armoured personnel carriers. A variant of the Soviet-designed BTR-152, the open-topped Al-Walid made its combat debut during the Six-Day War of 1967.

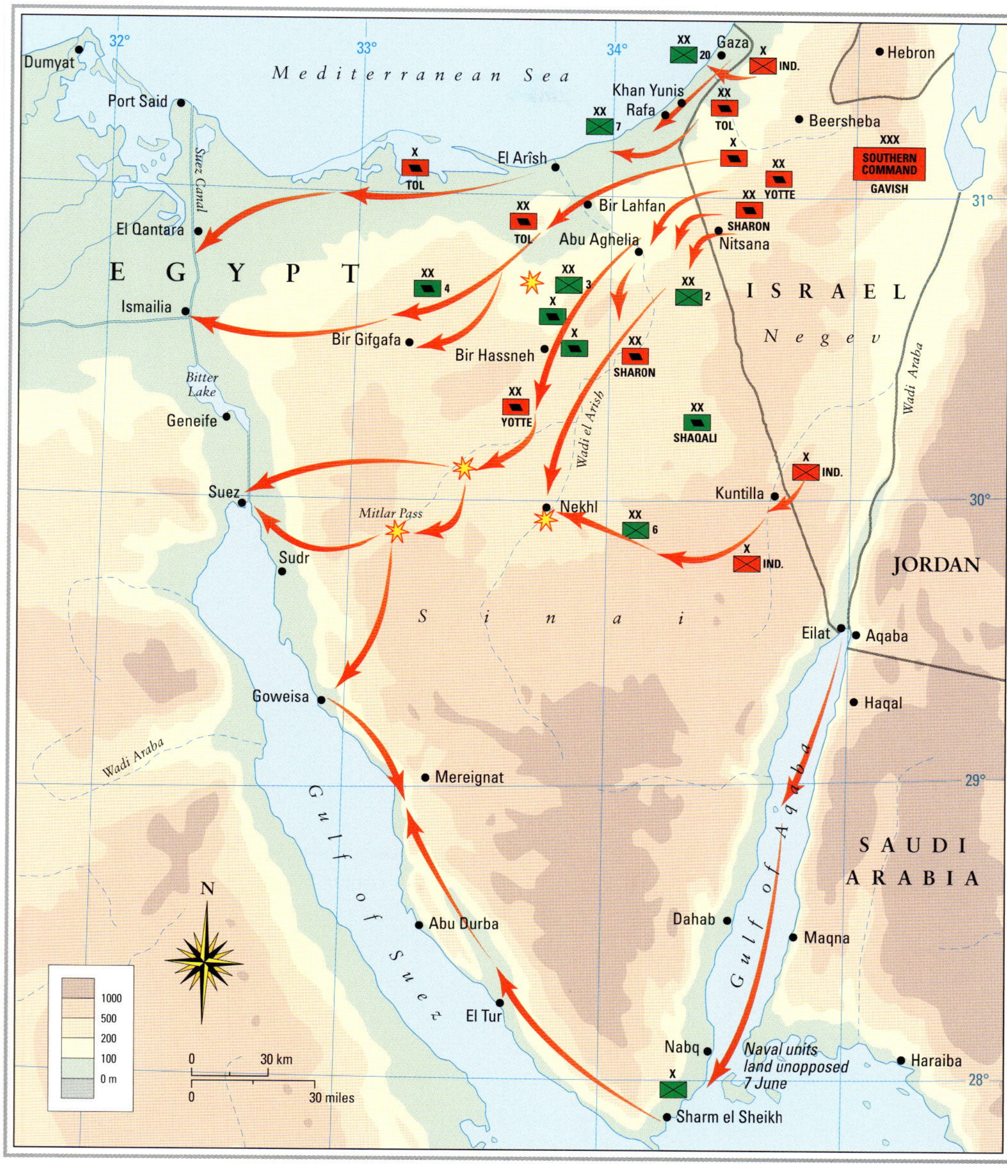

Israeli's Conquest of Sinai

5–8 June 1967

- Initial Egyptian positions
- Principal Israeli lines of advance
- Final destruction of Egyptian forces

CONQUERING SINAI
Seizing the initiative against the Egyptians, Israeli forces attacked at several key points and advanced swiftly to the banks of the River Nile near Suez. Thousands of Egyptian troops were trapped to the south and surrendered.

Israel counterattacks. With Israeli armoured units destroyed in these attacks, and the IAF smashed by new surface-to-air missile (SAM) assets, the two Arab states would insist on the return of the occupied lands as the price for Israel being allowed to end the war before the situation further deteriorated. It was a clever strategy, and such sophistication would be required in any future clash with the effective Israeli armed forces.

During early October 1973 Syrian and Egyptian forces amassed on the border amid the tightest security and under the cover of extensive deception schemes, including the guise of manoeuvres designed to initiate a fundamental reform of the services. Then at 14.00hrs on 6 October Syrian and Egyptian forces attacked the Israeli-held territory of the Golan and the Sinai. The onslaught essentially caught the Israelis by surprise, for they had only realized that an attack was imminent just 10 hours before it began. This was not long enough to mobilize Israel's military reserves, and so the Government eschewed on this occasion the option of pre-empting a seemingly imminent attack as it had done back in 1967. In an intricately planned and well-organized attack, five Egyptian infantry divisions crossed the Suez Canal and struck the Israeli defences of the Bar-Lev line. This was little more than a weak outpost line of 12 forts along the canal's eastern bank, each manned by around 500 men, supported by a total of 100 tanks and 30 guns. In total, the Egyptian campaign involved 500,000 troops, 2,200 tanks, 2,300 artillery pieces,

YOM KIPPUR ATTACK
Following months of planning, Egyptian troops crossed the Suez Canal in a surprise attack on Israeli defences. The coordination of ground troops, aircraft and air defence missiles initially resulted in great progress for the Egyptian offensive.

150 SAM batteries and 550 aircraft. This initial assault was achieved at relatively low cost – just 208 Egyptian personnel died, whereas their combat predictions had suggested that 10,000 might be killed in the initial crossing. In the 36 hours following the initial assault and the overrunning of these forts, the Egyptian forces moved on to establish a 16km (10-mile) deep bridgehead, before digging in; the whole bridgehead, therefore, remained under the protection of the static SAM batteries located on the western side of the canal. Into this bridgehead the Egyptians brought up anti-tank guns and 'Sagger' anti-tank missile teams to block the inevitable counterattacks, which were soon mounted by Israel's two reserve armoured divisions. Even though these divisions lacked many of their supporting arms (which were still mobilizing back in Israel proper), the tank regiments nonetheless charged headlong into the massed Egyptian anti-tank defences, while the Arab SAM umbrella prevented the IAF from providing effective close air support to its ground troops. In these repeated counterattacks, the Israel armour suffered appalling casualties, with Adan's division in the north losing 75 percent of its AFV strength in the first 72 hours.

At the same time as Egyptian forces assaulted across the Suez Canal, massed Syrian armoured units smashed their way through the Israeli defences on the strategically vital ground of the Golan Heights, while SAM batteries took their toll on the Israeli aircraft who tried to support the desperate ground defence. Despite facing fanatical resistance from many of the weak screening units that guarded the plateau, the numerically superior Syrian forces gradually made deep penetrations throughout the entire plateau. Nowhere was the fighting more intense than in a valley in the centre of the Heights that the defenders named the 'Vale of Tears' due to the number of comrades who fell there. By 7 June, the Syrian armour had closed on the vital Benot-Yaakov bridge over the River Jordan, the gateway into northern Israel proper. But by

SHERMAN LONGEVITY
The service life of the M4 Sherman tank extended for decades. In this photo, Israeli tank soldiers perform maintenance on their Shermans during the Suez operations in 1956. Israeli engineers modified some of their tanks to produce the Super Sherman.

Then Israeli reinforcements, newly mobilized within Israel itself, were reaching the line of the Jordan. Flung piecemeal into the fray as they arrived, these reinforcements helped halt the Syrian advance. Thereafter, the initiative passed to the Israelis, and they began their counterattacks in the southern and central sectors, aided by additional aircraft moved across from the Sinai front. By 18 October the Israelis had recaptured the whole of the plateau. Not satisfied with this, Israeli armoured units then began to advance northeast toward Damascus; by the ceasefire on 24 October, these units had advanced ten miles beyond the 1967 border to capture the key terrain of Mount Hermon.

In the Sinai, meanwhile, the front had remained static during 10-14 October, as the Israelis regrouped and the Egyptian forces strengthened their defences. The turning point of the war proved to be the point when Israel gained the initiative in the Golan, however. In response to Syria's desperate pleas for help, during 12-13 October the Egyptians moved their strategic armoured reserve of two divisions across the Suez Canal. On the 14th this armour struck east, deep into the Sinai. Having moved beyond the SAM umbrella, the Egyptian forces encountered well-balanced Israeli armoured forces that had relearned the all-arms battle and which were also well supported by the IAF. In the biggest tank battle since Kursk in 1943, the Egyptian armour was decimated by the Israeli counterattacks. Seizing the initiative, during 15-16 October the Israeli forces counterattacked west.

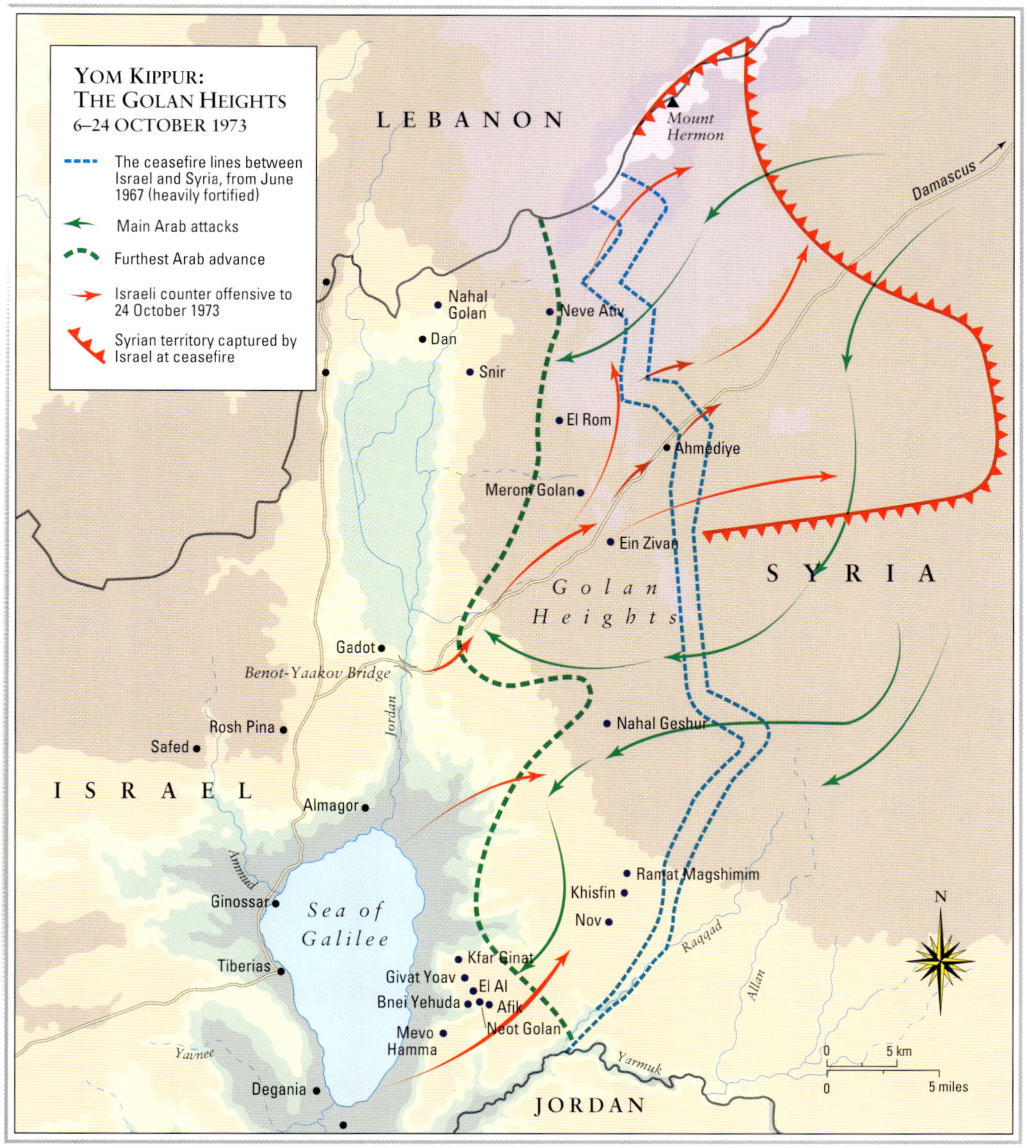

SYRIAN SURGE
Syrian tanks breached the thin Israeli defences along the Golan Heights, threatening to capture a key bridge across the River Jordan. However, the timely arrival of Israeli reinforcements eventually reversed the situation, and Israeli armour rolled toward Damascus.

This advance reached the canal just north of the Great Bitter Lake on the 16th. Over the ensuing 72 hours, Adan and Magan's divisions moved onto the western side of the canal and struck south deep into enemy territory down toward Suez. By the 24th, the Israeli advance had reached the Red Sea coast at Suez, trapping the Egyptian Third Army in a pocket located on the eastern side of the canal. The superpowers, fearing that they would be sucked into the conflict and trigger World War Three, both now exerted immense pressure on the belligerents to agree a ceasefire. On the 25th,

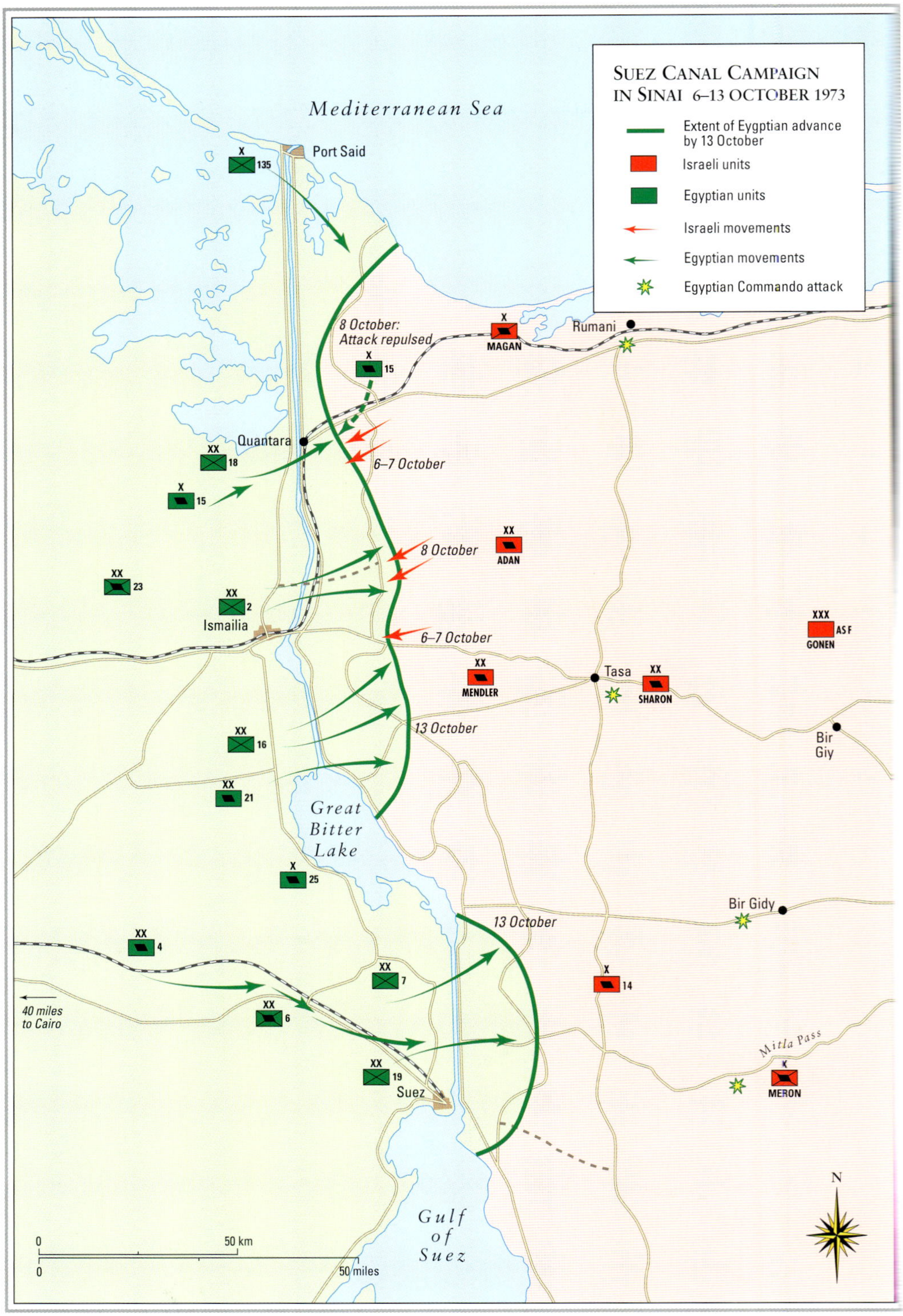

SUEZ CAMPAIGN
Committing their strategic armoured reserve to aid the Syrians, the Egyptian forces were subjected to coordinated Israeli air and ground attacks, which devastated their ranks, particularly during the largest tank-versus-tank battle since Kursk in World War II.

this was finally agreed. The conflict finally came to a permanent end in 1978, when, at the Camp David Accords, Egypt became the first Arab enemy to officially recognize Israel's existence, while in return, Israel agreed to withdraw in phases from the occupied Sinai Peninsula.

1982: The Invasion of Lebanon

During 1978 Israel had maintained a security screen across the southern Lebanon up to the River Litani to shield the north of the country from insurgent attacks by the Palestine Liberation Organization (P.L.O.); it then withdrew leaving a U.N.

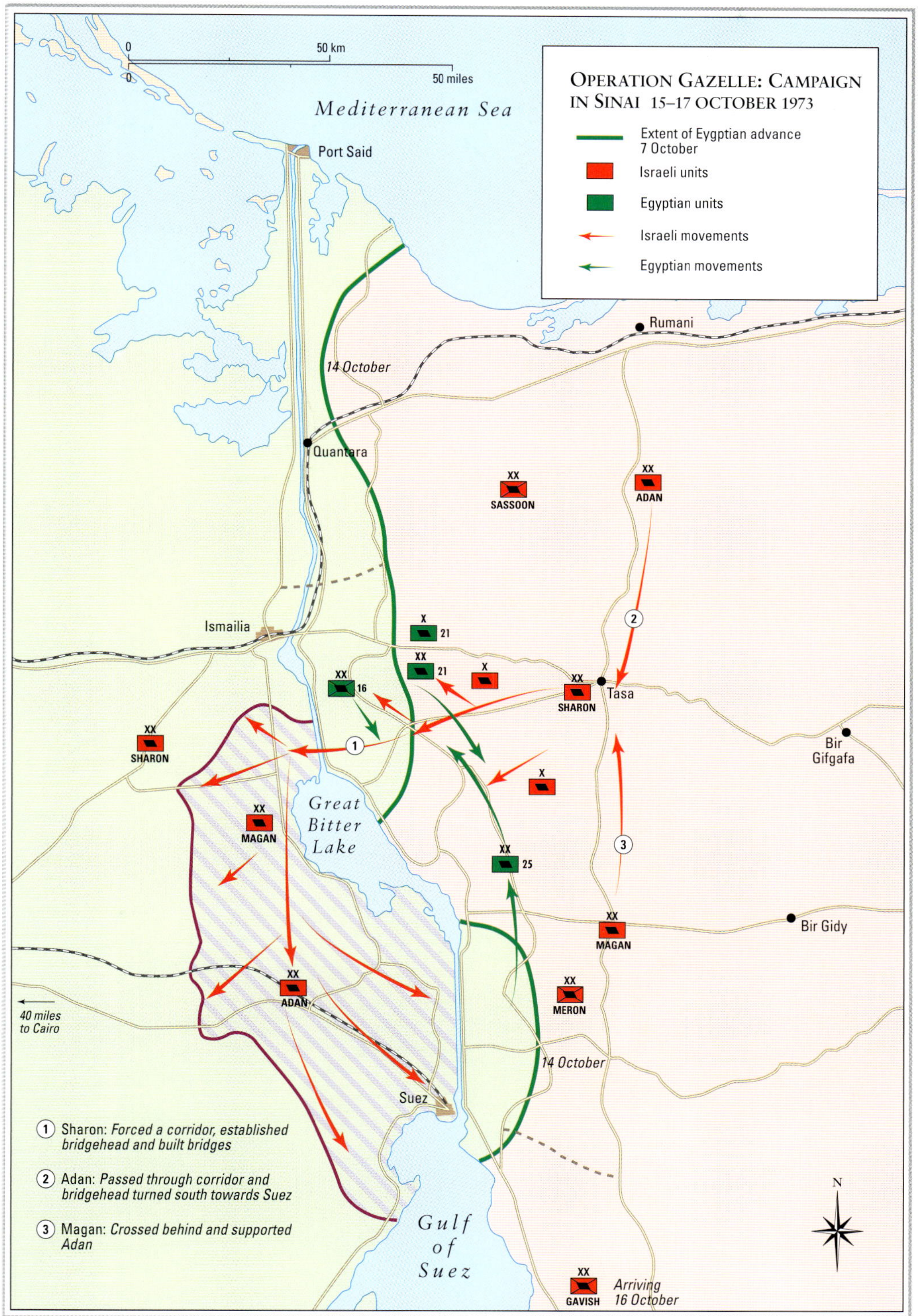

OPERATION GAZELLE
Israeli counterattacks pushed the Egyptian forces across the Sinai, and by late October, the Israeli spearheads reached the Red Sea near Suez. The Egyptian Third Army was trapped, and the super powers began urgently calling for a ceasefire.

MERKAVA MENACE
The Israeli Merkava tank, the first armoured vehicle of its kind designed and produced in Israel, entered service with the Israeli Defence Forces following more than a decade of development and proved a potent weapon.

peacekeeping force to monitor P.L.O. attacks on Israel and the friendly Haddad militia in control of the area immediately adjacent to the Israel's northern border; however, the P.L.O. soon began to increase its influence across southern Lebanon. During 1980-82 Israel became increasingly worried by P.L.O. attacks on northern Israel and by growing Syrian control of the Bekaa Valley in eastern Lebanon; here Syria had exploited the internecine strife of rival militias that afflicted the country.

In June 1982, in Operation Peace for Galilee, the Israelis invaded, although the extent of this incursion remained hotly debated within the Israeli government. Many ministers advocated a relatively limited military incursion against the P.L.O. along the west of the country up to the River Awali, or perhaps even as far as Beirut. But the 'hawks', led by Defence Minister Ariel Sharon (a combat veteran of Israeli wars since 1948), wanted an even larger conflict, which would encompass a fundamental struggle against Syria in eastern Lebanon. Gradually over the course of the first few days, the 'hawks' got their way.

Initially, in Operation Little Pines, 6-8 June 1982, four Israeli divisions attacked the P.L.O.'s positions in southern Lebanon, advancing along the western and central axis as far north as the outskirts of Beirut. In these operations, Israeli mobile divisions (spearheaded by the new Merkava main battle tank) conducted high-tempo advances, backed

MAGACH 7
Based on the design of the U.S. M60 Patton tank, the Magach 7 entered service with the Israeli Defence Forces in the early 1980s. It was used extensively in the fighting in Lebanon and remains deployed today.

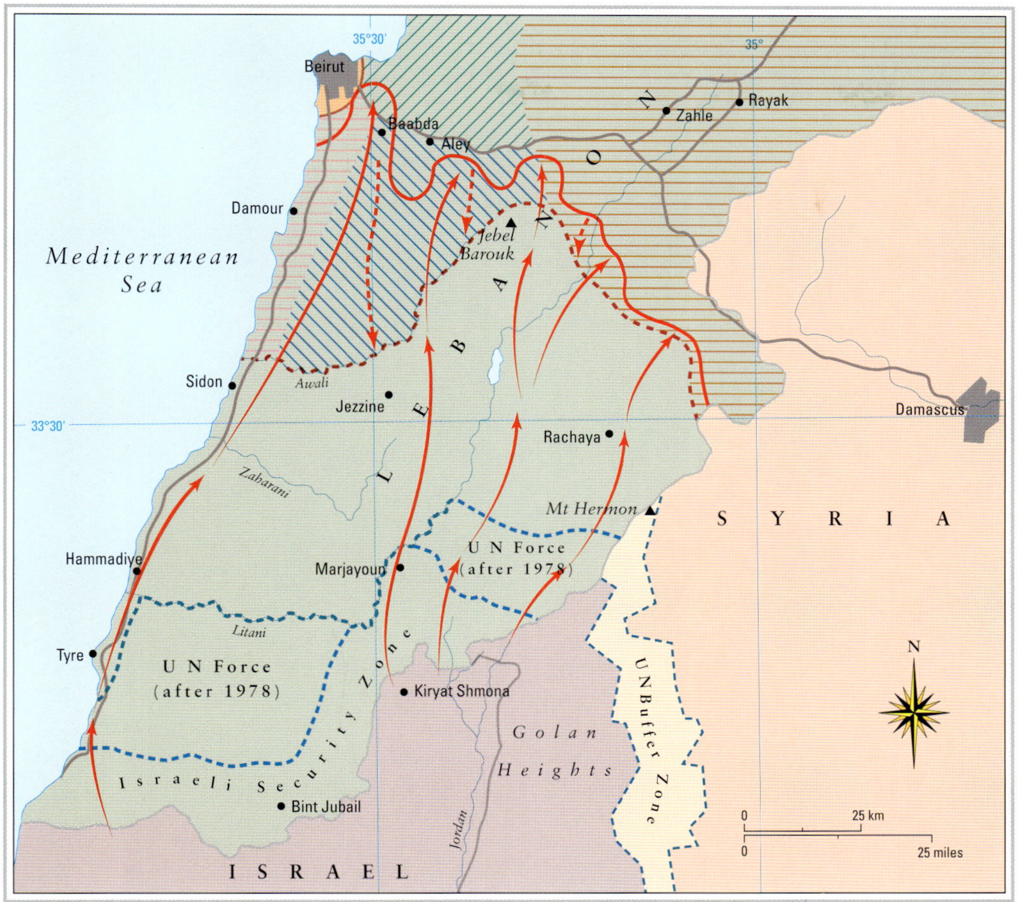

P.L.O. PUSH
In response to incursions and rocket attacks by the Palestinian Liberation Organization, Israeli forces invaded southern Lebanon in the spring of 1982, fighting the P.L.O. and the Syrians for control of Beirut prior to a United Nations brokered ceasefire.

by amphibious assaults along the coast, against the P.L.O.'s essentially non-conventional militias, which numbered 15,000 combatants. Next, during 9-11 June, at the instigation of the 'hawks', Israel widened the war. In this phase, two Israeli divisions deployed on the eastern axis advanced north and began to probe the southernmost Syrian forces located in front of the Bekaa valley. In particular, during 9 June, 188 Israeli aircraft, organized into carefully-selected dedicated force packages, destroyed 17 of the 19 Syrian SAM batteries in the Bekaa, with the remaining two being destroyed the next day. With the IAF now in control of the skies of Lebanon after this stunning success, the Syrian 1st Armoured Division suffered heavy casualties at the hands of IAF air strikes; simultaneously, Israeli ground forces pushed the Syrian forces back north toward the Beirut-Damascus Highway. By 11 June, the Israelis had been largely successful in driving the P.L.O. and Syrian threats away from northern Israel, sustaining only light casualties in the process.

During the ensuing six weeks, however, the IDF became bogged down in a long and bitter attritional battle with the P.L.O. for control of Beirut, which left much of the city devastated. Finally, on 12 August, in a U.N.-brokered ceasefire deal, the Israeli forces agreed to withdraw from Beirut while the P.L.O. forces withdrew by sea to Tripoli in Libya; this conflict resolution process was overseen by the deployment of the Multinational Force (MNF) in a traditional peacekeeping role. Although initially this process unfolded successfully, during 16-19 September it suffered a serious setback. IDF forces transported Lebanese Marionite Christian Phalange militia into the Sabra and Chatilla refugee camps to root out alleged P.L.O. para-militaries hiding there; the Phalange murdered at least 800 Palestinians, the vast majority innocent civilians, in an appalling atrocity, which did much to blacken Israel's international standing.

Despite the MNF's presence, sporadic fighting continued between the Israelis, the Lebanese forces and various local militias such as the Druze and the Phalange until, in May 1983, the new Lebanese president agreed a peace treaty with Israel, which saw the IDF withdraw back into the border region of southern Lebanon.

TANK DEVELOPMENT 1960–90

PROWLING LEOPARD
The Leopard 2 main battle tank was developed in the early 1970s and entered service with the West German Army in 1979. The tank has served in Kosovo and the Middle East.

As the Cold War dragged on, technology continued to advance, contributing to a new generation of combat hardware and refining the capabilities of armoured units of NATO, the Warsaw Pact and, in turn, non-aligned nations. Although some military tacticians asserted that the tank had seen its glory days, others believed in the continuing viability of firepower, speed and armour protection on the modern battlefield.

Resulting from a reassessment of the tank itself, a notable conceptual shift occurred. The specialization of tanks, some lightly armed and armoured for a reconnaissance role, others fitted with heavy guns and intended to engage enemy forces in tank-versus-tank showdowns, gave way to the concept of the main battle tank, or MBT. The British Centurion, the Soviet T-55 and the U.S. M48 Patton signalled the coming of a new era in modern armoured warfare. In concert with tank redevelopment, accompanying infantrymen were envisioned riding to battle in armoured fighting vehicles that could disgorge their human cargo and stand by to engage targets with direct fire support.

A new generation

During the 1960s, the Soviets continued to improve their armoured capabilities with the T-62, followed a decade later by the further improved T-72. Meanwhile, the British Chieftain entered service in 1961, and its contemporary the West German Leopard 1 debuted following the collapse of a joint development project with France. This was followed by the improved Leopard 2 in the mid-1970s, and by 1980 the United States had deployed the M1 Abrams. Further enhancements to the M1 resulted in the improved M1A1 in the mid-1980s and later the M1A2, which saw action during the Gulf War in 1990–91 and Operation Iraqi Freedom in 2003, earning lasting fame. In the hotbed of the Middle East, Israeli engineers recognized the need to produce their own main battle tank. The result was the Merkava, an enduring design that continues to anchor armoured units of the Israeli Defence Forces.

Armed with main guns of 120mm (4¾in) or 125mm (5in) and suites of supporting machine guns each tank mounted progressively more sophisticated offensive and defensive systems, taking into account the improvements realized in anti-tank weapons, including shoulder-fired and vehicle-mounted missiles, as well as the augmented penetrating power of specially designed tank-killing ordnance.

BRITISH WARRIOR
The British Warrior infantry fighting vehicle is capable of delivering up to seven combat infantrymen to battle and supporting them with its 30mm (1in) Rarden cannon and a pair of machine guns. The Warrior entered service in 1988.

Technological wonder

Numerous design and tactical deployment enhancements set ever-higher standards for combat effectiveness in the main battle tank from 1960 to 1990. Specialized turrets housing larger-calibre guns were mounted on heavier chassis that provided more stabilized firing platforms. Automatic loading systems were perfected and increased performance and rate of fire, while providing for a division of

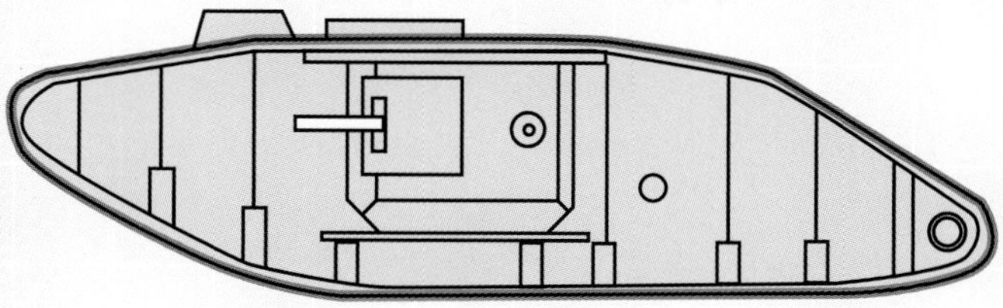

Mk. IV

Manufactured	**1915**
No. built	150
Weight	28 tonnes
Length	9.94 m
Crew	8
Armour	6–12 mm
Main Armament	2 x 6 pdr QF
Secondary Armament	4 x .393 MG
Speed	4 mph (6.5 kph)

Tiger

Manufactured	**1942**
No. built	1,347
Weight	62.71 tons
Length	6.29 m
Crew	5
Armour	25–120 mm
Main Armament	88 mm KWK 36 L/56
Secondary Armament	4 x 7.9 mm MG
Speed	24 mph (38.5 kph)

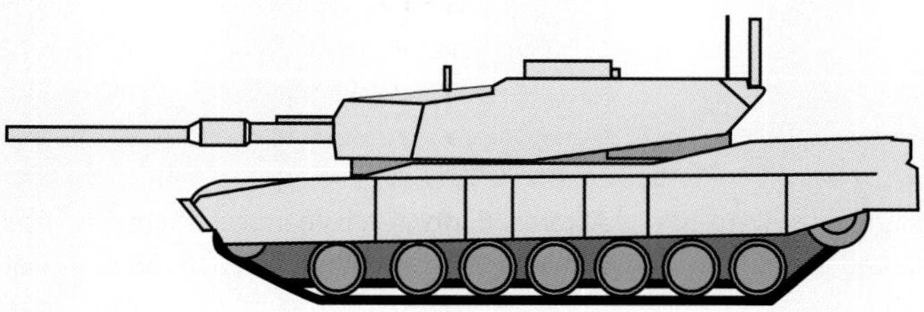

Abrams

Manufactured	**1980**
No. built	9,000 +
Weight	60.4 tons
Length	7.93 m
Crew	4
Armour	1620–940 Comp. rifled
Main Armament	105 mm or 120 mm smooth bore
Secondary Armament	1 x 12.7 HMG 2 x 7.62 MG
Speed	42 mph (67.5 kph)

ARMOURED EVOLUTION
Three iconic incarnations of the tank in modern times include the British Mk IV of World War I, the German Tiger of World War II, and the U.S.-designed M1 Abrams main battle tank that entered service in the early 1980s.

labour that allowed a tank crew to operate with greater efficiency in the heat of battle. Diesel powerplants were joined by innovative gas turbine designs, providing impressive speed and quiet running.

Infrared vision systems and laser range-finding equipment pierced the veil of the night and rendered inclement weather inconsequential, easily compensating for atmospheric conditions such as varied wind speeds. Targets could now be acquired more rapidly than with the naked eye. Later variants of proven main battle tanks include stabilization systems that permit the tank to fire on the move. Battlefield command systems allow a tank commander to track several targets simultaneously, while quickly identifying contacts as friend or foe.

On the defensive side, the British innovation of composite Chobham armour, designed to minimize the damaging impact of incoming ordnance, has been complemented by modular and explosive reactive armour. The perennial problem of storing explosive ammunition has resulted in the redesign of interior space and the installation of components that focus the force of an explosion outward in the event of a direct hit. Nuclear, biological and chemical (NBC) defensive systems increase survivability under such adverse conditions, and sophisticated equipment warns of the presence of radiation from nuclear weapons.

Low-tech threat

Despite advancing technology, the continuing threat of improvised explosive devices (IED), land mines, and other rather low technology has persisted, while the confines of urban warfare present continuing hazards as well. In response, the generation of main battle tanks emerging in the late 1980s began to exhibit improved armour protection in vulnerable areas such as the underside, exhaust and ventilation systems, and crew compartment.

INDOCHINA AND VIETNAM

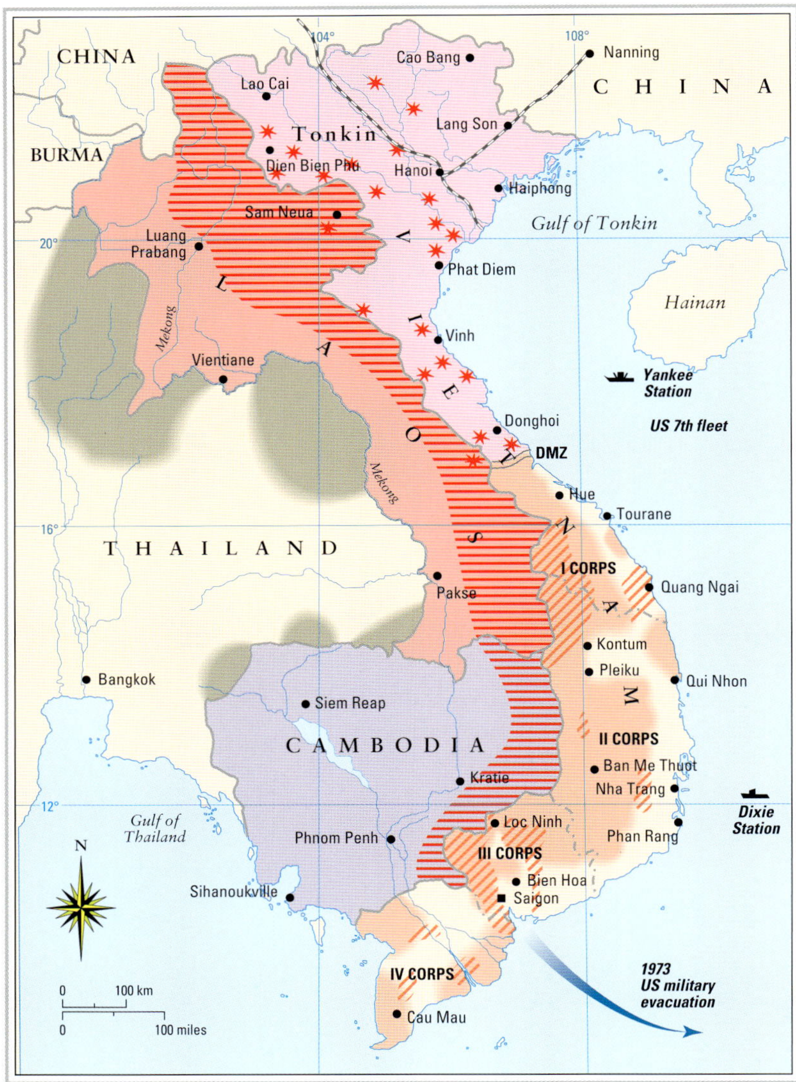

VIETNAM WAR 1959–75

- Communist-controlled area in Laos and Cambodia 1950–75
- Communist-held area January 1973 "ceasefire"
- Controlled by Khmer Rouge c. 1975
- Controlled by Pathet Lao c. 1975
- Area of Communist guerrilla activity c. 1975
- U.S. Corps command area
- North Vietnam subject to air attack

The Indochina War ended in 1954 with a French withdrawal from their former colonial possessions, but it left many local issues unresolved. North and South Vietnam were established under local governments and separated by a demilitarized zone (DMZ) that was monitored by international forces. The Communist regime in North Vietnam was content, at first, to await developments in the south.

The South Vietnamese government was expected to collapse under internal pressures, and the Communists were poised to take control. Instead, South Vietnam emerged as a reasonably stable state, absorbing many refugees from the north.

Northern intervention

To hasten the demise of South Vietnam, the North launched a guerrilla campaign in 1959. Some personnel were already in place, while others were able to enter South Vietnam across the long frontiers via Cambodia and Laos. Some also infiltrated across the DMZ or by sea, landing on the long eastern coastline.

The North Vietnamese had extensive experience of guerrilla operations, having recently ejected the French from their territory. Some personnel had also fought in the Chinese Civil War. Conversely, the Army of the Republic of Vietnam (ARVN) was inexperienced and ill-prepared to deal with an insurgency. Organized and equipped along conventional lines, ARVN forces were optimized for meeting a conventional invasion across the DMZ by similarly equipped North Vietnamese forces.

These forces did exist. The North Vietnamese Army (NVA) was quite capable of engaging in conventional operations and possessed tanks and artillery. This threat had to be countered, and throughout the conflict that ensued, ARVN forces remained fixated on the DMZ. Weakening this frontier to deal with relatively minor guerrilla threats was not a viable option.

To counter the guerrilla threat, South Vietnam created lightly equipped defensive forces, originally named the Civil Guard and later, the Regional Forces. This might have been effective, but they were poorly trained and lacked confidence. The result was that while the DMZ remained secure, South Korea gradually lost control of the countryside.

AUSTRALIAN TANKS IN VIETNAM The Centurion tank arrived too late to see action in World War II, but became the main battle tank of the British and Australian forces for many years afterward.

THE WAR IN SOUTH-EAST ASIA (Map opposite). The defence of Vietnam was complicated by a long land frontier with nations more sympathetic to the Communist North than the pro-Western South.

M50 ONTOS TANK DESTROYER (bottom left) The Ontos was an experimental tank destroyer design using six 106mm (4in) recoilless rifles. Although it underperformed in its intended role, it proved useful as a fire support platform.

MULTI-ROLE AMPHIBIOUS APC The U.S. LVT-5 featured a stabilized 75mm (3in) gun, to allow accurate firing when in the water. It served as a transport and fire support vehicle, with some examples also outfitted for combat engineering.

Some aid was provided by the U.S. and other countries, but it was not until U.S. vessels were attacked by North Korean naval forces in the Gulf of Tonkin in 1964 that full intervention took place. By this time the Vietnamese Communists (or Viet Cong, as the guerrilla forces were now known) were able to defeat even regular ARVN units. Their tactics were familiar from the earlier war against France, using the jungle to move into the target area, launching an attack and then vanishing again. Often an attack on an outpost was used to draw a response which was then in turn ambushed.

U.S. forces in Vietnam

Large-scale U.S. intervention began in 1965, with the arrival of Marine units. Troop strength was gradually increased thereafter to counter the intensified Communist attacks. U.S. forces concentrated on dealing with NVA units which had begun to enter South Vietnam, while South Vietnamese forces took back control of the countryside from the Viet Cong.

U.S. tactics emphasized firepower, with mutually supporting firebases able to deliver artillery fire on any hostiles detected in the area. These bases naturally invited attack, which suited American tactics. Defended by wire and mines, as well as dug-in armoured vehicles and infantry, a firebase served as a honey-pot to concentrate enemy forces in a predictable location where they could be shelled or would be broken on the defences. Meanwhile, 'search and destroy' operations inflicted steady casualties on the NVA and Viet Cong.

Amphibious armoured vehicles proved highly useful in the swampy terrain of South Vietnam, and for crossing the country's many rivers. AMTRACs (Amphibious Tractors), first developed to support amphibious invasions in World War II, were often used as road transport, which increased the wear and tear on the vehicles, but provided useful mobility. Their light weapons might have been handicapped in a battle against an enemy that deployed tanks, but within South Vietnam the enemy had none.

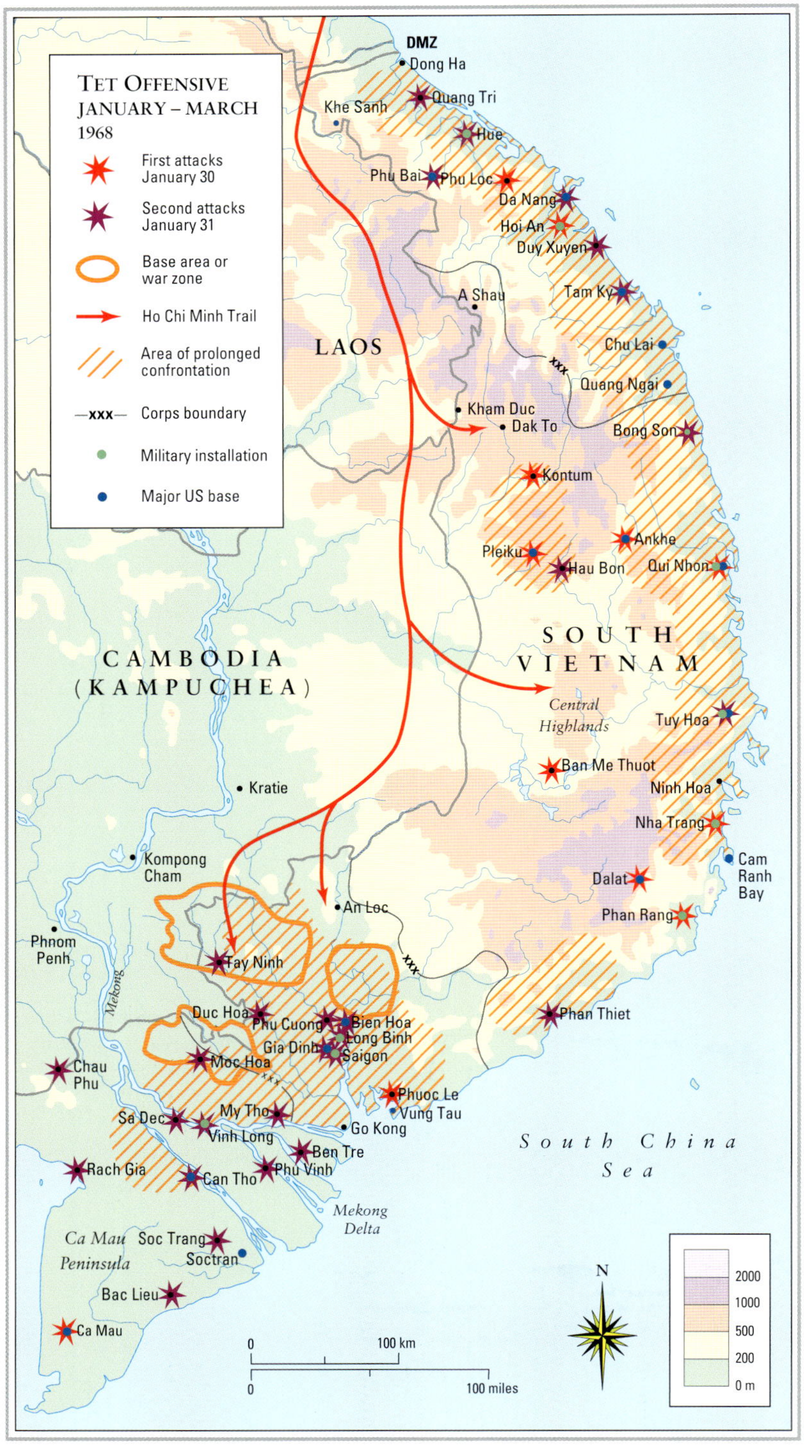

TET OFFENSIVE, 1968
The Tet Offensive involved coordinated strikes against regional capitals and other major targets as well as a conventional attack across the DMZ. The Viet Cong exhausted its strength for little gain, leaving the NVA to shoulder the burden of the war.

The only North Vietnamese tanks were in the hands of regular NVA formations deployed along the DMZ, and saw little action during the American involvement in Vietnam. U.S. tanks served mainly as mobile strongpoints and as infantry support weapons, able to bring heavy firepower to bear on an enemy position, or to blast snipers or machine-gunners out of their cover. Tank operations were heavily constrained by the terrain, which made rapid movement impossible. Thus tanks were primarily deployed in 'penny packets' to support infantry.

The Tet Offensive

By early 1968 the situation had been stabilized and U.S. forces were largely concentrated on the borders. Nearly half a million American troops were deployed and every major action was won by U.S. firepower. Although NVA units did slip into South Vietnam to join the Viet Cong there, the North was most definitely not winning. In an effort to reverse this, a major offensive was launched by the Communists. The Tet Offensive, named after the Tet new year holiday when it began, was a simultaneous attack on many points by NVA and Viet Cong forces within South Vietnam, while other troops attacked across the DMZ.

It began well enough for the North Vietnamese, with surprise attacks on installations and regional capitals, but most were quickly beaten off. At the city of Hue, however, fighting went on for a month before the North Vietnamese forces were ejected from the city. U.S. and ARVN reinforcements moving up to the city had to fight past enemy blocking positions on the road before joining the battle proper.

Tet was part of a wider series of operations and large-scale actions launched in late 1967. It was followed by a siege of the U.S. base at Khe Sanh in January 1968, which lasted from January to April, although the base was never in serious danger of being overrun. Resupplied and reinforced by air, Khe Sanh was eventually relieved by an airmobile/ground advance which broke through the opposition at several points to lift the siege.

By this time, the Tet Offensive was long over. Tet was a decisive military defeat for the North Vietnamese. The Viet Cong suffered such huge casualties that they were thereafter unable to contribute greatly to the conflict, and had to be replaced by lightly equipped NVA formations, which slipped into South Vietnam to continue the guerrilla conflict. However, the political implications for the U.S.A. were significant, with the American public and their leaders shocked by the Communists' ability to launch such devastating attacks. Moves towards disengagement began in 1969.

The U.S. adopted a strategy of strengthening South Vietnam and preparing its forces to take control of the country. Even though the U.S. troop commitment had been halved by 1971, South Vietnamese government control steadily improved.

However, peace negotiations imposed conditions on American forces which limited their ability to respond to situations as they developed.

In March 1972, the NVA launched an invasion across the DMZ. Despite having recently been rearmed with Soviet equipment including T-55 tanks and new artillery, the NVA was repelled after hard fighting. U.S. air power was instrumental in supporting the defence, but it seemed that ARVN forces were now capable of defending their nation. The U.S. ended its military involvement in January 1973, withdrawing the last of its forces.

Fall of South Vietnam

The NVA had taken severe losses in the 1972 offensive, not least due to poor infantry-armour cooperation. However, with the Americans gone it was able to rearm until, by 1975, its conventional forces outnumbered and outgunned those of the south by a broad margin. Conversely, the withdrawal of U.S. support was accompanied by a reduction in other forms of aid, which severely limited the capabilities of ARVN units.

South Vietnam was placed in an impossible strategic position. Forced to maintain heavy forces along the DMZ to counter the threat of invasion, it was open to an invasion and a drive on the capital, Saigon, through Cambodia and Laos. Weakness at either point invited defeat; strength at both was impossible. Reliance on U.S. aid limited South Vietnam's strategic options and ensured that the initiative rested firmly with the North.

In March 1975, the North launched an offensive that aimed to secure a route for a later, more widespread assault. Low morale and conflicting orders caused the ARVN defence to collapse, enabling the offensive to be expanded. Despite stubborn defence at Hue and a two-week stand at Saigon, the South could not withstand the offensive.

The final advance into Saigon was led by NVA armoured forces which had played a relatively small part in the war up until this point. However, once the standoff on the DMZ was broken, it was an armoured advance that finally sealed the fate of South Vietnam.

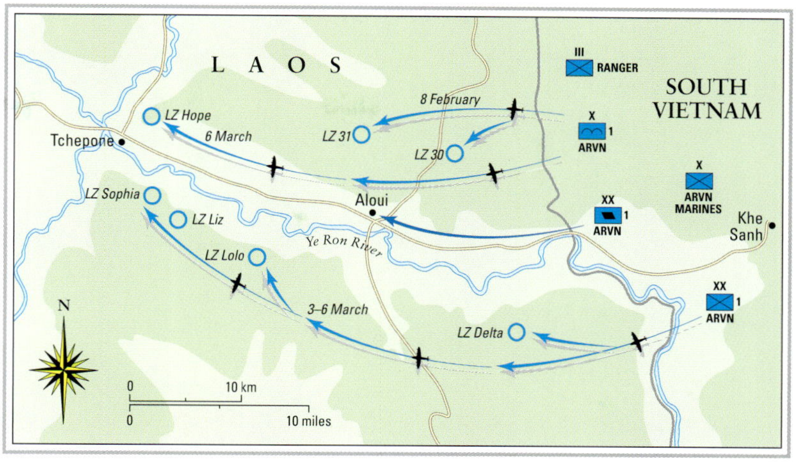

OPERATION LAM SON 719
FEBRUARY – MARCH 1971

Tanks attack route

Aircraft attack route

Main Landing Zones

OPERATION LAM SON 719
Intended as a spoiling attack to prevent further NVA incursions via Laos, Operation Lam Son 719 involved South Vetnamese ground troops with U.S. air support. The operation failed in the face of Laotian resistance.

M113 APC
The M113 is numerically the most successful APC design in history. Thousands have been converted to specialist roles such as air defence or fire support vehicles.

THE TANK ON THE NUCLEAR BATTLEFIELD

The Cold War was a stand-off between NATO and the Warsaw Pact. Had it ever gone 'hot', the fate of Western Europe would have rested upon the ability of NATO forces to resist a massive armoured advance across the North European Plain.

A difference in doctrines

Western armoured forces emphasized quality over quantity, and were well suited to a defensive stance. Their tanks' higher turrets allowed them to take hull-down positions behind hard cover such as walls or ridges, while their excellent targeting systems enabled them to fire accurately at long range, hopefully eliminating enemy tanks before they could close to their effective combat range.

Conversely, the Warsaw Pact preferred designs optimized for the assault role, with a low silhouette to make a hard target and rounded armour for improved defence. Large forces of relatively unsophisticated tanks could absorb heavy casualties and still break through the enemy's positions, forcing a fluid battle where the Warsaw Pact vehicles should have the advantage.

The nuclear dimension

Both sides agreed that a war in Europe might well 'go nuclear', and, massive retaliation with nuclear weapons was a clearly stated defensive response. A full-scale nuclear exchange would probably make tank warfare irrelevant, but limited use of nuclear weapons might not trigger the use of strategic assets. Nuclear weapons might be used to attack enemy troop concentrations or deny a route to the enemy by contaminating it with radioactive fallout.

U.S. ARMOURED DIVISION
In addition to the fighting battalions, an armoured division requires an array of support formations from armoured recovery and resupply units to military police and artillery support. NBC defence formations were added once nuclear warfare became a real threat.

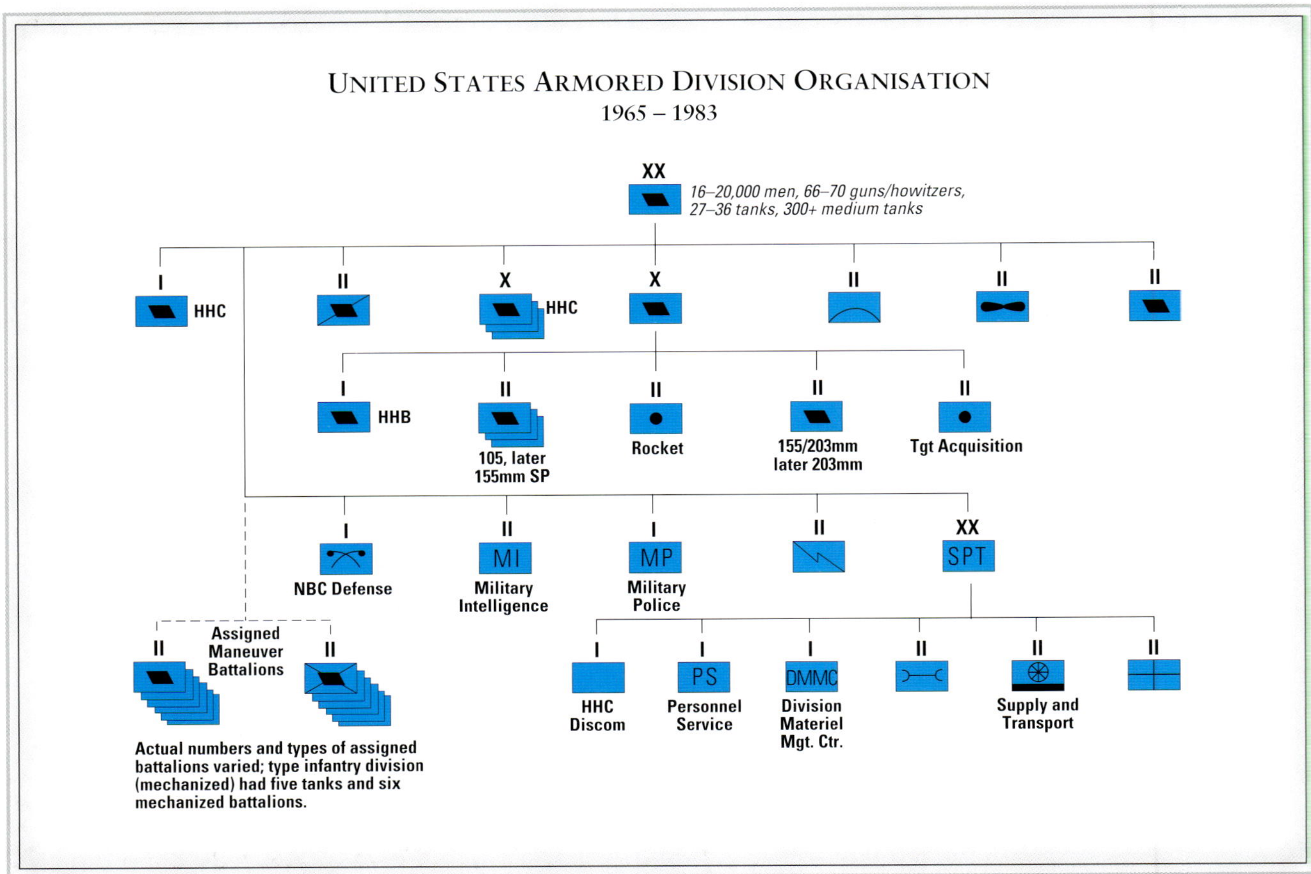

The best defence against nuclear attack is to be somewhere else, so doctrines were developed for operating in a nuclear threat environment. While concentration of force is a key military concept, concentrating high-value assets like armoured forces invited attack, so armoured units were trained to break down into smaller sub-units and disperse when necessary. This reduced fighting power, but also made the formation a less-inviting target. It also ensured that some elements of the formation would survive even if it were attacked with nuclear weapons.

The ability to function in the nuclear environment was built into Cold War tank designs. While nothing would survive at ground zero in a nuclear attack, an armoured vehicle could perhaps protect its crew from the blast and heat produced by the nuclear fireball of a strike, even at fairly short distances. Air filtration units, for example, were used to prevent radioactive fallout penetrating the vehicle's environmental systems.

Armoured formations were given a nuclear defence element. This included nuclear reconnaissance vehicles tasked with finding areas of contamination and ensuring that other units were directed around them or transited quickly under full protective measures, and did not stop in the contaminated area. Mobile decontamination units, able to wash radioactive materials off other vehicles, allowed the crew of a contaminated vehicle to leave it without being harmed and thus reduced secondary casualties.

Ironically, perhaps, these measures made nuclear weapon use less likely. With the benefits of using nuclear weapons greatly reduced by effective countermeasures, there was less to gain in return for the immense risk of escalation to full-scale strategic nuclear warfare.

AVENUES OF ATTACK
Terrain and the location of key targets made the axis of advance for an invasion of Western Europe predictable. NATO had years to plan a defensive strategy, and the Warsaw Pact had as long to find a way to defeat it.

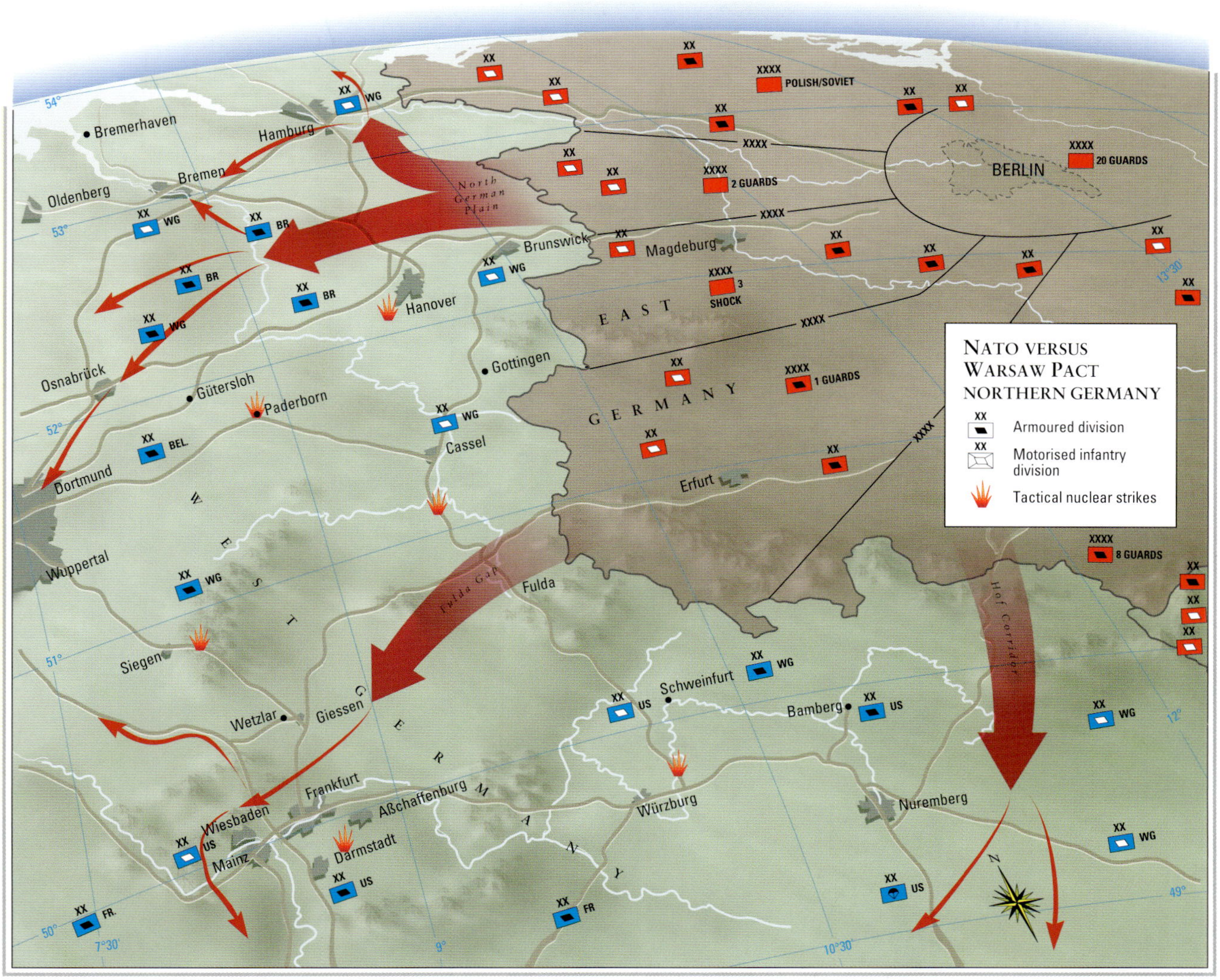

THE FIRST GULF WAR 1990–91

The First Gulf War was fought to liberate Kuwait from Iraqi occupation, and although Coalition forces did enter Iraq this was in the context of defeating the Iraqi Army and eliminating the threat to Kuwait rather than ousting the Iraqi government.

Extensive preparations

Building up sufficient force to remove the powerful Iraqi Army from Kuwait took a considerable time during the winter of 1990, during which Coalition air forces attacked Iraqi command and control facilities, logistics and infrastructure assets, as well as combat units. By the time the ground offensive opened in January 1991, the ability of the Iraqi military to make an effective and coherent response had been severely degraded.

During the build-up period, Iraqi forces launched an attack into Saudi Arabia, aimed at capturing the strategically important city of Khafji. Had this succeeded, Khafji could have been used as a forward base for further operations that would have threatened the Coalition flank and perhaps disrupted the ground offensive.

The advance on Khafji was made by armoured units advancing along several routes. One component of the Iraqi force was able to reach Khafji and briefly occupy the city, but the rest were turned back by Coalition troops after coming under air attack. The city was retaken two days later by Saudi and Qatari troops, assisted by Coalition air power. Not only was the threat to Khafji eliminated, but the Iraqi Army also lost a significant proportion of its armoured assets.

INITIAL DISPOSITIONS
Iraqi dispositions in and around Kuwait assumed that the desert protected their right flank. GPS navigation allowed Coalition forces to launch a rapid outflanking manoeuvre through this terrain, which might otherwise have proven impossible.

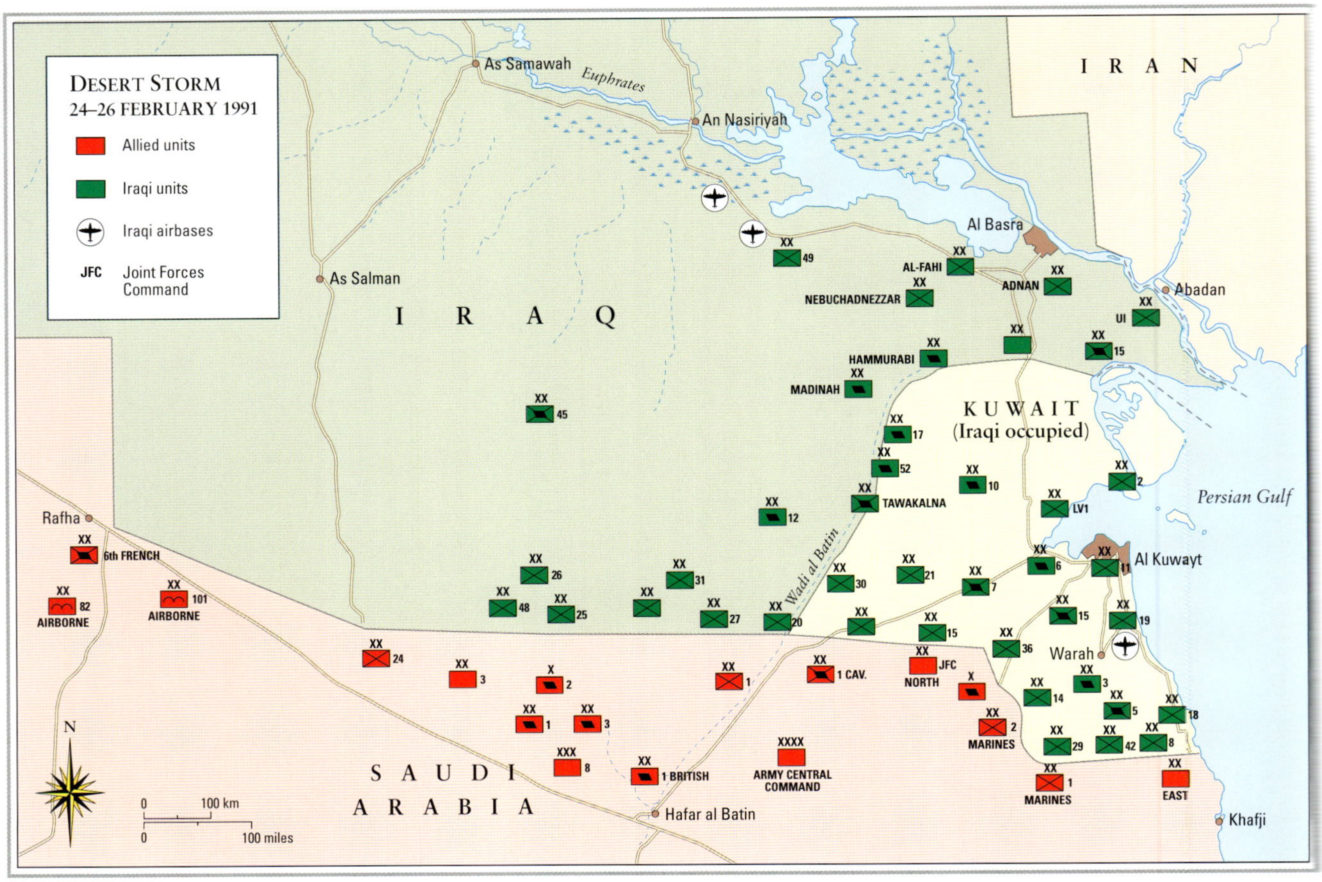

The ground offensive

Once preparations were complete, Coalition forces drove into Kuwait from the south, meeting relatively light opposition. Iraqi units began pulling back into home territory, which would have preserved the formidable Iraqi ground forces for future operations. To prevent this, Coalition armoured forces launched a 'left hook' into Iraq, intended to cut off the line of retreat.

Taking advantage of GPS navigation, the Coalition forces were able to move rapidly through the desert and attack Iraqi units from an unexpected direction. Even without the benefits of air support and reconnaissance, the Coalition's advanced armoured forces had a massive advantage in fighting power over the obsolescent post-Soviet weaponry of the Iraqis.

The ground campaign lasted 100 hours, during which time it had become a pursuit rather than a battle. Heavy casualties were inflicted on those units that stood and fought, and those that fled came under heavy air attack on what became known as the 'highway of death'. With Kuwait liberated and likely to stay that way, the operation was halted.

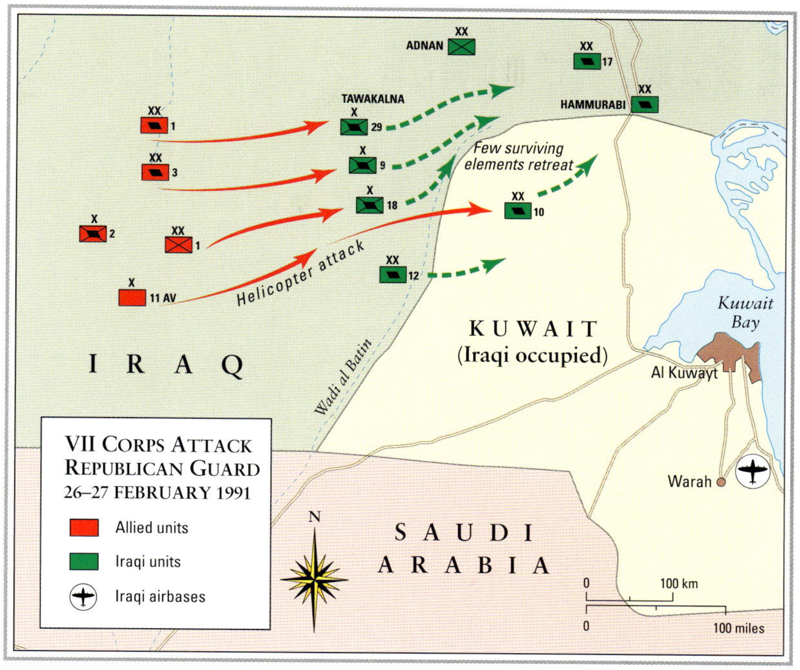

THE COALITION GROUND OFFENSIVE
Coalition forces hooked deep into the enemy rear, ensuring the destruction of the Iraqi Army as it retreated from Kuwait.

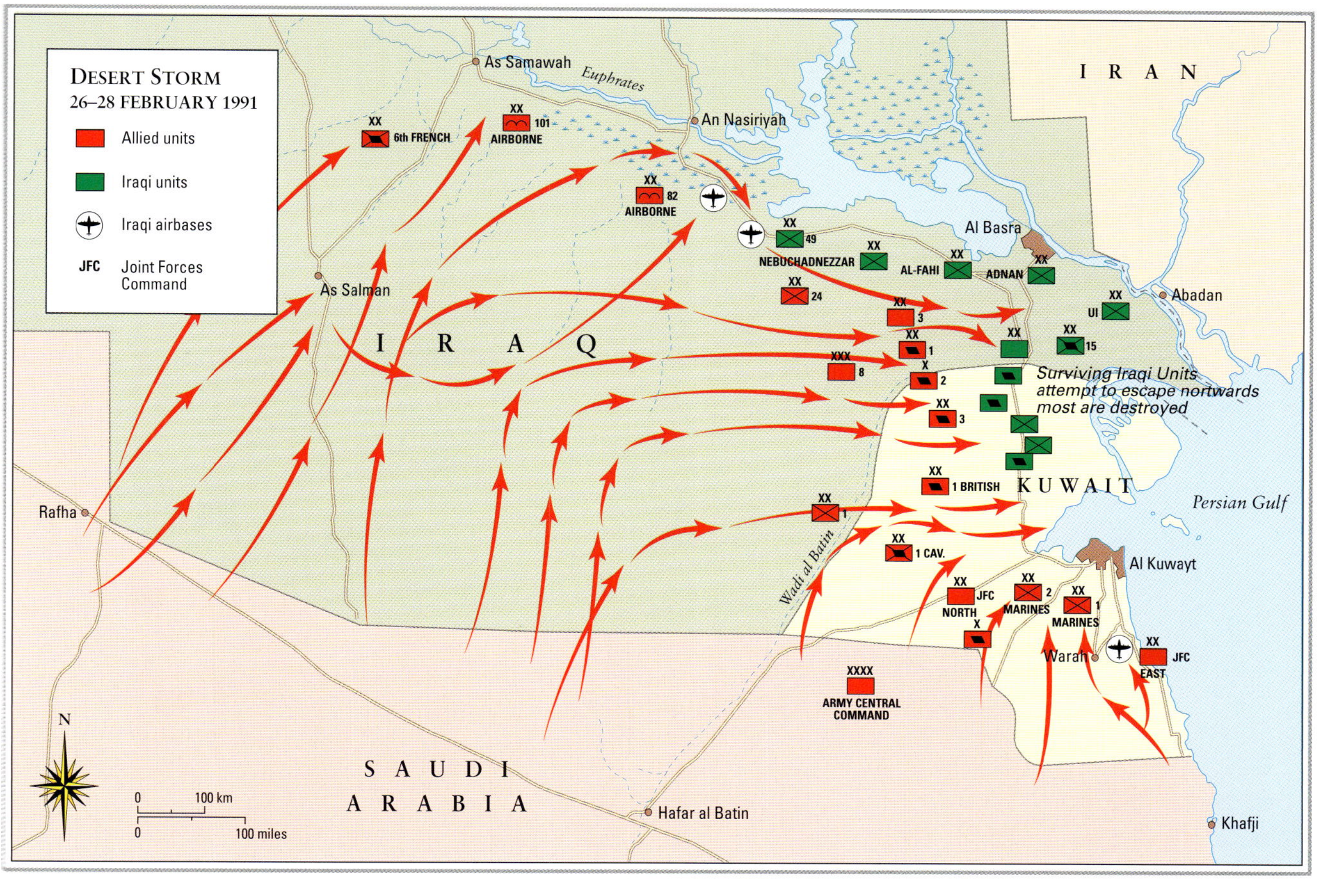

THE SECOND GULF AND AFGHANISTAN WARS

There were those who predicted, at the end of the First Gulf War, that a second campaign against Iraq would be necessary sooner or later if Saddam Hussein remained in power. However, there was nothing that could be done at the time – the United Nations' mandate for the First Gulf War allowed only for the liberation of Kuwait.

After the 1991 Gulf War, Saddam Hussein was able to rebuild the Iraqi armed forces and to crush several uprisings. Thus, by the time of the Second Gulf War in 2003, the international Coalition faced what certainly seemed to be a powerful foe.

No repeat of 1991

It may have appeared that the stage was set for a repeat performance of the 1991 conflict, but there were significant differences in 2003. The Coalition was much smaller, consisting almost entirely of U.S. and British forces, and its goal was different. In 1991 the aim was simply to liberate Kuwait. In 2003 the Coalition was intent on invading Iraq and deposing its government as part of the 'War on Terror'.

On the Iraqi side, the situation was also somewhat different. Although some of the losses incurred in 1991 had been made good, the Army was in poor

OPENING POSITIONS
During the Second Gulf War U.S. Special Forces secured key objectives and assisted Kurdish rebels before the ground offensive opened, creating a distraction. However, it was not possible to conceal the main Coalition axis of advance.

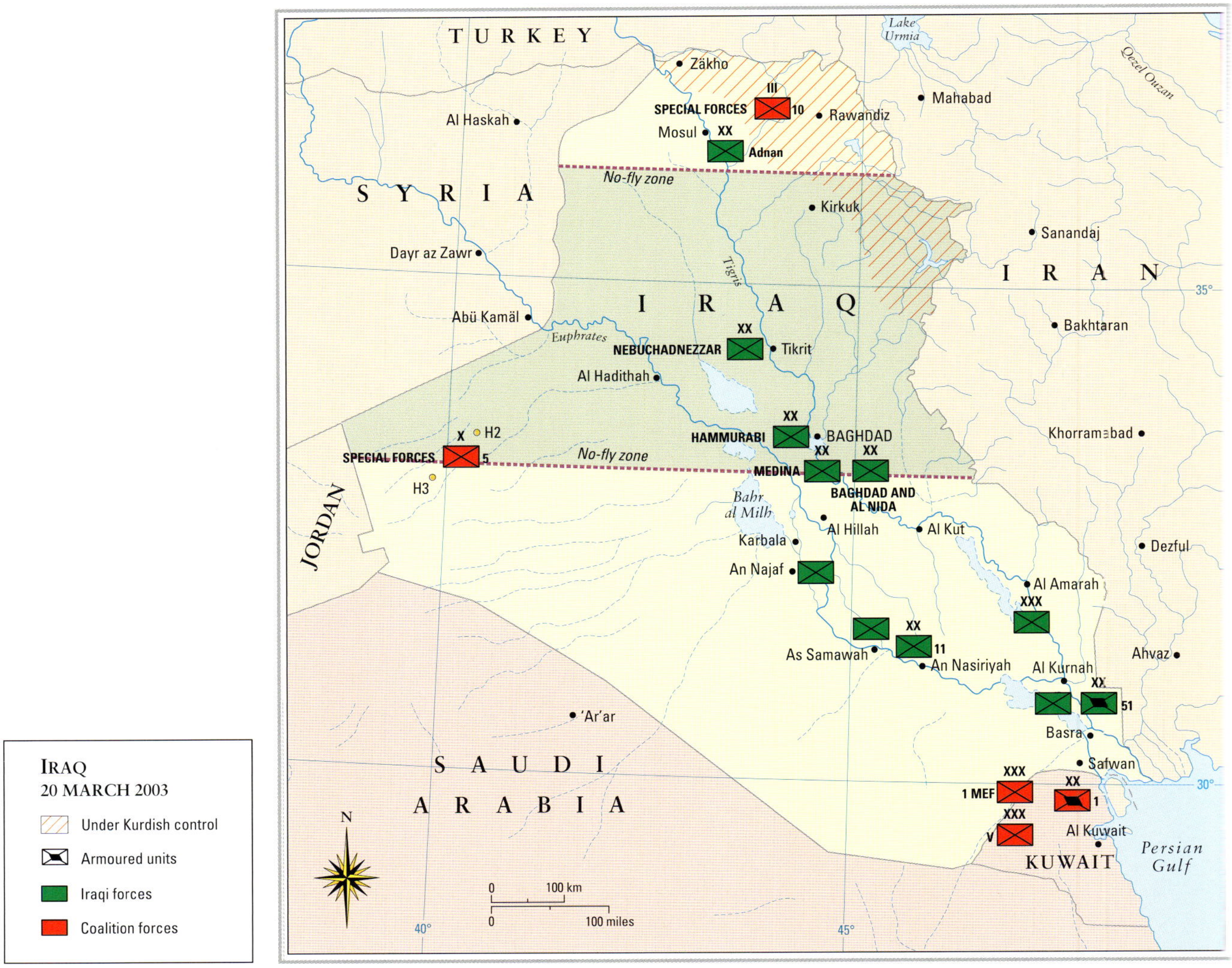

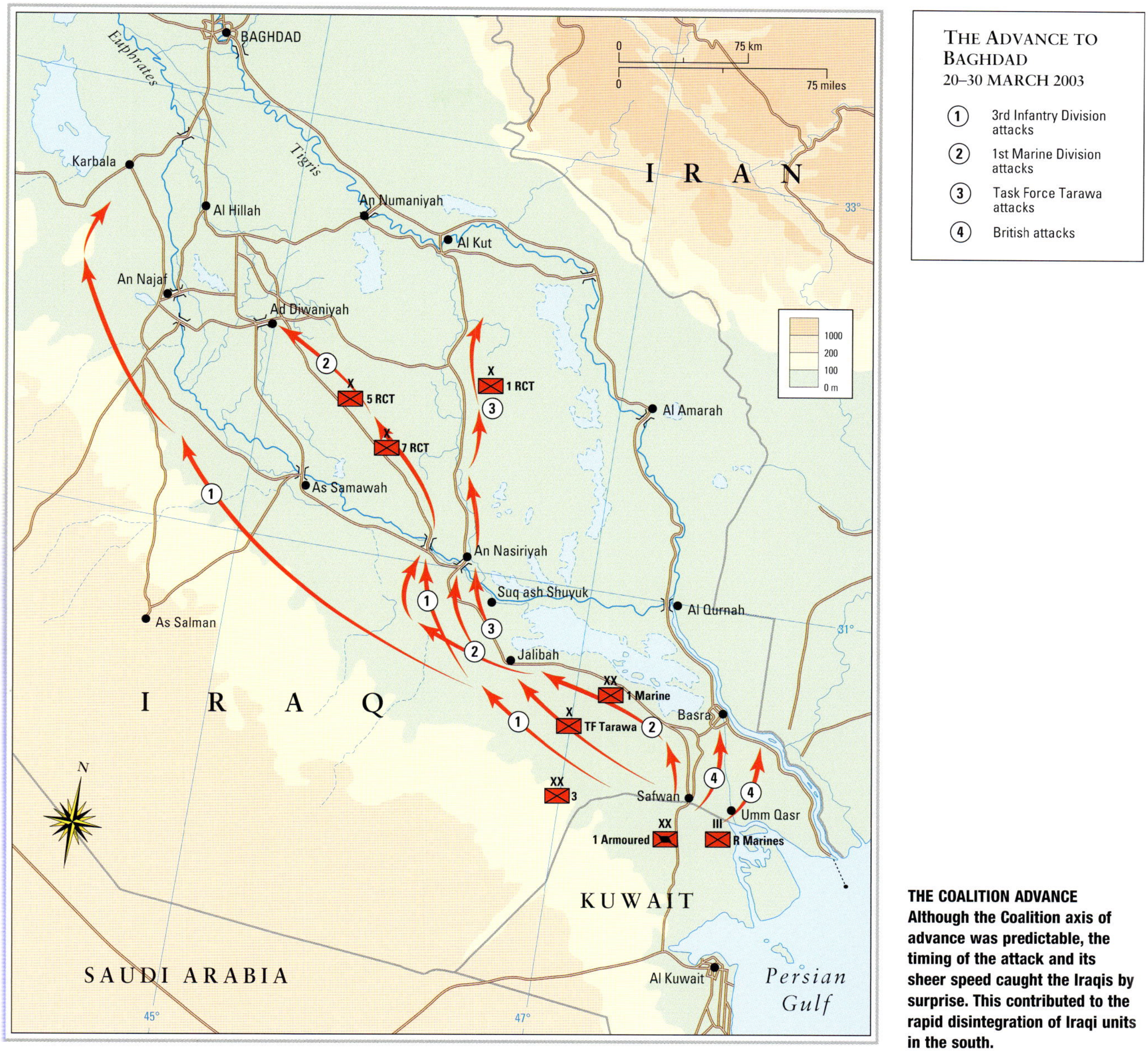

THE COALITION ADVANCE
Although the Coalition axis of advance was predictable, the timing of the attack and its sheer speed caught the Iraqis by surprise. This contributed to the rapid disintegration of Iraqi units in the south.

shape in other ways. Morale and training levels were low among regular units, and there was little popular support among the general population. After years of brutalization by the Saddam Hussein regime, the people from whom the Army was recruited showed little support for the government. There were those who were more willing to fight, of course. These were mostly found among the Republican Guard and the volunteer Fedayeen fighters, who were more interested in fighting against the West than for Saddam Hussein.

Opening moves

The Iraqi leadership expected a lengthy air campaign to soften up their defences, but instead the Coalition launched a rapid ground offensive from the south almost immediately. From the north, Kurdish rebels supported by U.S. Special Forces and aircraft flying out of bases in Turkey began an advance southward that ultimately led to the capture of Tikrit.

Aircraft and missiles attacked key command and control facilities to fragment the Iraqi response, while attack helicopters eliminated border defences. The Coalition's armoured forces were thus able to advance rapidly into Iraq. The initial objective was the port of Umm Qasr and the Al Faw peninsula, which were secured by amphibious forces, before the U.S. contingent began the long advance on the capital, Baghdad.

The primary objective for the British contingent

M2 BRADLEY
The days of the armoured personnel carrier as a 'battle taxi' are long over. Today's infantry fighting vehicle can support its personnel with a 25mm (1in) cannon and has limited anti-tank capability.

was Iraq's second city, Basra, which was reached within hours. Rather than advancing directly into the city, which might have resulted in massive civilian casualties, the British contingent halted and allowed noncombatants to leave the city. After defeating Iraqi armoured forces in the area, the British troops pushed into the city, taking control after a week of urban combat in which armoured forces supported infantry as they cleared the city street by street.

The U.S. advance

U.S. forces met resistance throughout the advance on Baghdad, but this was for the most part fragmented for all its ferocity. The Marine Corps

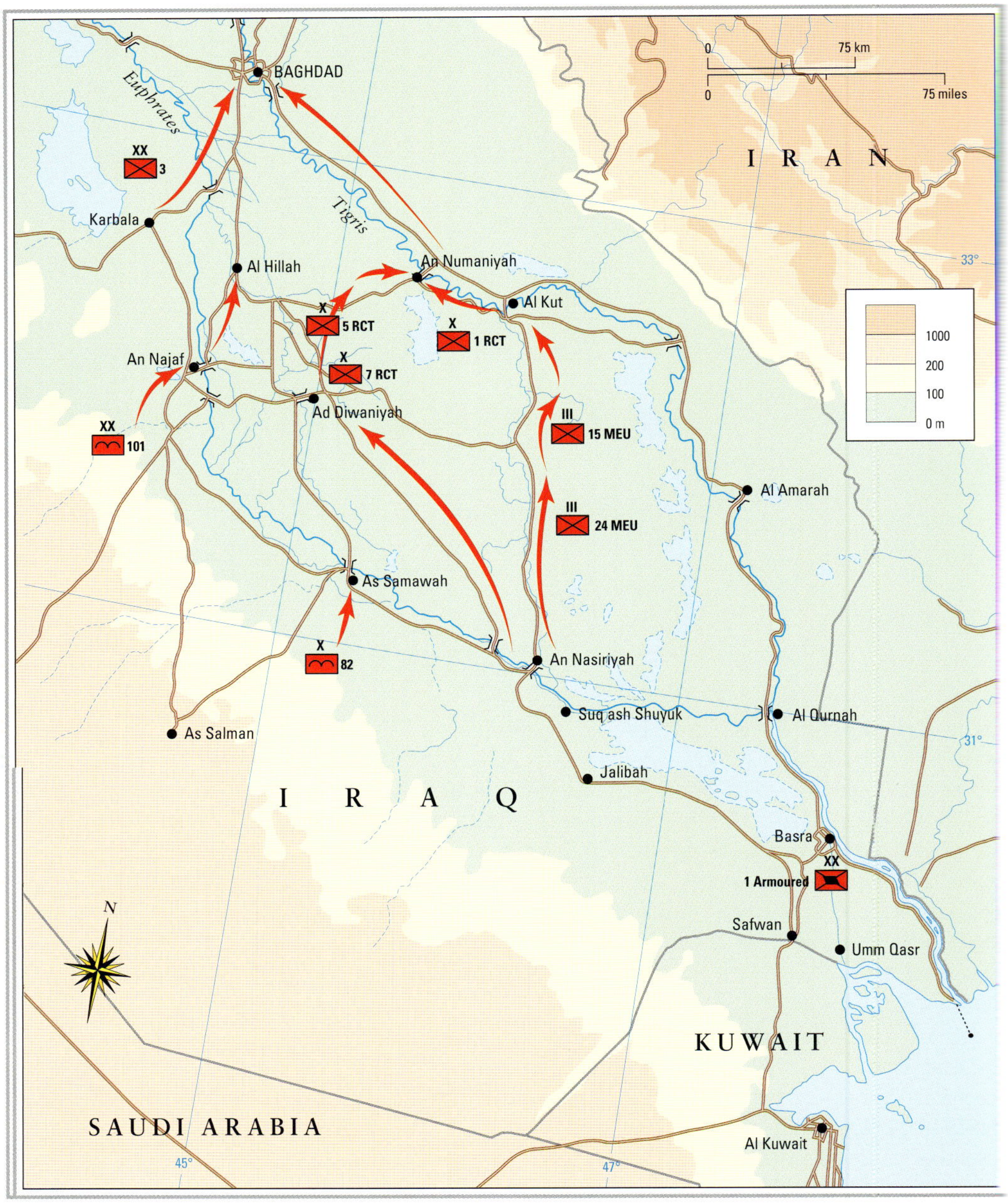

STRATEGY OF SPEED
The Coalition offensive towards Baghdad was conducted at breakneck pace. Urban areas and key bridges that might be held as an obstacle were seized to facilitate the advance, but other centres of resistance were simply bypassed.

THE ADVANCE TO BAGHDAD
30 MARCH – 12 APRIL 2003
Coalition attacks

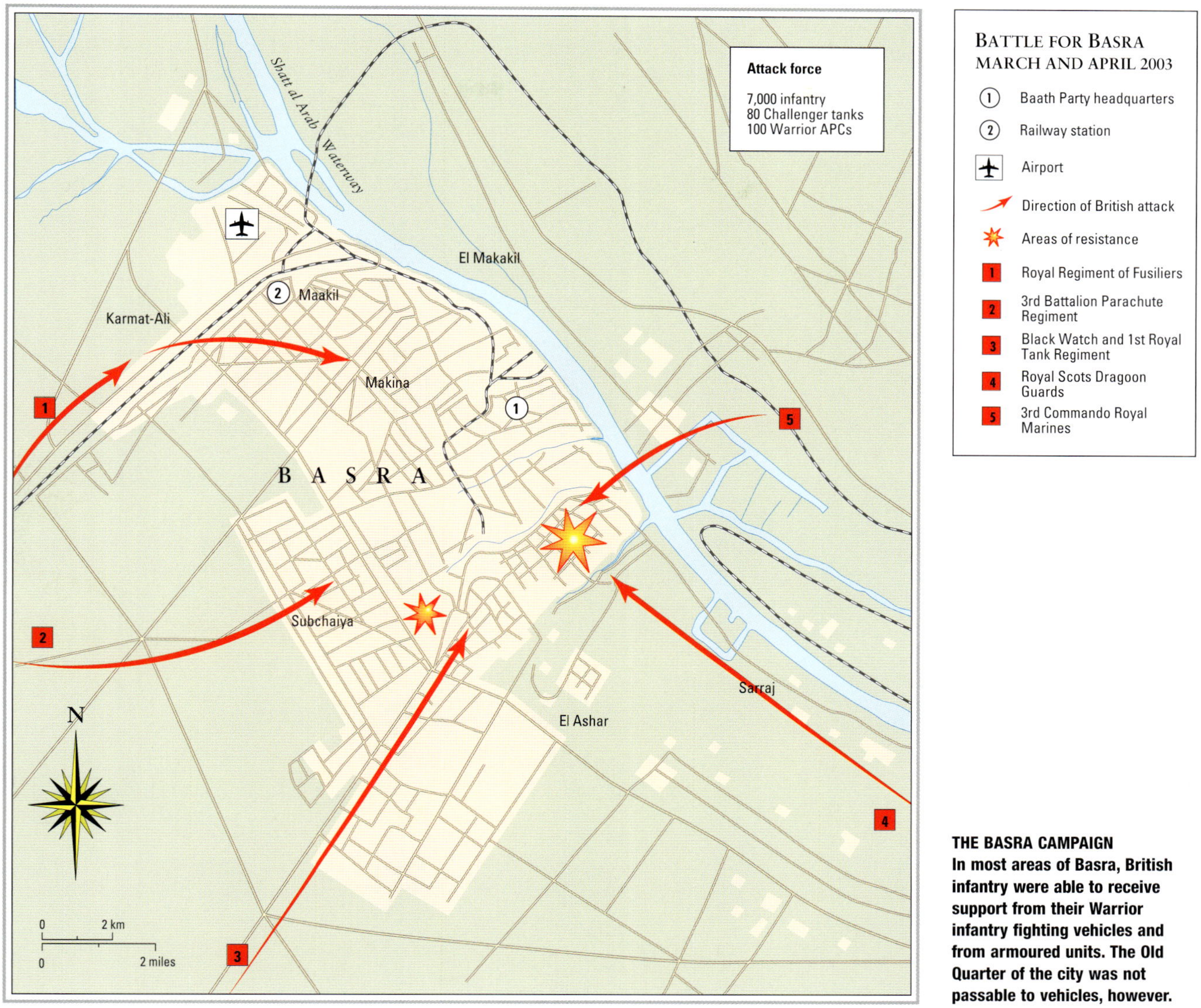

THE BASRA CAMPAIGN
In most areas of Basra, British infantry were able to receive support from their Warrior infantry fighting vehicles and from armoured units. The Old Quarter of the city was not passable to vehicles, however.

contingent deployed as Regimental Combat Teams (RCTs), which were essentially a Marine infantry regiment to which had been added a force of armour, artillery and supporting units to allow it to function as an independent all-arms formation. U.S. Army units were similarly integrated, though on a larger scale. Combined-arms operations allowed most pockets of resistance to be efficiently eliminated or scattered.

Many areas were contested by Iraqi infantry or irregulars with few heavy weapons, against which the well-supported U.S. forces had a huge advantage. Each obstacle was overcome as it was encountered and there was no coherent counter-attack, but all the time the U.S. supply line was lengthening. Logistic issues threatened to slow or even stop the advance, and although the front-line units had punched through the defenders, they were not eliminated. Supply convoys were harassed throughout the route, with the city of An Nasiriyah becoming a centre for Iraqi resistance.

Organized resistance began to firm up after the first days of the advance, with the Republican Guard putting up a stiff fight, in contrast to the regular army, which was more inclined to surrender than face overwhelming force. The city of An Najaf was encircled and eventually cleared to protect the flank of the advance, while other potential centres of resistance were bypassed.

Capture of Baghdad

Although the regular Iraqi Army had largely collapsed from low morale and crippled command and control, the Republican Guard prepared to defend Baghdad.

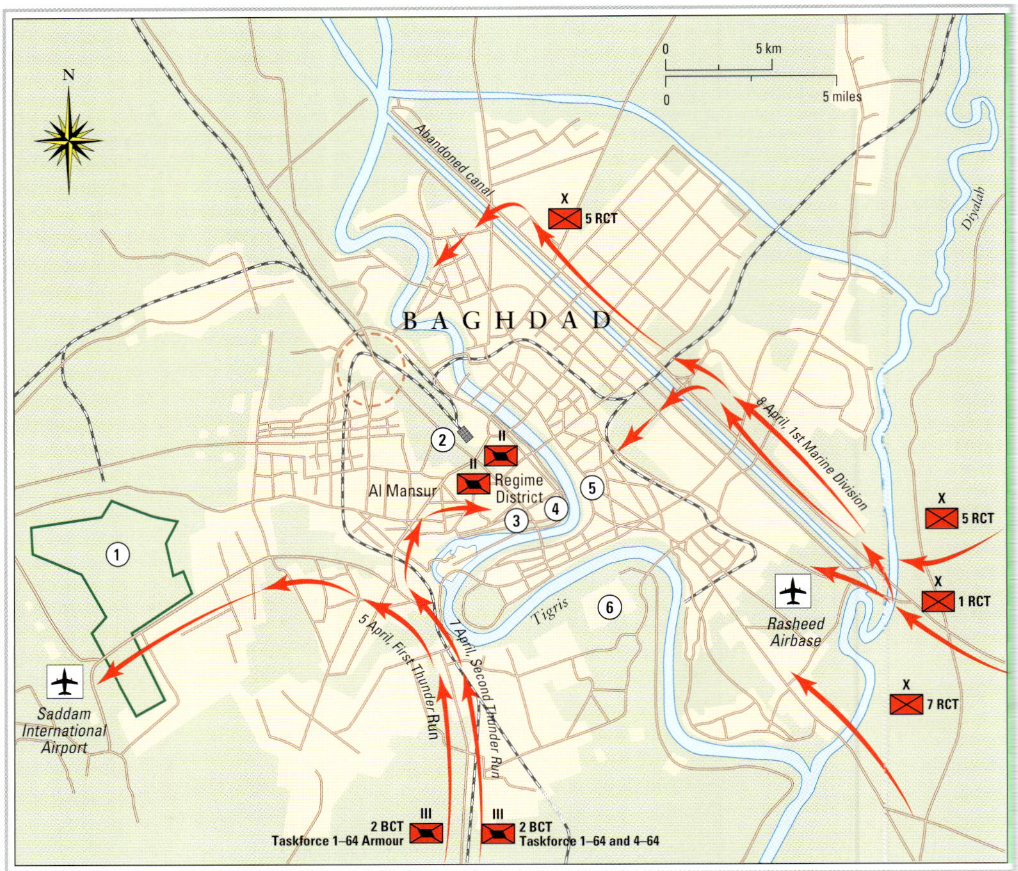

ENTRY INTO BAGHDAD
As Marine Corps units pushed into the city from the east, U.S. armoured forces made two extremely rapid advances into Baghdad. They punched through the defences to secure the administrative centre of the country, decapitating the Iraqi war effort.

AFGHAN ARMOUR
The U.S.S.R. deployed large numbers of BTR armoured personnel carriers during their invasion and occupation of Afghanistan during the 1980s. Some were taken into service with the Afghan National Army after the Soviet withdrawal.

As U.S. commanders hoped, the Republican Guard tried to make a stand in front of the city rather than retreat into the streets. Hammered by air attack, they were smashed by the U.S. forces, opening the way for an advance into the capital.

Rather than mount a slow street-by-street advance through Baghdad, U.S. forces instead launched two audacious rapid assaults directly into the city. The first of these attacks, known as Operation Thunder Run, was more of a reconnaissance in force than an attempt to capture the city, although it succeeded in breaking through to Saddam International Airport and linking up with U.S. forces which had secured it.

The second Thunder Run, again spearheaded by armoured vehicles, was a charge straight into the administrative area of the city. Despite ferocious resistance, Operation Thunder Run was successful and key objectives were secured within hours. Pacifying the rest of the city took a little longer, but the strike effectively disabled the Iraqi government and military command apparatus. This rapid assault was characteristic of the war as a whole, in which the swiftness of the Coalition advance was a factor in keeping casualties down.

Security operations in Iraq and Afghanistan

The fall of Baghdad was a symbolic end to the Iraq campaign, but resistance did not end there.

The work of rebuilding Iraq as a democratic state required a long counterinsurgency campaign in which armoured forces could offer support, but could not take the place of infantry patrols and checkpoints.

Similarly, armoured forces can only achieve so much in a campaign like that in Afghanistan. If concentrations of insurgents can be located or they can be induced to stand and fight, armoured units can quickly overrun them. It has long been said that the best terrain for tanks is any terrain with few anti-tank weapons, and modern armour has proven all but invulnerable to most weapons fielded by insurgents in Iraq and Afghanistan. However, the main problem facing the armoured forces in Afghanistan is that while the terrain has few anti-tank weapons, it also lacks targets suitable for armoured assault.

T-72
The Russian-supplied T-72 was the best tank available to the Iraqi Army. Although a good vehicle in its day, it was no match for the far more advanced armour fielded by the Coalition.

M1A1
The M1A1 Abrams proved almost invulnerable to Iraqi anti-tank weapons, though several were immobilized. Sophisticated gunlaying systems allowed long-range first-shot kills against enemy tanks and armoured vehicles, even on the move.

THE FUTURE ARMOURED FORCE

Modern armoured forces are the product of a century of evolution, during which tanks developed from short-ranged infantry support vehicles to the arm of decision in fluid breakthrough-and-exploitation battles. In order to survive into the future, the tank must continue to evolve to meet the requirements of the modern battlespace.

One of the main challenges facing the heavily armoured combat vehicle is justifying its immense cost. If helicopters and missiles can kill enemy tanks, and lightly armoured vehicles can effectively support infantry against most likely opponents, does the tank still have a role to play?

Armoured warfighting

The capabilities of light vehicles, remotely controlled drones and helicopters are impressive, but the ability of a tank to bring massive firepower to bear where it is needed, and to protect its crew while doing so, remains unrivalled. This allows the tank force to act as a spearhead for less well protected units and to advance rapidly even against heavy opposition.

Tanks justify their cost by being able to bring massive force to bear at the critical point. The ability to smash through an enemy force and overrun its command and logistics assets makes a costly attritional battle unnecessary, saving both money and lives. Thus tanks remain a cost-effective investment even though cheaper alternatives might look attractive.

Tank forces are becoming ever more effective, partly through advances that make each tank individually more capable. Other force-multipliers include computerized data-sharing systems which are fed information from reconnaissance drones or other units. These allow for increased cooperation between the armoured spearhead and other units such as artillery or helicopters. This allows the tank force to 'see what is over the next hill' and to engage it in the most effective manner possible.

The light armoured role

There are some roles for which the main battle tank is not well suited, and which can be fulfilled

TOWARDS A LIGHTER MBT
The CV90120 is part of a family of advanced armoured vehicles. It mounts a 120mm (44/5in) gun in a lighter package than many previous battle tank designs.

by cheaper vehicles. In today's conflicts, armoured vehicles are used more often in support of infantry attempting to secure an area than in the traditional 'tank battle' role. A much lighter vehicle can serve as a heavy weapons platform and still provide excellent protection to its crew. Several of these lighter vehicles can be purchased and maintained in the field for the cost of a single tank. Several armoured vehicles can obviously cover more ground and have a better chance of bringing their firepower to bear than a single vehicle.

Thus tank forces are complemented by lighter vehicles optimized for security operations and infantry support. Infantry Combat Vehicles (or Infantry Fighting Vehicles) have largely replaced the 'battle taxi' or Armoured Personnel Carrier. Equipped with weapons for infantry support, these vehicles can be effective in open warfare and in security operations. However, there are some threats that they simply cannot tackle, which makes cooperation with tank forces essential.

Future armoured forces

The future of armoured forces lies as part of an integrated combat team which shares tactical information and trains for mutual support. This is particularly important in urban combat, where fields of vision are limited and tanks can be ambushed by infantry tank-hunting teams. Tank forces must be able to enter cities where enemy forces have taken refuge, confident that they can work closely with infantry, air assets and light armoured vehicles to bring their firepower to bear where it is most urgently needed.

3D BATTLESPACE
Any future conflict will take place in a complex three-dimensional 'battlespace' where air power must always be considered.

NEXT-GENERATION APC
The Armadillo is the latest vehicle in the CV90 family. Its weapons are remote-controlled from inside the vehicle.

INDEX

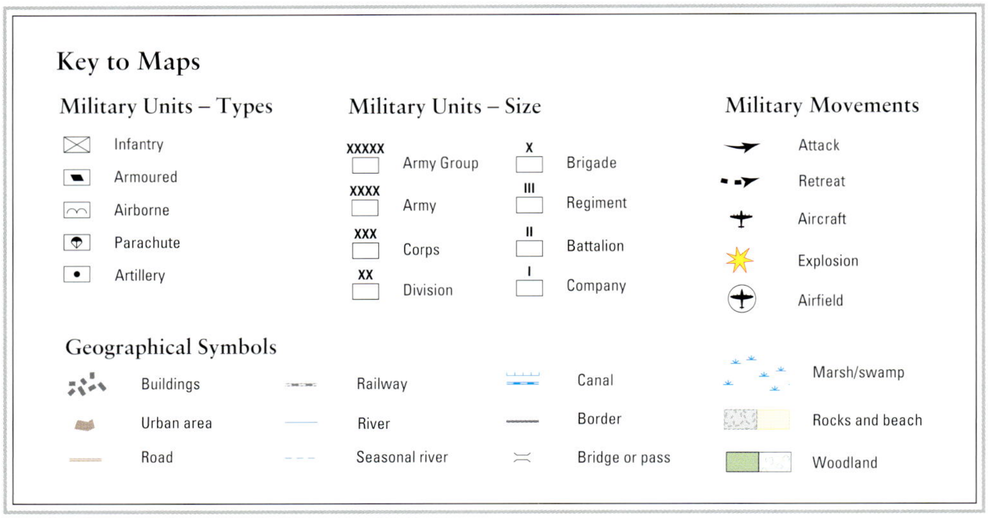

Entries in *italics* refer to captions for photographs, illustrations or maps

INDEX

INDEX

PICTURE CREDITS

Alamy: 54 bottom (Mirrorpix)

Art-Tech/Aerospace: 9, 12, 13, 15 bottom, 20-24 all, 26 bottom, 32, 45, 54 top, 57 top, 67 both, 68 both, 93, 104 top, 106 top, 110 bottom, 112-118 all, 126-130 all, 134 top, 141 top, 142, 144, 148-154 all, 160-165 all, 170 bottom, 174-181 all, 194, 200 bottom, 202 both, 205 both

Art-Tech/MARS: 8, 34, 36, 41 bottom, 74 bottom, 207

BAE Systems: 218, 219

Cody Images: 10-11 all, 15 top, 17-19 all, 26 top, 28, 29, 38, 41 top, 42, 46, 48, 50 top, 51, 53 top, 56 both, 58, 61 both, 64 bottom, 66, 72 top, 74 top, 82-85 all, 87, 88 top, 89, 94-103 all, 104 bottom, 106 bottom, 109 top, 110 top, 122-124 all, 141 bottom, 172, 193, 196, 200 top

Corbis: 75 (Underwood & Underwood), 86 (Hulton)

Nik Cornish/Stavka: 77, 90, 91, 146 both, 184 both

Mary Evans Picture Library: 72 bottom (Robert Hunt Collection)

Krigsarkivet: 7 top, 14, 30, 33, 62 both, 64 top 65

Tank Museum: 57 bottom, 69, 70, 88 bottom, 92

Ukrainian State Archive: 7 bottom, 78-81 all, 1[illegible]9-120 all, 134 bottom, 136-138 all, 157, 158

U.S. Department of Defense: 6, 50 bottom, 53 bottom, 109 bottom, 170 top, 204, 214-217 all

INDEX OF MAPS